CARDIOVASCULAR IMPLICATIONS OF STRESS AND DEPRESSION

Edited by

PAUL D. CHANTLER
Division of Exercise Physiology, Robert C. Byrd Health Sciences Center, School of Medicine, West Virginia University, Morgantown, WV, United States

KEVIN T. LARKIN
Department of Psychology, West Virginia University, Morgantown, WV, United States

ELSEVIER

ACADEMIC PRESS
An imprint of Elsevier

Academic Press is an imprint of Elsevier
125 London Wall, London EC2Y 5AS, United Kingdom
525 B Street, Suite 1650, San Diego, CA 92101, United States
50 Hampshire Street, 5th Floor, Cambridge, MA 02139, United States
The Boulevard, Langford Lane, Kidlington, Oxford OX5 1GB, United Kingdom

Library of Congress Cataloging-in-Publication Data
A catalog record for this book is available from the Library of Congress

British Library Cataloguing-in-Publication Data
A catalogue record for this book is available from the British Library

ISBN 978-0-12-815015-3

For information on all Academic Press publications
visit our website at https://www.elsevier.com/books-and-journals

Publisher: Stacy Masucci
Acquisition Editor: Katie Chan
Editorial Project Manager: Carlos Rodriguez and Aleksandra Packowska
Production Project Manager: Poulouse Joseph
Cover Designer: Miles Hitchen

Typeset by SPi Global, India

Contents

CARDIOVASCULAR IMPLICATIONS OF STRESS AND DEPRESSION

Contributors

Julie A. Brefczynski-Lewis
Department of Neuroscience, West Virginia University, Morgantown, WV, United States

Leah A. Brown
Department of Psychology, West Virginia University, Morgantown, WV, United States

Matthew M. Burg
Section of Cardiovascular Medicine, Department of Internal Medicine, Yale School of Medicine, New Haven; VA Connecticut Healthcare System, West Haven; Department of Anesthesiology, Yale School of Medicine, New Haven, CT, United States

Casey E. Cavanagh
Section of Cardiovascular Medicine, Department of Internal Medicine, Yale School of Medicine, New Haven; VA Connecticut Healthcare System, West Haven, CT, United States

Paul D. Chantler
Division of Exercise Physiology, Robert C. Byrd Health Sciences Center, School of Medicine, West Virginia University, Morgantown, WV, United States

Christine L. Foxx
Department of Integrative Physiology and Center for Microbial Exploration, University of Colorado Boulder, Boulder, CO, United States

Sierra Cronan
Department of Psychology and Health Psychology Program, University of Utah, Salt Lake City, UT, United States

Jefferson C. Frisbee
Department of Medical Biophysics, Schulich School of Medicine & Dentistry, Western University, London, ON, Canada

Allison E. Gaffey
Section of Cardiovascular Medicine, Department of Internal Medicine, Yale School of Medicine, New Haven; VA Connecticut Healthcare System, West Haven, CT, United States

Jeffrey L. Goodie
Department of Medical & Clinical Psychology, Uniformed Services University, Bethesda, MD, United States

Hilary M. Gould
Psychology Service (116B), VA San Diego Healthcare System, San Diego, CA, United States

Angela J. Grippo
Department of Psychology, Northern Illinois University, DeKalb, IL, United States

Phillip Kroke
Department of Medical & Clinical Psychology, Uniformed Services University, Bethesda, MD, United States

Joshua Landvatter
Department of Psychology and Health Psychology Program, University of Utah, Salt Lake City, UT, United States

Kevin T. Larkin
Department of Psychology, West Virginia University, Morgantown, WV, United States

Christopher A. Lowry
Department of Integrative Physiology and Center for Microbial Exploration, University of Colorado Boulder, Boulder, CO, United States

Michael Papadakis
Department of Kinesiology, University of Utah, Salt Lake City, UT, United States

Lindsey Rosman
Section of Cardiovascular Medicine, Department of Internal Medicine, Yale School of Medicine, New Haven; VA Connecticut Healthcare System, West Haven, CT, United States

Thomas Rutledge
Psychology Service (116B), VA San Diego Healthcare System, San Diego; UC San Diego Department of Psychiatry, La Jolla, CA, United States

Emily Scott
Department of Psychology and Health Psychology Program, University of Utah, Salt Lake City, UT, United States

Alaina G. Tiani
Department of Psychology, West Virginia University, Morgantown, WV, United States

Bert N. Uchino
Department of Psychology and Health Psychology Program, University of Utah, Salt Lake City, UT, United States

Kevin Wilfong
Department of Medical & Clinical Psychology, Uniformed Services University, Bethesda, MD, United States

Susan K. Wood
Pharmacology, Physiology and Neuroscience, University of South Carolina School of Medicine, Columbia, SC, United States

CHAPTER 1

Stress, depression, and cardiovascular disease

Kevin T. Larkin[a], Paul D. Chantler[b]
[a]Department of Psychology, West Virginia University, Morgantown, WV, United States
[b]Division of Exercise Physiology, Robert C. Byrd Health Sciences Center, School of Medicine, West Virginia University, Morgantown, WV, United States

Contents

Cardiovascular disease (CVD) has held the unenviably position of being the leading cause of death in industrialized nations for almost a century (Heron and Anderson, 2016). Today, CVD accounts for 1 out of every 4 deaths for men and women in the United States (Centers for Disease Control and Prevention, 2017), and even higher mortality rates worldwide (Mendis et al., 2011). Coupled with cerebrovascular disease (stroke), the third leading cause of death, diseases of the circulatory system account for considerably more morbidity and mortality in the world than any other health condition. Not surprisingly, these diseases represent one of the world's largest health care concerns, leading to significant costs to the health care system, lost productivity among workers, and serious restrictions to life activities and mobility for those affected by them. Despite gains made in the assessment and treatment of diseases of the circulatory system over the latter half of the 20th century that resulted in reductions in morbidity and mortality, we are again noticing an increased incidence of CVD (Heron and Anderson, 2016).

The vast majority of cases of diseases of the circulatory system are associated with the underlying process of atherosclerosis. Atherosclerosis is the progressive blockage of blood flow through arteries that occurs over decades as the body responds to regular injuries to endothelial cells comprising artery walls caused by perturbed blood flow, heightened arterial pressure, and/or exposure to toxins circulating in the bloodstream. Based upon our understanding of atherosclerosis, it is not surprising that the earliest risk

Cardiovascular Implications of Stress and Depression
https://doi.org/10.1016/B978-0-12-815015-3.00001-5

factors for CVD identified were variables that influenced the atherosclerotic process. For example, the diagnosis of essential hypertension presumably promotes endothelial injury through the impact of having chronically elevated arterial pressures throughout daily life. High levels of circulating blood lipids (i.e., hypercholesterolemia), result in coating the endothelium with fatty streaks and eventually fatty plaques, also promoting cellular injury and an impaired bodily response to that injury. Finally, recognizing that tobacco smoke contains many chemicals toxic to the endothelium, one can easily identify biologically plausible mechanisms through which these risk factors for CVD exert their negative health effects by promoting atherosclerosis and eventually the onset of CVD.

Unfortunately, smoking tobacco products, having high cholesterol, and exhibiting high blood pressure accounts for less than half of the variance in predicting onset of CVD (Centers for Disease Control and Prevention, 2017). In this regard, many nonsmokers with perfectly normal cholesterol and blood pressure levels can succumb to a myocardial infarction (i.e., heart attack) and/or stroke, and many individuals who have extensive smoking histories accompanied by untreated hypertension and hypercholesterolemia show no evidence of CVD, even at advanced ages. Our inability to predict onset of CVD accurately from recognized risk factors has led to a search for additional variables that may indeed capture some unique explanatory variance in making these sorts of predictions.

The influence of psychological factors on medical health conditions has been a fertile ground for research for some time, so it was only natural that scientists began examining potential psychological risk factors for CVD. In fact, one could trace theories linking medical conditions to temperaments back to the Greek physician, Hippocrates (460–375 BC), who proposed that medical conditions resulted from an imbalance of various humors in the body that were each associated with specific personality profiles. Although contemporary personality theory traces its roots back to some of the beliefs of early scientists like Hippocrates, adoption of the belief that personality type influenced specific medical disease progression was never fully adopted, partly because a biologically plausible mechanism for such associations was lacking.

In the early part of the 20th century, physiologist Walter Cannon (1915) and endocrinologist Hans Selye (1955) made significant contributions in outlining how exposure to various environmental stimuli, mainly those that elicited strong emotional reactions, led to a cascade of physiological responses that affected most organ systems in the body. Borrowing the terms "stress" and "strain" from mechanical engineers, they described how exposure to environmental events (i.e., stress) was met with an expected and predictable physiological response of the organism (i.e., strain). The importance of their work cannot be understated in providing clues to the potential mechanisms for explaining how stressful events encountered in life could alter the physiological processes that affected blood flow and ultimately the process of atherosclerosis. If indeed the brain transduces environmental events into the physiological sequelae of atherosclerosis, much like it transduces sound or light waves to neural signals, organisms exposed to certain types of toxic environments may be at greater risk for developing CVD. Consequently, if individual differences existed in the magnitude or pattern of these physiological responses to stress, then these specific personality or behavioral profiles could be assessed and identified as potential psychological risk factors for CVD.

One of the initial forays into the search for specific psychological risk factors for CVD was done by cardiologists Friedman and Rosenman (1959) who coined the term Type A Behavior Pattern (TABP) to refer to a constellation of highly prevalent behaviors observed among their cardiac patients. This pattern of behaviors was characterized by a chronic sense of time urgency and impatience, showing competitiveness even in non-competitive situations, and easily-aroused hostility, and was detected using a structured interview designed to evoke behavioral displays of these traits (Rosenman, 1978). Although showing initial promise as a recognized behavioral risk factor, subsequent trials failed to corroborate initial reports (e.g., Dimsdale et al., 1978; Johnston et al., 1987; Shekelle et al., 1985) and TABP was never uniformly accepted as an established risk factor for predicting CVD onset. Subsequent investigations focused on other psychosocial constructs that might enhance our ability to comprehend how exposure to stress increased risk for CVD and functioned as a behavioral risk factor for CVD. Among those that were examined, depression emerged as the one that is most strongly associated with onset of CVD among initially healthy adults as well as for subsequent cardiac events among patients diagnosed with coronary heart disease (for meta-analytic reviews, see Rugulies, 2002; Barth et al., 2004, respectively). In fact, it has been argued that presence of a diagnosis of depression or its less severe counterpart, dysphoric mood, was as strongly associated with CVD as having high blood pressure, high cholesterol, or smoking tobacco. In recognition of this status, the American Heart Association published a statement recommending that depression be considered an established risk factor for poor prognosis among patients with CVD in 2014 (Lichtman et al., 2014).

Despite evidence linking depression to CVD, it is still not given the same status as treating high blood pressure and high cholesterol or encouraging smoking cessation in most medical settings. One reason that depression screening has not been readily adopted in these health care settings pertains to our lack of knowledge pertaining to how exposure to environmental stress can lead to the neuropsychiatric sequelae of depression or perturbations in blood flow that lead to atherosclerosis within the cardiovascular system. If the mechanisms linking stress with depression and CVD were established and understood, it might result in greater adoption of strategies to assess psychological risk factors for CVD in contemporary health care settings. This edited volume aims to provide readers with a comprehensive appraisal of the literature linking stress and depression with CVD, with special attention to the various physiological mechanisms through which stress exerts its negative health effects on both the nervous and circulatory systems.

Basic definitions

Prior to introducing conceptual models of potential relations among these variables, it is important to consider basic definitions of key terms. Although we have already introduced a brief description of the process of atherosclerosis as the underlying mechanism for CVD, it is important to recognize that there are other cardiovascular problems that occur that also lead to significant health consequences, including death. These include disturbances in the pacing of the heart's contractions (e.g., arrhythmias), compromised mechanical functioning of the four valves of the heart, infections that invade the pericardium that surrounds the heart, and genetic physical abnormalities of the heart itself. Because atherosclerosis underlies the majority of cases of coronary heart disease, stroke, and peripheral vascular disease, we will use the term CVD synonymously with cardiovascular problems caused by atherosclerosis. When other cardiovascular conditions are considered and discussed throughout this volume, we will refer to them by name rather than lumping them together with cardiovascular conditions caused by atherosclerosis.

As noted above, Selye (1955) differentiated stress from strain by referring to stress as the environmental precipitant of the disease and strain as the physiological processes that occur as the body adapted to or attempted to adapt to the precipitating event or circumstance. In this regard, stress is the stimulus and strain is the response. We will make the same distinction here. It needs to be said, however, that not all researchers make this distinction, as many measures of stress used in clinical and research settings actually measure the magnitude of the response to the stress (i.e., strain) or one's perception of the severity of the stressful event. For purposes of this volume, the term "stress" will refer to the specific stimuli or alterations in environmental contexts to which an organism is exposed, much like Selye's usage of this term. Although our use of the term "stress" typically refers to stimuli external to the organism, it is true that internal stimuli can also trigger physiological responses comparable to those evoked by exposure to external stressors. For example, experiencing an angina attack in the middle of the night may elicit a physiological response comparable to those that occur when confronting a bear in the woods. In this regard, both external and internal stimuli can justifiably be considered sources of stress.

In contrast to CVD that refers to restricted blood flow in the circulatory system and stress that most commonly occurs in the external environment, depression and depressive symptoms presumably largely occurs in the brain. In fact, there is a substantial literature documenting the changes in neural functioning that occurs in the central nervous system and neuroendocrine system associated with depression (e.g., Davidson et al., 2002; Thase et al., 2015). Depression itself is a mood state characterized by a range of cognitive (reduced concentration, thoughts of hopelessness or worthlessness, suicidal thinking), emotional (e.g., feeling sad or empty, loss of interest in previously pleasurable activities), behavioral (e.g., inactivity, social withdrawal), and physical symptoms (e.g., fatigue, restlessness, sleep disturbance, weight loss or gain, altered appetite). When the mood state persists for at least 2 weeks and is associated with a reduction in daily functioning, but with no instances of mania or hypomania, a Major Depressive Disorder is diagnosed (American Psychiatric Association, 2013). The prolonged experience of depressive mood states for over 2 years duration in adults meets diagnostic criteria for Persistent Depressive Disorder or Dysthymia (American Psychiatric Association, 2013). Both Major Depressive and Persistent Depressive Disorders must be distinguished from Bipolar Disorder that involves altered mood rhythms that vacillate between depressive moods and periods of extreme energy and activation. For purposes of this book, the term "depression" will refer to the diagnosed conditions of either Major Depressive or Persistent Depressive Disorders, but not the diagnosis of Bipolar Disorder, and the term "depressive symptoms" will refer to the underlying cognitive, behavioral, emotional, and physical symptoms upon which these diagnoses are made. Although humans can be diagnosed with depressive disorders, animals cannot. However, animals can display behaviors that resemble depressive symptoms (e.g., social withdrawal, restlessness, reduced grooming, inactivity). Therefore, research on animals can shed light on relations between stress, depressive symptoms, and CVD outcomes, but are of lesser value in relating stress and CVD with the complex symptomatology of clinically diagnosed depressive disorders.

Conceptual models linking stress with depression and cardiovascular disease

Although bi-directional pathways exist between stress and CVD, of primary importance here is the causal association between stress as the stimulus and CVD as the pathogenic response. This is not to say that experiencing a heart attack or stroke does not constitute a source of stress, but that we are mostly interested in how a broad range of stressful life events leads to CVD over time. In other words, if exposure to stress functions as a behavioral risk factor for CVD, it must precede the onset of disease. What is unclear in this model is whether depression is directly involved in this causal pathway or whether it serves as simply a correlated outcome. These two conceptual models are depicted in Fig. 1.1. In Panel A, exposure to environmental

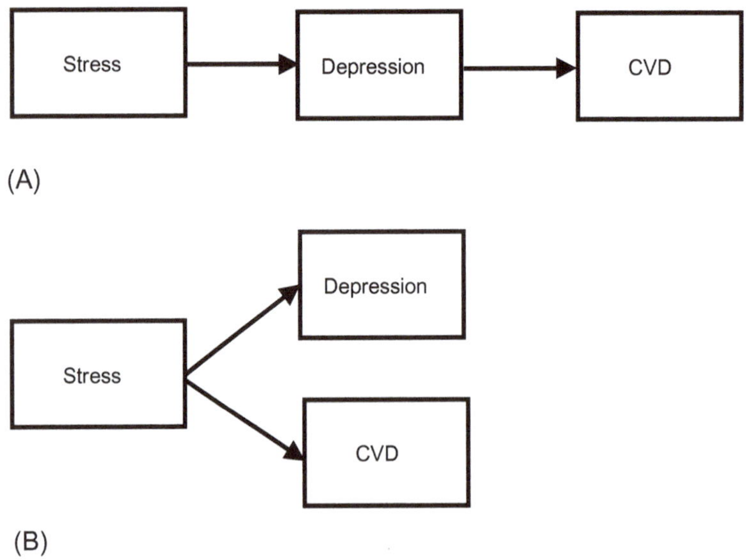

Fig. 1.1 (A) Depression mediates the relation between stress and CVD and (B) Depression as a corollary of the relation between stress and CVD.

stress influences brain functioning that causes depression and/or depressive symptoms. Over time, the brain changes that occur with experiencing a depressed mood lead to physiological perturbations in the circulatory system that lead to atherosclerosis.

In contrast, the corollary model depicted in Panel B reveals independent links between stress and both the brain changes that occur with experiencing depression or depressive symptoms and the restrictions in blood flow caused by atherosclerosis. The corollary model assumes no causal role for depression, although it still functions as a credible marker for CVD through their shared lineage.

In Section One of this book, evidence supporting these two conceptual models will be examined in order to gain a comprehensive understanding of the role of depression in explaining how stress leads to CVD. Evidence for relations among stress, depression, and cardiovascular disease has accumulated in three separate bodies of literature: human epidemiologic studies, translational research that emanates from animal models, and human intervention research using therapeutic strategies known to influence these health conditions. In Chapter 2, Rutledge and Gould review the body of evidence from prospective epidemiological trials on humans that examines relations between stress and depression with onset of CVD or experiencing subsequent cardiac events among patients with pre-existing CVD. Next, in Chapter 3, Grippo reviews the extant animal literature linking depressive behaviors of animals with cardiovascular consequences, with a special focus on translational studies extending animal research to human pre-clinical and clinical studies. Experimental methods for inducing depressive symptoms in animals by exposing them to stress is a highlight of this work. Finally, in Chapter 4, Cavanagh, Gaffey, Rosman, and Burg review clinical trials aimed at treating either depression or CVD using pharmacotherapeutic agents or psychological interventions, including stress management, cognitive behavior therapy, and nutritional and exercise interventions. By examining the evidence from these three bodies of literature, we can begin to determine the proper role for depression in explaining the relation between exposure to stress and CVD outcomes. Although the differentiation between depression as a causal psychological phenomena and as a corollary phenomenon may not be as important for establishing depression as a behavioral risk factor for CVD, it is critical for developing intervention and prevention programs that target reducing behavioral risk for CVD. If depression is not part of the causal pathway linking stress with CVD, it is unlikely that the numerous evidence-based interventions we have for treating depression will influence subsequent incidence of CVD.

Mechanisms linking stress with cardiovascular disease

As mentioned previously, one of the reasons that behavioral risk factors may receive a lower priority than the established risk factors of smoking tobacco, having high blood pressures, or exhibiting hypercholesterolemia in clinical settings is the lack of clarity surrounding the behavioral and physiological mechanisms linking stress to CVD. Because stress, for the most part, results from one's surrounding environment and atherosclerosis occurs in the circulatory system, it is important to consider biologically plausible mechanisms regarding how stress infects the organism and causes the dramatic changes in the cardio-vasculature that lead to onset of CVD. Clearly, there is no direct portal through which events like loss of a loved one seep through the skin into the vascular bed leading to reductions in arterial flow. In this regard, experts in this field agree that the linkage between environmental stress and atherosclerosis relies on the nervous system, the organ system responsible for transducing environmental stimuli (e.g., sound waves; light spectra) into neural phenomena that enable the brain and other organ systems in the body to develop appropriate responses to these stimuli. In brief, the transduction of stressful events occurring in the environment to pathogenic processes in the cardio-vasculature requires a functional nervous system that is capable of detecting these events, appraising them as being threatening or presenting a challenge for the organism to overcome, and orchestrating a physiological response capable of coping with or handling the event in the most efficient way. The purpose of Section Two of this volume is to examine several behavioral and/or physiological mechanisms that have been proposed to explain how exposure to stress might lead to restricted blood flow that occurs in CVD. This is not a new question, as it was proposed shortly after the association between depression and CVD was discovered and reported. In an early review examining potential mechanisms linking depression and CVD, Carney et al. (1995) identified increased sympathetic nervous system activity and reduced parasympathetic activity as physiological mechanisms that potentially explained how onset of depression led to an increased risk for CVD. These authors also suggested that reductions in adherence to cardiac treatment regimens or less engagement in healthy behaviors (e.g., consuming healthy diets, exercising regularly, limiting consumption of alcohol, and not using tobacco) may be other avenues through which depression increased risk for CVD. They dismissed the hypothesis that the depression-CVD linkage could be associated with cardio-toxic side effects of anti-depressant medication, mainly because strong associations between depression and CVD were evident among depressed patients who had not been treated with anti-depressant medications.

Almost a decade later, Lett et al. (2004) reviewed the literature again and added several new biologically plausible mechanisms through which depression might increase risk for CVD. These included increased platelet aggregation, hypothalamic pituitary adrenal (HPA) dysregulation, and immune system alterations via inflammatory responses markers. Skala et al. (2006) confirmed these findings in a review a few years later, and added new information pertaining to imbalances among blood clotting mechanisms for explaining how depression increases risk for CVD, specifically the balance between prothrombotic and antithrombotic processes. Dysfunction within the endothelium itself was also introduced as a potential mechanism based on evidence showing reduced flow-mediated dilation among cardiac patients with higher scores on measures of depressive symptoms (Sherwood et al., 2005). Additionally, shared genetic substrates as well as plasma concentrations of Omega-3 polyunsaturated fatty acids were reported as showing promise as underlying mechanisms for explaining the depression-CVD association (Skala et al., 2006). Given that over a decade has passed since these initial forays examining underlying mechanisms linking depression with CVD were published, the time is ripe to examine the progress made in understanding how stress gets under the skin and into the circulatory system, ultimately leading to atherosclerotic complications. The remaining chapters of this volume serve this purpose and lead to a greater understanding of the biologically plausible mechanisms that link stress with CVD for both scientists examining how stress and depression lead to increased risk for CVD and health care providers who will likely pay closer attention to these modifiable risk factors in hopefully reducing the consequences of the disabling array of diseases that affect the cardiovascular system.

In Fig. 1.2, several biologically plausible behavioral and physiological mechanisms for linking stress and depression with CVD that have been historically identified are presented and serve as the organization for the remainder of this book. These include the magnitude and patterning of autonomic and neuroendocrine responses to stress (Chapter 5), cardiac changes (e.g., electrical conductance, cardiac remodeling) and vascular alterations (peripheral arterial disease; arterial stiffness and atherosclerosis) that occur in response to stress (Chapter 6), central nervous system involvement (e.g., brain signaling; cerebrovascular functioning) in response to stress (Chapter 7), inflammatory system responses to stress (Chapter 8), physiological influences of social support and lack thereof (Chapter 9), alterations in health behaviors, including dietary and exercise behaviors, smoking tobacco and use of other substances (such as electronic cigarettes

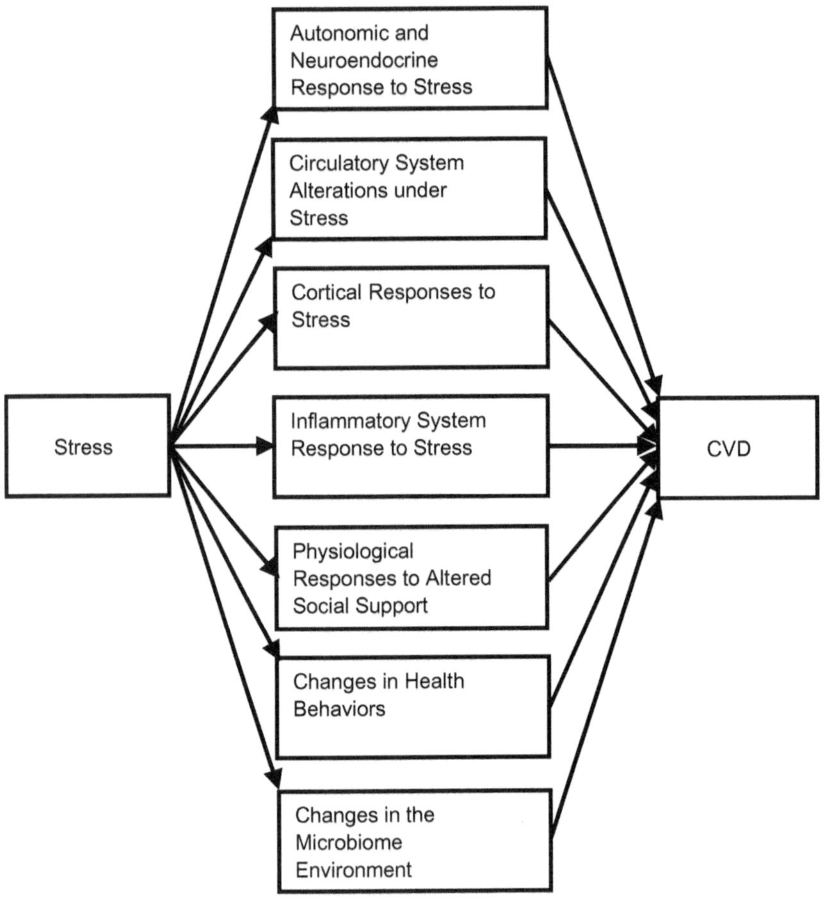

Fig. 1.2 Potential mechanisms through which exposure to stress leads to CVD onset.

with or without nicotine), and engagement with health care providers (Chapter 10), and alterations in the brain-microbiome-gut interaction in response to stress and depression (Chapter 11). Each chapter considers (a) how exposure to stress influences the proposed behavioral or physiological mechanism, and (b) whether the proposed mechanism has been shown to be associated with increased incidence of CVD. Additionally, the association of the proposed mechanism with the presence of depression and/or depressive symptoms is also introduced to gain an understanding of how depressed mood fits into the hypothesized causal pathways depicted in Fig. 1.1.

Conclusion

As the disease encumbered with the status of being the leading cause of death and disability around the world, CVD continues to present challenges to scientists who study it and health care providers who treat patients who suffer from it. Psychological factors, primarily exposure to stress and depression, hold much promise in advancing our understanding and predicting risk for CVD, but have not been uniformly adopted as risk factors by the medical community. Consequently, psychological factors are not often thoroughly assessed as often as the traditional risk factors of hypertension, hypercholesterolemia, and tobacco use. We argue that the status of psychological risk factors would be more widely adopted if we understood the biologically plausible mechanisms responsible for explaining psychosocial risk for CVD. By thoroughly examining the current state of the literature examining these mechanisms as they relate to stress and depression, we will be in a better position to focus on the most promising avenues for future research. Based on the evidence presented in the chapters that follow, we aim to achieve the ultimate purpose of expediting the development of effective strategies for reducing risk for CVD associated with both exposure to stress and the experience of depression and other negative mood states.

References

American Psychiatric Association, 2013. Diagnostic and Statistical Manual of Mental Disorders (DSM-5), fifth ed. American Psychiatric Association Publishing, Washington, DC.

Barth, J., Schumacher, M., Herrmann-Lingen, C., 2004. Depression as a risk factor for mortality in patients with coronary heart disease: a review and meta-analysis. Psychosom. Med. 66 (6), 802–813.

Cannon, W.B., 1915. Bodily Changes in Pain, Hunger, Fear and Rage: An Account of Recent Researches into the Function of Emotional Excitement. Appleton.

Carney, R.M., Freedland, K.E., Rich, M.W., Jaffe, A.S., 1995. Depression as a risk factor for cardiac events in established coronary heart disease: A review of possible mechanisms. Ann. Behav. Med. 17 (2), 142–149.

Centers for Disease Control and Prevention, National Center for Health Statistics, 2017. Underlying Cause of Death 1999-2016 on CDC WONDER Online Database, released December, 2017. Data are From the Multiple Cause of Death Files, 1999-2016, As Compiled from Data Provided by the 57 Vital Statistics Jurisdictions through the Vital Statistics Cooperative Program.

Davidson, R.J., Pizzagalli, D., Nitschke, J.B., Putnam, K., 2002. Depression: Perspectives from affective neuroscience. Annu. Rev. Psychol. 53 (1), 545–574.

Dimsdale, J.E., Hackett, T.P., Hutter, A.M., Block, P.C., Catanzano, D.M., 1978. Type A personality and extent of coronary atherosclerosis. Am. J. Cardiol. 42 (4), 583–586.

Friedman, M., Rosenman, R.H., 1959. Association of specific overt behavior pattern with blood and cardiovascular findings; blood cholesterol level, blood clotting time, incidence of arcus senilis, and clinical coronary artery disease. J. Am. Med. Assoc. 169 (12), 1286–1296.

Heron, M., Anderson, R.N., 2016. Changes in the Leading Cause of Death: Recent Patterns in Heart Disease and Cancer Mortality. NCHS data brief, no 254, National Center for Health Statistics, Hyattsville, MD.

Johnston, D.W., Cook, D.G., Shaper, A.G., 1987. Type A behaviour and ischaemic heart disease in middle aged British men. Br. Med. J. 295 (6590), 86–89.

Lett, H.S., Blumenthal, J.A., Babyak, M.A., Sherwood, A., Strauman, T., Robins, C., Newman, M.F., 2004. Depression as a risk factor for coronary artery disease: evidence, mechanisms, and treatment. Psychosom. Med. 66 (3), 305–315.

Lichtman, J.H., Froelicher, E.S., Blumenthal, J.A., Carney, R.M., Doering, L.V., Frasure-Smith, N., Freedland, K.E., Jaffe, A.S., Leifheit-Limson, E.C., Sheps, D.S., Vaccarino, V., Wulsin, L., On behalf of the American Heart Association Statistics Committee of the Council on Epidemiology and Prevention and the Council on Cardiovascular and Stroke Nursing, 2014. Depression as a risk factor for poor prognosis among patients with acute coronary syndrome: Systematic review and recommendations. Circulation 129 (12), 1350–1369.

Mendis, S., Puska, P., Norrving, B. (Eds.), 2011. Global Atlas on Cardiovascular Disease Prevention and Control. World Health Organization, Geneva.

Rosenman, R.H., 1978. The interview method of assessment of the coronary-prone behavior pattern. In: Dembroski, T.M., Weiss, S., Shields, J., Haynes, S.G., Feinleib, M. (Eds.), Coronary-Prone Behavior. Springer-Verlag, New York, pp. 55–69.

Rugulies, R., 2002. Depression as a predictor for coronary heart disease: a review and meta-analysis. Am. J. Prev. Med. 23 (1), 51–61.

Selye, H., 1955. Stress and disease. Science 122 (3171), 625–631.

Shekelle, R.B., Hulley, S.B., Neaton, J.D., Billings, J., Borhani, N.O., Gerace, T.A., Jacobs, D.R., Lasser, N.L., Mittlemark, M.B., Stamler, J., For the MRFIT Research group, 1985. The MRFIT behavior pattern study. II. Type A behaviour and incidence of coronary heart disease. Am. J. Epidemiol. 122 (4), 559–570.

Sherwood, A., Hinderliter, A.L., Watkins, L.L., Waugh, R.A., Blumenthal, J.A., 2005. Impaired endothelial function in coronary heart disease patients with major depression. Am. J. Cardiol. 46 (4), 656–659.

Skala, J.A., Freedland, K.E., Carney, R.M., 2006. Coronary heart disease and depression: a review of recent mechanistic research. Can. J. Psychiatr. 51 (12), 738–745.

Thase, M.E., Hahn, C., Berton, O., 2015. Neurobiological aspects of depression. In: Gotlib, I.H., Hammen, C.L. (Eds.), Handbook of Depression, third ed. Guilford Press, New York, pp. 187–217.

SECTION 1

Evidence linking stress with depression and cardiovascular disease

CHAPTER 2

Epidemiological evidence linking stress and depression with CVD

Thomas Rutledge[a,b], **Hilary M. Gould**[a]
[a]Psychology Service (116B), VA San Diego Healthcare System, San Diego, CA, United States
[b]UC San Diego Department of Psychiatry, La Jolla, CA, United States

Contents

The epidemiological literature describing associations between stress, depression, and cardiovascular disease (CVD) is extensive, encompassing hundreds of scientific articles, involving participants from dozens of countries, and spanning more than a half century of research (Dimsdale, 2008; Hemingway and Marmot, 1999). Given the sheer size of this literature, and the continued proliferation of research questions and methodologies, high quality systematic and narrative reviews have become an invaluable resource for scientists in the field. The most impactful reviews of epidemiological relationships involving stress, depression, and CVD easily accumulate citation numbers exceeding 1000 (e.g., Musselman et al., 1998; Barth et al., 2004). New topic reviews appear regularly (e.g., Carney and Freedland, 2017; Huffman et al., 2013; Steptoe and Kivimaki, 2013), and the American Psychological Association has published two versions of the cardiac psychology volume, Heart and Mind (in 1996 and 2012, respectively), containing epidemiological research summaries in multiple areas of the field as resources for clinicians and researchers. This chapter will integrate findings from these many prior reviews and resources, in addition to offering novel themes and suggestions for future epidemiological research based on the most recent science.

Cardiovascular Implications of Stress and Depression
https://doi.org/10.1016/B978-0-12-815015-3.00002-7

Methodological considerations

In discussing epidemiological findings from the stress, depression, and CVD literature, it is important to initially address several of the most noteworthy methodological concerns that raise interpretive challenges. The first is that there remains no clear consensus regarding the optimal measurement format of either stress or depression. Variability in measurement is the rule in these literatures rather than the exception. Stress is not recognized as a psychiatric condition itself, but is considered a symptom of or a contributor to many psychiatric conditions. Lacking a diagnostic standard (posttraumatic stress disorder [PTSD] representing arguably the psychiatric disorder most closely associated with stressful life experiences), scientists define stress in epidemiological studies multifariously using measures of work stress, home stress, financial stress, posttraumatic stress, and using subjective (e.g., harassment or loneliness) and objective (e.g., natural disasters, loss of loved ones) stressful life events. Self-report measures of stress in the CVD literature are usually brief—sometimes as short as a single item—and this feature contrasts sharply with measures of depression that sometimes comprise dozens of items or that require lengthy diagnostic interviews to identify.

Depression measures in CVD research are arguably even more diverse than those found in the stress and CVD literature, including multiple diagnostic interviews, brief measures of depression treatment status (e.g., binary questions assessing current or prior treatment for depression), and more than a dozen validated depression questionnaires (Davidson, 2012). The selection of depression measurement format forces methodological tradeoffs between practicality (e.g., brief and easy to administer questionnaires) and validity (diagnostic interviews representing the gold-standard measurement for depressive disorders). Depression questionnaires differ not only in length, but also in content (e.g., measures such as the Beck Depression Inventory [BDI] that include substantial "somatic" symptom content, versus measures such as the Hospital Anxiety and Depression Scale [HADS] that minimize somatic symptom content).

The result of this measurement heterogeneity is that readers of the epidemiological science reporting stress and depression associations with CVD should consider that many studies may be measuring these factors dissimilarly enough to reduce the congruity of research findings. In practice, researchers have observed methodologies consistent enough across different stress and depression research to aggregate them in most meta-analyses and systematic reviews, with perhaps two periodic modifications: (1) Research comparing depression measured as a psychiatric diagnosis (i.e., meeting criteria for major depressive disorder) consistently show stronger relationships with CVD relative to questionnaire measures of depressive symptoms and are often described separately in quantitative reviews (e.g., Barth et al., 2004; Rugulies, 2002). Whether diagnostic depression is more robustly related to CVD relative to depressive symptom questionnaires because the diagnosis captures greater depression severity, greater duration of symptoms, or both, is unknown; and (2) There is some evidence from epidemiological studies of depression and CVD that separating somatic and cognitive symptom content can improve the strength of depression relationships with CVD (Carney and Freedland, 2012).

A second challenge from the epidemiological literature concerns interpreting the clinical significance of observational relationships between stress, depression, and CVD. Epidemiological designs lack the randomization feature inherent to randomized controlled trials and do not possess the degree of environmental control and standardization present in laboratory research designs. These latter qualities increase the internal validity of research by minimizing the risk of pre-existing group differences and regulating the level of exposure (e.g., exposure to a controlled laboratory stressor vs self-reported stressful life events). Beyond their advantages for inhering causal relationships compared to epidemiological design, these same design features lend themselves more directly to statistics such as numbers needed to treat and to clearer interpretations of the temporal direction of relationships useful for appraising the real-world implications of research results. Epidemiological studies of stress, depression, and CVD, in contrast, typically rely heavily on risk ratios and hazard ratios that require additional information (such as base rates of depression, stress, or CVD events) to properly interpret. The same hazard ratio, for example, can have markedly different clinical implications for events that vary in frequency or severity. For example, a 1.5 risk ratio for stress or depression with a more common CVD outcome could translate to many additional events in a large hospital—and thereby create considerable medical and financial implications—whereas the same 1.5 risk ratio with a more rarely occurring CVD outcome might translate into only a small number of additional CVD outcomes in the same setting and possibly have less clinical significance. Regarding outcome severity, the same risk or hazard ratio for fatal CVD events has greater clinical significance than for non-fatal CVD events. These distinctions are important to understanding observational studies of stress and depression with CVD because most studies mix categories of CVD events with differing frequencies and severities into their primary outcomes to improve statistical power. Observational studies such as INTERHEART (Yusuf et al., 2004) represented an exception to the pattern of reliance on odds, risk, and hazard ratios in this literature. INTERHEART investigators also calculated attributable risk values for depression as a predictor of myocardial infarction that could be compared to the attributable risk values for standard CVD risk factors.

Of final note regarding the challenge of interpreting clinical significance, the meaning of even the same risk or hazard ratio value related to stress or depression has changed over time in the epidemiological CVD literature as treatment advances have reduced rates of CVD events. As overall CVD event rates reduce, the same risk or hazard ratio between more stressed or more depressed vs normal patients reflects increasingly smaller differences in actual event numbers between the groups, and thereby potentially smaller clinical significance. Because lower CVD event rates also reduce statistical power, one of the undesirable methodological results is that researchers are increasingly compelled to combine across categories of CVD outcomes in their studies to obtain statistically significant findings.

Two final challenges to interpreting the epidemiological literature regarding stress, depression, and CVD relationships are: (1) difficulties disentangling stress and depression associations with CVD from overlapping psychological characteristics; and (2) the absence of repeated measurements to evaluate the stability of stress and depression over study follow-up intervals. Many reviews of depression and CVD, for example, also discuss relationships between CVD and psychological factors such as anger, anxiety, and social isolation that frequently co-occur and correlate highly with measures of stress and depression (Hare et al., 2014). Researchers have adopted different approaches to this challenge, most often attempting to statistically control for related psychological factors, sometimes treating them as independent CVD predictors in separate statistical models, and sometimes exploring ways to statistically combine them into omnibus psychological predictors of CVD outcomes (e.g., Whittaker et al., 2012). Unfortunately, the most widespread practice in the epidemiological literature regarding this challenge is to not measure or report on related psychological characteristics at all; this practice artificially simplifies the task of studying stress, depression and CVD relationships at the expense of construct validity and interpretation clarity.

The general absence of repeated measures of stress and depression is a problem of particular concern for epidemiological studies of CVD because these studies frequently involve years of follow-up over which psychological factors can (and probably do) change. Clinical observation studies show, for example, that measures of depression show substantial variability over time (Whiteford et al., 2013). Without confidence that stress or depression characteristics remain stable over time, it becomes more difficult to interpret prospective relationships with CVD; the longer the follow-up period, the greater this potential concern affects interpretation. Table 2.1 summarizes the above four methodological limitations for epidemiological studies of stress, depression, and CVD,

Table 2.1 Primary methodological limitations of epidemiological studies of stress and depression with cardiovascular disease (CVD) outcomes and potential remedies.

Limitation and influence	Methodological options
1. Variable measurement formats for stress and depression	a. Include multiple measurement formats to advance understanding of measurement differences with CVD outcomes and permit direct empirical comparisons of formats b. Standardize the reporting of the strengths and weaknesses of each measurement format to improve results interpretation
2. Difficulties evaluating clinical significance and the direction of relationships between stress, depression, and CVD	a. Standardize the reporting of CVD event base rates when reporting odds, risk, or hazard ratios b. Consider the reporting of effect sizes to complement odds, risk, and hazard ratio reporting (e.g., number needed to treat values can be calculated using event rates from depressed and non-depressed groups [Zapletal et al., 1996])
3. Overlap between stress and depression, and between stress, depression, and other important psychological factors such as anger, anxiety, and social isolation	a. Reliably include other major psychosocial factors associated with CVD in stress and depression studies (i.e., particularly anger/hostility, anxiety, and social isolation) b. Consider multiple statistical approaches within epidemiological studies, such as examining bivariate stress and depression relationships with CVD in addition to multivariate or data reduction (e.g., factor analyses) methods for comparison
4. Absence of stress and depression measures over follow-up to demonstrate their duration or stability over time	a. Include stress and depression measures at one or more follow-up intervals to provide statistical basis for evaluating reliability b. Consider use of technology formats (internet, smartphones) to facilitate tracking of psychosocial changes over time

and offers recommendations for future research that could reduce the adversity of their impact.

Stress and cardiovascular disease

Serendipity is a common source of discovery in many scientific disciplines. Although the field of medicine is replete with examples of serendipitous factors contributing to the discovery of important treatments and causes of disease, our current understanding of stress and CVD also benefited from opportunistic sources of observation. Daily variations in cardiovascular deaths first noticed in epidemiological data, for example, fueled interest in and later provided empirical support for work stress associations with CVD by revealing relatively higher CVD rates on Mondays than other days of the week (e.g., Evans et al., 2000). These daily variations in CVD events were later observed to be even more pronounced following changes in daylight savings time (Janszky and Ljung, 2008) in CVD data from the Swedish registry, offering a possible example of stress resulting from sleep deprivation. Individual case studies of mysterious deaths that accumulated in the early-to-mid-20th century were the source of what came to be known as "voodoo death," describing the phenomenon of intense psychological experiences potentially resulting in sudden death, usually from CVD causes (Cannon, 1942). Consistent increases in CVD events following natural disasters represent among the most compelling arguments for the adverse CVD effects of stress, yet the body of knowledge we possess regarding this relationship was produced almost entirely by scientists taking skillful advantage of uncontrolled data sources to identify the associations (Dimsdale, 2008). Finally, observations that marital partners were more likely to die from CVD causes shortly after the deaths of their spouses may have originated from the statistician, Karl Pearson, who noticed unusually close patterns of dates on the tombstones of husbands and wives (Jauhar, 2018). This form of stress-CVD relationship is now well-documented in epidemiological research (Elwert and Christakis, 2008) and described more generally as an example of takotsubo cardiomyopathy (Akashi et al., 2008). In short, epidemiological data represent the historical foundation for current stress-CVD research, even as the field has evolved to include controlled laboratory stress models, research focus on identifying potential stress-CVD mechanisms, and intervention studies examining the effects of stress treatments.

In recent decades, the results from numerous prospective and case-control studies have supported relationships between different forms of psychological stress with CVD. Table 2.2 summarizes some of the specific forms of stress associated with CVD in epidemiological studies.

Table 2.2 Forms of psychological stress potentially associated with cardiovascular disease (CVD) from epidemiological research.

Form of stress	Definition and examples
Work stress	Psychological stress resulting from employment circumstances. Common examples include job strain and burnout
Financial stress	Psychological stress related to income or financial status
Marital stress	Psychological stress resulting from marital or romantic partner relationships
Environmental stress	Psychological stress related to environmental events. Natural disasters are the most common types of environmental stressors associated with CVD
Posttraumatic stress	Psychological stress associated with a traumatic event. Traumatic events include events ranging from assaults, accidents, and major health events, with the definition requiring that the trauma was life threatening or very severe
Stressful life events	Psychological stress related to events generally deemed stressful. This category can include a range of stressors from traumatic events to daily hassles such as traffic
Home stress	Psychological stress related to the home environment. This may include marital stress, family stress, caregiver stress, and socioeconomic stress, among others
Socioeconomic stress (SES)	Includes psychological, physical, and social sources of stress related to SES. Studies commonly define SES using measures of income, employment, and education. Adult and childhood SES status may be differentiated
Childhood stress	Psychological stress associated with childhood events. These may include traumatic events, childhood SES, and abuse, as common examples
Caregiver stress	Psychological stress resulting from involvement in caregiving relationships. Examples include caregiving for family members or friends with dementia, HIV, and cancer
Stress from abuse	Psychological stress reported from experience of abuse. This category can include adult or childhood events as well as physical, sexual, and emotional forms of abuse
Interpersonal stress	Psychological stress related to general interpersonal factors. In addition to marital stress, this category includes social isolation, loneliness, poor social support, and grief as examples
Stress from racism or sexism	Psychological stress related to various forms of discrimination. The frequency and intensity of racism and sexism are the most common examples
Emotional stress	Psychological stress, more broadly, related to non-psychiatric emotions. Examples include anger, hostility, and type A behavior pattern characteristics
Acute and chronic stress	Psychological stress measured in relation to short-term (acute) events, such as natural disasters or grief, versus long-term (chronic) events such as job stress or marital stress

In a detailed 2013 review on stress and CVD (Steptoe and Kivimaki, 2013), the authors summarize a large and diverse literature linking various forms of stress to acute CVD events and CVD development. For example, the authors reported meta-analytic relationships between social isolation and CAD risk (relative risk [RR] = 1.5, 95% CI = 1.2–1.9), between work stress and CAD (RR = 1.3, 95% CI = 1.2–1.5), and between stress and acute coronary syndrome (RR = 2.5, 95% CI = 1.8–3.5). The INTERHEART study (Yusuf et al., 2004) employed a case-control design to compare more than 10,000 patients suffering a first myocardial infarction to age and sex-matched control patients without an MI history. The INTERHEART protocol included multiple measures of stress (financial stress, ranging from little or none to severe; work stress, home stress, and general stress [either work or home] quantified in terms of frequency from none to permanent; and the number of stressful life events in the past year). There were significant dose-response relationships with MI risk across each of the stress measures, and these associations remained stable when examined across subgroups of both men and women. The relationships between factors such as job strain, home stress, and financial stress with CVD may partly explain the robust epidemiological relationship between lower socioeconomic status (SES) and higher CVD and total mortality risk (e.g., Kaplan and Keil, 1993). With SES most commonly defined in some combination of occupation, income, and education, it likely captures multiple forms of psychological stress simultaneously. Lower SES, however, also likely reflects physical (e.g., violence) and environmental stressors (e.g., pollution); for this reason, SES should be interpreted as a form of psychological stress with caution regarding its status as a predictor of CVD.

Relationships between natural disasters (a potent source of psychological stress in addition to the physical hazards) and cardiovascular events have been summarized in several reviews (e.g., Dimsdale, 2008; Steptoe and Kivimaki, 2013). Because of their capricious nature and often life-threatening severity, natural disasters rank among the most severe sources of stress and would be expected to provoke a rapid increase in CVD events if stress was a causal mechanism. Observational data from multiple populations affected by natural disasters, in fact, confirms precisely this expected pattern. Beyond natural disasters, other naturalistic sources of stress also show associations with CVD, albeit sometimes in divergent patterns. For example, CVD events rise short-term not only in response to major stressors—natural disasters such as earthquakes or following deaths of loved ones—but also in response to comparatively "minor" stressful events such as the results of popular sports contests (e.g., major American football and European soccer matches [Kloner et al., 2011; Wilbert-Lampen et al., 2008]). Similarly, epidemiological research indicates that stressful events show associations with both acute increases in CVD events (e.g., in the immediate days and weeks following a natural disaster) and long-term increases in CVD events (e.g., years or even decades later among survivors of war or adult victims of childhood abuse); the latter suggests that sufficiently stressful events may affect CVD risk long after the stressor has ended. The associations between acute episodes of intense stress and long-term CVD risk should be distinguished from examples of prolonged or chronic stress (e.g., low SES, job or marital stress that may be present for long periods) predicting increased CVD risk in epidemiological research (Edmondson and von Känel, 2017; Steptoe and Kivimaki, 2013); the latter forms of stress may involve different biopsychosocial mechanisms to explain CVD associations.

One of the newest and fastest growing areas of stress research in relation to CVD risk is posttraumatic stress disorder (PTSD; Gander and von Känel, 2006). According to the most recent version of the Diagnostic and Statistical Manual of Mental-Disorders-5 (DSM-5), PTSD contains both an objective component (the Criterion A standards of experiencing or witnessing an event risking death, serious injury, or sexual violation) and a subjective component (Criterion's B-E involving different cognitive, affective, and behavioral symptoms). With the prevalence of PTSD being high among certain portions of the population—such as U.S. military Veterans, among whom 15–20% report PTSD—relationships between PTSD and CVD could have important research and clinical implications due to the millions of persons at risk (Seal et al., 2009). An increasing volume of epidemiological research suggests a greater risk for CVD among those with PTSD. Among the multiple cohort and case-control studies available from the last decade, several are noteworthy regarding methodological strengths. For example, Vaccarino et al. (2013) examined the relationship between PTSD and CVD risk using discordant twins (i.e., twin pairs where one twin had PTSD and the other did not) from the Vietnam Era Registry. The incidence of CHD was more than twice as high among the twins with PTSD (odds ratio = 2.2, 95% CI = 1.2–4.1). Impressively, this relationship was observed *after* adjusting for lifestyle factors, established CHD risk factors, and depression.

Probably the largest epidemiological study of PTSD and CVD risk was published by Cohen et al. (2009) in the *Journal of the American Medical Association*. The authors used a VA electronic medical record database to assess PTSD relationships with CVD risk factors among more than 300,000 Veterans. Relative to those without mental health diagnoses, Veterans with PTSD with or without other mental health conditions had significantly higher rates of hypertension, smoking, diabetes, obesity, and hyperlipidemia after adjusting for demographic factors.

A third recent PTSD-CVD epidemiological study noteworthy for methodological characteristics was the report from the Nurses' Health Study II including nearly 50,000 women. In a field otherwise reporting PTSD analyses drawn from predominantly male samples, Sumner et al. (2016) reported a twofold increased relationship between PTSD (defined as either PTSD symptom severity or trauma exposure history) and venous thromboembolism events that were robust to covariate adjustment.

Another distinctive area of PTSD research related to CVD risk is comprised of research suggesting that PTSD may often be a consequence of CVD (e.g., PTSD resulting from a heart attack) rather than a precursor. Spendler and Pedersen (2005) authored one of the most detailed reviews of this epidemiological literature, summarizing results from 18 independent studies, with the most consistent evidence for PTSD developing in the aftermath of CVD among patients experiencing myocardial infarction (i.e., the evidence for CVD-induced trauma symptoms is weaker for other forms of CVD relative to MI). This evidence of bidirectional relationships between PTSD and CVD parallels similar findings observed for depression that may also both precede and follow CVD events. Edmondson and von Känel (2017) provide one of the highest quality and up to date reviews of the PTSD-CVD field, including epidemiological results from prior meta-analyses and systematic reviews and describing some of the pending results from ongoing studies in the area likely to yield important findings in the years ahead.

If PTSD is among the newest and most rapidly growing areas of epidemiological stress-CVD research, then stress related to the presence, size, nature, and quality of interpersonal relationships is perhaps the most senior predictor of CVD and overall mortality risk. For example, Holt-Lunstad et al. (2010) identified 148 prospective studies meeting criteria for their review on social relationships and mortality, observing a 50% greater risk of premature mortality among participants reporting relatively weaker interpersonal relationships. Similarly, Schulz and Beach's (1999) prospective results reported from the Caregiver Health Effects Study data were the basis for a now burgeoning field of research examining associations between caregiver strain, CVD and other health consequences (e.g., Adelman et al.,

2014). Whereas stressful or limited numbers of social relationships are associated with increased CVD risk, the corresponding pattern is also observed in epidemiological research; a larger social network and higher quality relationships (e.g., greater social support or marital quality) predict lower CVD risk, perhaps by reducing stress (Steptoe and Kivimaki, 2013).

Depression and cardiovascular disease

The study of epidemiological relationships between depression and CVD gained widespread momentum following a 1987 review of psychological predictors of coronary heart disease (CHD; Booth-Kewley and Friedman, 1987). Booth-Kewley and Friedman's review described a surprisingly strong and consistent depression-CHD relationship—across a then relatively small literature of observational studies—that raised scientific interest in depression and CVD at an opportune time; during the same period, research concerning the Type A behavior pattern as a CVD predictor was in decline (Kuper et al., 2002).

In comparison to the primarily cross-sectional and lower-quality cohort study designs available at the time of this 1987 review, Musselman et al. (1998) summarized epidemiological findings in 1998 on a rapidly emerging depression-CVD field that had grown measurably in size and methodological rigor. Even narrowed only to prospective studies measuring depression with diagnostic interviews and validated depressive symptom questionnaires, Musselman's review reported more than a dozen studies meeting criteria, among which the majority supported a depression-CVD relationship. From the multiple important epidemiological papers published during this period, Frasure-Smith et al. (1993) findings reported in the *Journal of the American Medical Association* were arguably the most impactful on the field. This paper described a more than fourfold increase in post-MI mortality over 6-months of follow-up among patients with major depressive disorder, as diagnosed using a structured clinical interview.

Research concerning epidemiological relationships between depression and CVD had also diversified in the 1990's, with separate areas of research focused on depression as a predictor of incident CVD, and assessing depression as a predictor of CVD events and progression among patients with established CVD. Beyond the research implications of these two areas, there were equally important clinical implications. For example, establishing depression and as independent predictor of CVD development could support the addition of depression to the list of established CVD risk factors; whereas an independent association between depression and CVD progression could support measures such as screening for depression among patients with CVD and incorporating depression treatment into cardiac rehabilitation and other CVD treatment protocols. Whether evidence would demonstrate that depression treatment improved objective CVD outcomes was the empirical question that then remained unanswered towards these latter goals. In 1996, ENRICHD—the largest clinical trial to date testing depression treatment among patients with CVD—had begun recruitment to yield an answer to this question (Berkman et al., 2003). The rationale for the ENRICHD study was based heavily upon the compelling epidemiological depression-CVD literature. Meanwhile, the number of prospective studies supporting the depression-CVD relationship continued to increase.

After several additional years of continuing growth, the journal, Psychosomatic Medicine, published two meta-analytic reviews (Barth et al., 2004; van Melle et al., 2004) in 2004 analyzing depression relationships with CVD and overall mortality among patients with established CHD. Each meta-analysis included data from at least 20 prospective cohort studies in their analyses. For example, van Melle's review (2004) included 22 observational studies, showing that depression present among patients' post-myocardial infarction (MI) was associated with a 2.6-fold greater odds of cardiac mortality over follow-up (95% CI = 1.8–3.8) relative to post-MI patients without depression. These latter reviews, focused on depression-CVD outcomes in patients *with existing CVD*, had followed a separate 2002 meta-analysis (Rugulies 2002) quantifying depression as a predictor of CHD development among CVD-free adults based on data from 11 prospective studies (showing that the relative risk for CHD development was 1.6 [95% CI = 1.3–2.1] among depressed vs non-depressed participants). Despite the separate focus of these quantitative reviews, each reported statistically significant relationships between depression and CVD development and outcomes that were robust to adjustment for demographic and biomedical covariates, along with evidence of a dose-response relationship such as that "clinical depression" defined by diagnostic standards or clinically significant elevations was a stronger CVD predictor relative to less severe depressive symptoms.

With the proliferation of epidemiological studies during this period—employing many different measures of depression—there emerged a growing consensus that rates of depressive disorders approached 20% among patients with CVD (more than twofold the rates observed among community samples of adults) and that 30–50% of patients with CVD showed evidence of "subclinical" depression (i.e., elevated depressive symptoms not meeting diagnostic criteria; Davidson, 2012). During the same time frame, the scope of epidemiological depression-CVD research expanded beyond CHD to include stroke (e.g., Everson et al., 1998) and congestive heart failure (e.g., Rutledge et al., 2006), multinational populations (Yusuf et al., 2004), healthcare costs (e.g., Sullivan et al., 2002), and to explore gender differences as changes in NIH guidelines shifted the sex composition of CVD studies to enable a stronger focus on epidemiological predictors of CVD in women (e.g., Merz et al., 1999). Based on the consistency of relationships and magnitude of associations shown in several high profile epidemiological studies of depression and CVD, there was also growing momentum to include depression among the list of major CVD risk factors. Findings from the 2004 INTERHEART study (Yusuf et al., 2004), for example, combined brief measures of depression and stress into a psychosocial factor in their analyses, finding that the attributable risk for psychosocial factors (32.5%) exceeded the values of established CVD risk factors such as hypertension (17.9%), diabetes (9.9%), and regular physical activity (12.2%) for predicting MI events in their case-control design.

The year 2008 marked the next major development related to accumulating epidemiological depression-CVD research with the publication of the American Heart Association endorsement for routine depression screening among CHD patients (Lichtman et al., 2008). These recommendations drew from more than 60 prospective studies of depression and CHD available at that time and from the conclusions of more than 100 then-existing narrative and quantitative reviews showing elevated rates of depression among CHD patients and consistent relationships between depression and CHD development, events, and mortality. Initially, the AHA recommendations seemed to forecast an even larger role for depression and mental health in the study and treatment of CVD. However, the endorsement quickly became controversial due to mixed statistical evidence regarding the value of routine depression screening (e.g., Thombs et al., 2008) and to the failure of early clinical trials to demonstrate event reduction from depression treatment among patients with CVD (Berkman et al., 2003).

Historically, the decade from 2000 to 2010 observed a turning point in the interpretation of depression-CVD relationships that continued to be reported in new epidemiological research. In contrast to the momentum from prospective research in the initial years of the 21st century that, among other products, fueled NIH support for ambitious clinical trials such as ENRICHD (Berkman et al., 2003) and SADHART (Glassman et al., 2002), the series of negative findings from these first era of depression-CVD clinical trials had resulted in a more cautious and complex perspective on the role of depression in CVD. For example, epidemiological research on depression and CVD from the mid-2000's forward began to focus on depression subtypes and specific symptoms (Davidson et al., 2005) and increasingly to consider the depression-CVD relationship as bi- or multi-directional (e.g., de Jonge and Roest, 2012). Similarly, studies focused on biopsychosocial mechanisms of depression-CVD relationships were surpassing simpler epidemiological designs assessing depression-CVD event relationships as the most important focus of depression-CVD research (e.g., Joynt et al., 2003). By the release of the next official AHA statement concerning depression and CVD in 2014, it was now a greater subject of debate whether depression merited status as a major risk factor for CVD (Lichtman et al., 2014) despite, objectively, an even larger body of observational data available to support depression-CVD associations relative to 2008. Established CVD risk factors (e.g., smoking, hypercholesterolemia, hypertension), after all, also showed the prospective relationships with CVD that were observed with depression; however, research also demonstrated that effective interventions for these established CVD risk factors reduced CVD risk and showed more convincing evidence regarding pathophysiological mechanisms to explain their link to CVD development and progression. Fig. 2.1 offers a chronological overview of some major epidemiological studies and reviews from the 1980's to present.

The most recent decade of epidemiological research addressing depression-CVD relationships may be viewed as undergoing an important course correction that will improve the quality of the science going forward. Some of the largest observational studies to date supporting associations between depression and CVD, for example, have been published since 2010 (e.g., Daskalopoulou et al., 2016). Similarly, some of the most recent depression-CVD reviews (e.g., Hare et al., 2014) offered more tempered descriptions of depression as a predictor of or risk marker for CVD (rather than risk factor), presenting more sophisticated biopsychosocial models to explain complex depression-CVD relationships, and discussing depression-CVD pathways through mechanisms such as psychiatric comorbidities with overlapping conditions such as anxiety (Huffman et al., 2013). Other recent reviews propose more nuanced epidemiological questions concerning depression and CVD, such as whether the presence of depression is more strongly associated with CVD outcomes when it precedes versus follows an acute coronary event (e.g., a 2012 meta-analysis showed stronger support for post-CVD event depression as an outcome predictor; Leung

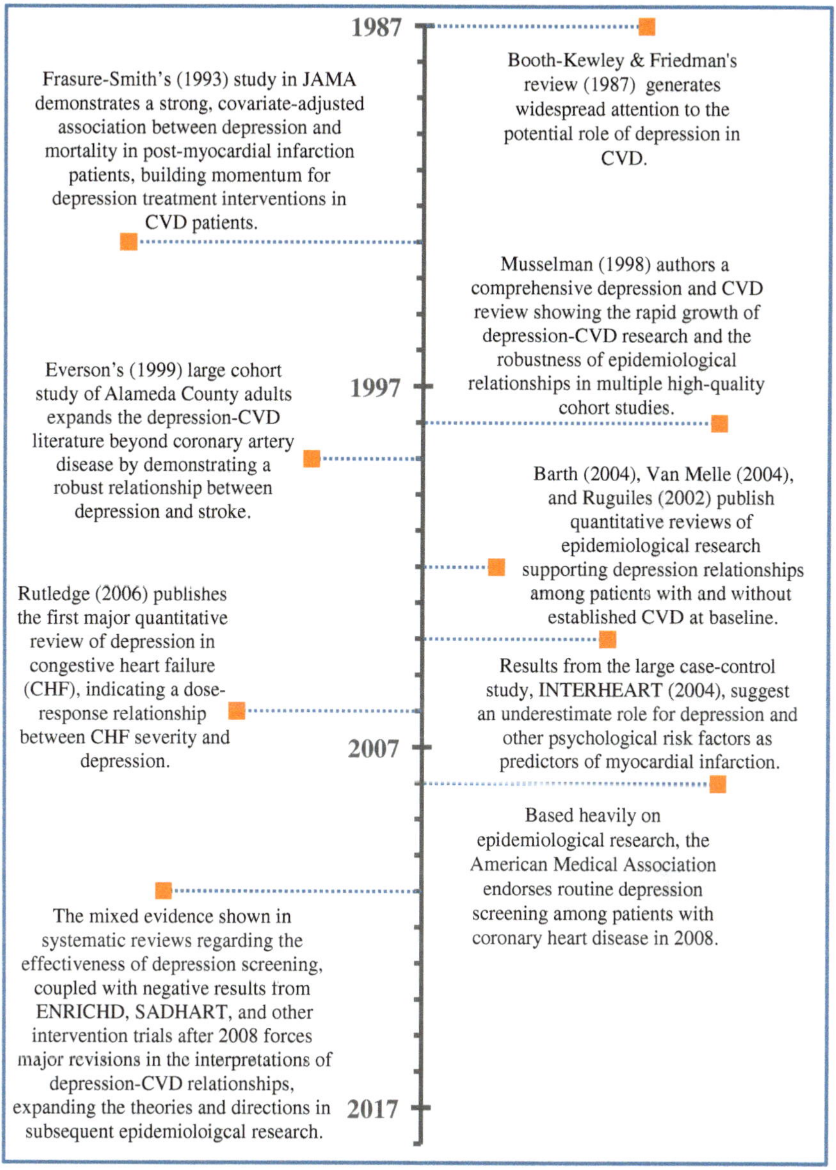

Fig. 2.1 An abbreviated timeline of major developments in epidemiological research assessing depression and cardiovascular disease (CVD).

et al., 2012). Lastly, as described in Carney and Freedland's (2017) review of depression and CHD, the consistent and still growing epidemiological evidence favoring a relationship between depression and CHD incidence and progression remains important in supporting continued research concerning topics such as: (a) potential higher risk depression subtypes that may more reliably predict CVD outcomes; and (b) the development of more effective depression treatments for CVD patients to deploy in the next phase of clinical trials.

In summary, epidemiological research evaluating depression and CVD relationships now spans half a century. After the promising initial decades of cross-sectional and prospective studies suggested a potentially direct depression-CVD relationship, more recent research reflects an association that is multidirectional and considerably more complex. As the epidemiological literature concerning depression and CVD continues to mature in the decades ahead, it is likely to evolve in parallel with research advances in the areas of depression mechanisms and depression treatment. After serving as the initial catalyst for depression and CVD research, epidemiological designs are likely to serve a valuable secondary role in the future, offering cost-effective methods of: (a) identifying prospective depression-CVD mechanisms that could be verified in controlled laboratory studies; (b) assessing long-term relationships of depression treatments with CVD events, costs, quality of life, and other clinical targets; (c) serving as an important source of data for the growing genetic literature on depression and CVD (Hare et al., 2014) and; (d) helping to identify relationships between depression subtypes or specific depressive symptoms that could be integrated into treatment studies (e.g., Davidson et al., 2005).

Conclusion

After decades of epidemiological research, stress and depression remain reliable predictors of CVD. Although each area of study has improved in methodological rigor over time, data from these higher quality cohort and case-control designs continue to support CVD relationships. For example, even though stress and depression show consistent associations with major CVD risk factors, recent systematic and quantitative reviews suggest that relationships with CVD remain even after statistical adjustment for both established CVD risk factors and demographic variables. The robustness of these epidemiological literatures, however, should not be interpreted to suggest a static understanding of stress and depression associations with CVD. In contrast, stress and depression research from the past two decades has compelled scientists to adopt an evolving model for how these psychological factors may contribute to CVD risk. This evolution is reflected in the current thinking of epidemiological researchers, for example, in the now well-established bidirectional relationships between stress, depression, and CVD; in the increasingly complex biological, behavioral, and psychosocial mechanisms believed to account for relationships with CVD that are important to include in current and future cohort studies to advance the understanding of stress, depression, and CVD associations; and in the increasing attention to genetic and other individual level factors that epidemiological designs can capture to contribute to "precision medicine" era efforts to clarify individual-level risks and identify individual-level targets to be potentially explored in interventional research.

References

Adelman, R.D., Tmanova, L.L., Delgado, D., Dion, S., Lachs, M.S., 2014. Caregiver burden: a clinical review. J. Am. Med. Assoc. 311 (10), 1052–1060.

Akashi, Y.J., Goldstein, D.S., Barbaro, G., Ueyama, T., 2008. Takotsubo cardiomyopathy: a new form of acute, reversible heart failure. Circulation 118, 2754–2762.

American Psychological Association, 1996. In: Allen, R., Fisher, J. (Eds.), Heart and Mind: The Practice of Cardiac Psychology, second ed. American Psychological Association, Washington, DC.

American Psychological Association, 2012. In: Allen, R., Fisher, J. (Eds.), Heart and Mind: The Practice of Cardiac Psychology, second ed. American Psychological Association, Washington, DC.

Barth, J., Schumacher, M., Herrmann-Lingen, C., 2004. Depression as a risk factor for mortality in patients with coronary heart disease: a review and meta-analysis. Psychosom. Med. 66 (6), 802–813.

Berkman, L.F., Blumenthal, J., Burg, M., Carney, R.M., Catellier, D., Cowan, M.J., Czajkowski, S.M., DeBusk, R., Hosking, J., Jaffe, A., Kaufmann, P.G., Mitchell, P., Norman, J., Powell, L.H., Raczynski, J.M., Schneiderman, N., Enhancing Recovery in Coronary Heart Disease Patients Investigators (ENRICHD), 2003. Effects of treating depression and low perceived social support on clinical events after myocardial infarction: The enhancing recovery in coronary heart disease patients (ENRICHD) randomized trial. JAMA 289 (23), 3106–3116.

Booth-Kewley, S., Friedman, H.S., 1987. Psychological predictors of heart disease: a quantitative review. Psychol. Bull. 101, 343–362.

Cannon, W.B., 1942. Voodoo death. Am. Anthropol. 44 (2), 169–181.

Carney, R.M., Freedland, K.E., 2012. Are somatic symptoms of depression better predictors of cardiac events than cognitive symptoms in coronary heart disease? Psychosom. Med. 74 (1), 33–38.

Carney, R.M., Freedland, K.E., 2017. Depression and coronary artery disease. Nat. Rev. Cardiol. 14, 145–155.

Cohen, B.E., Marmar, C., Ren, L., 2009. Association of cardiovascular risk factors with mental health diagnoses in Iraq and Afghanistan war veterans using VA health care. J. Am. Med. Assoc. 302 (5), 489–492.

Daskalopoulou, M., George, J., Walters, K., Osborn, D.P., Batty, G.D., Stogiannis, D., Rapsomaniki, E., Pujades-Rodriguez, M., Denaxas, S., Udumyan, R., Kivimaki, M., Hemingway, H., 2016. Depression as a risk factor for the initial presentation of twelve cardiac, cerebrovascular, and peripheral arterial diseases: data linkage study of 1.9 million women and men. PLoS One 11, e0153838.

Davidson, K.W., 2012. Depression and coronary artery disease. ISRN Cardiol. 2012, 743813.

Davidson, K.W., Rieckmann, N., Rapp, M.A., 2005. Definitions and distinctions among depressive syndromes and symptoms: Implications for a better understanding of the depression-cardiovascular disease association. Psychosom. Med. 67 (1), s6–s9.

de Jonge, P., Roest, A.M., 2012. Depression and cardiovascular disease: the end of simple models. Br. J. Psychiatry 201 (5), 337–338.

Dimsdale, J., 2008. Psychological stress and cardiovascular disease. J. Am. Coll. Cardiol. 51, 1237–1246.

Edmondson, D., von Känel, R., 2017. Post-traumatic stress disorder and cardiovascular disease. Lancet Psychiatry 4 (4), 320–329.

Elwert, F., Christakis, N.A., 2008. The effect of widowhood on mortality by the causes of death of both spouses. Am. J. Public Health 98 (11), 2092–2098.

Evans, C., Chalmers, J., Capewell, S., Redpath, A., Finlayson, A., Boyd, J., Pell, J., McMurray, J., Macintyre, K., Graham, L., 2000. "I don't like Mondays"-day of the week of coronary heart disease deaths in Scotland: study of routinely collected data. Br. Med. J. 320 (7229), 218–219.

Everson, S.A., Roberts, R.E., Goldberg, D.E., Kaplan, G.A., 1998. Depressive symptoms and increased risk of stroke mortality over a 29-year period. Arch. Intern. Med. 158 (10), 1133–1138.

Frasure-Smith, N., Lesperance, F., Talajic, M., 1993. Depression following myocardial infarction. Impact on 6-month survival. J. Am. Med. Assoc. 270 (15), 1819–1825.

Gander, M.L., von Känel, R., 2006. Myocardial infarction and post-traumatic stress disorder: frequency, outcome, and atherosclerotic mechanisms. Eur. J. Cardiovasc. Prev. Rehabil. 13 (2), 165–172.

Glassman, A.H., O'Connor, C.M., Califf, R.M., Swedberg, K., Schwartz, P., Bigger Jr., J.T., Krishnan, K.R., van Zyl, L.T., Swenson, J.R., Finkel, M.S., Landau, C., Shapiro, P.A., Pepine, C.J., Mardekian, J., Harrison, W.M., Harrison, W.M., Barton, D., Mclvor, M., 2002. Sertraline treatment of major depression in patients with acute MI or unstable angina. J. Am. Med. Assoc. 288, 701–709.

Hare, D.L., Toukhsati, S.R., Johansson, P., Jaarsma, T., 2014. Depression and cardiovascular disease: a clinical review. Eur. Heart J. 35 (21), 1365–1372.

Hemingway, H., Marmot, M., 1999. Evidence based cardiology: psychosocial factors in the aetiology and prognosis of coronary heart disease. Systematic review of prospective cohort studies. Br. Med. J. 318, 1460–1467.

Holt-Lunstad, J., Smith, T.B., Layton, J.B., 2010. Social relationships and mortality risk: a meta-analytic review. PLoS Med. 7 (7), e1000316.

Huffman, J.C., Celano, C.M., Beach, S.R., Motiwala, S.R., Januzzi, J.L., 2013. Depression and cardiac disease: epidemiology, mechanisms, and diagnosis. Cardiovasc. Psychiatry Neurol. 2013, 695925.

Janszky, I., Ljung, R., 2008. Shifts to and from daylight savings time and incidence of myocardial infarction. N. Engl. J. Med. 359, 1966–*1968*.

Jauhar, S., 2018. Heart: A History. Farrar, Straus and Giroux, New York.

Joynt, K.E., Whellan, D.J., O'Connor, C.M., 2003. Depression and cardiovascular disease: mechanisms of interaction. Biol. Psychiatry 54 (3), 248–261.

Kaplan, G.A., Keil, J.E., 1993. Socioeconomic factors and cardiovascular disease: a review of the literature. Circulation 88 (4), 1973–1998.

Kloner, R.A., McDonald, S.A., Leeka, J., Poole, W.K., 2011. Role of age, sex, and race on cardiac and total mortality associated with super bowl wins and losses. Clin. Cardiol. 34, 102–107.

Kuper, H., Marmot, M., Hemingway, H., 2002. Systematic review of prospective cohort studies of psychological factors in the etiology and prognosis of coronary heart disease. Semin. Vasc. Med. 2 (3), 267–314.

Leung, Y.W., Flora, D.B., Gravely, S., Irvine, J., Carney, R.M., Grace, S.L., 2012. The impact of premorbid and postmorbid depression onset on mortality and cardiac morbidity among patients with coronary heart disease: meta-analysis. Psychosom. Med. 74, 786–801.

Lichtman, J.H., Bigger Jr., J.T., Blumenthal, J.A., Frasure-Smith, N., Kaufmann, P.G., Lespérance, F., Mark, D.B., Sheps, D.S., Taylor, C.B., Froelicher, E.S., 2008. Depression and coronary heart disease: recommendations for screening, referral, and treatment—a science advisory from the American heart association prevention committee of the council on cardiovascular nursing, council on clinical cardiology, council on epidemiology and prevention, and interdisciplinary council on quality of care and outcomes research. Circulation 118 (17), 1768–1775.

Lichtman, J.H., Froelicher, E.S., Blumenthal, J.A., Carney, R.M., Doering, L.V., Frasure-Smith, N., Freedland, K.E., Jaffe, A.S., Leifheit-Limson, E.C., Sheps, D.S., Vaccarino, V., Wulsin, L., On Behalf of the American Heart Association Statistics Committee of the Council on Epidemiology and Prevention and the Council on Cardiovascular and Stroke Nursing, 2014. Depression as a risk factor for poor prognosis among patients with acute coronary syndrome: systematic review and recommendations. Circulation 129 (12), 1350–1369.

Merz, C.N., Kelsey, S.F., Pepine, C.J., Reichek, N., Reis, S.E., Rogers, W.J., Sharaf, B.L., Sopko, G., 1999. The Women's ischemia syndrome evaluation (WISE) study: protocol design, methodology, and feasibility report. J. Am. Coll. Cardiol. 33 (6), 1453–1461.

Musselman, D.L., Evans, D.L., Nemeroff, C.B., 1998. The relationship of depression to cardiovascular disease: epidemiology, biology, and treatment. Arch. Gen. Psychiatry 55, 580–592.

Rugulies, R., 2002. Depression as a predictor for coronary heart disease: a review and meta-analysis. Am. J. Prev. Med. 23 (1), 51–61.

Rutledge, T., Reis, V., Linke, S., Greenberg, B., Mills, P., 2006. Depression in heart failure: a meta-analytic review of prevalence, intervention effects, and associations with clinical outcomes. J. Am. Coll. Cardiol. 48, 1527–1537.

Schulz, R., Beach, S.B., 1999. Caregiving as a risk factor for mortality: the caregiver health effects study. J. Am. Med. Assoc. 282 (23), 2215–2219.

Seal, K.H., Metzler, T.J., Gima, K.S., Bertenthal, D., Maguen, S., Marmar, C.R., 2009. Trends and risk factors for mental health diagnoses among Iraq and Afghanistan veterans using Department of Veterans Affairs health care, 2002–2008. Am. J. Public Health 99, 1651–1658.

Spendler, H., Pedersen, S.S., 2005. Posttraumatic stress disorder in the wake of heart disease: prevalence, risk factors, and future research directions. Psychosom. Med. 67, 715–723.

Steptoe, A., Kivimaki, M., 2013. Stress and cardiovascular disease: an update on current knowledge. Annu. Rev. Public Health 34, 337–354.

Sullivan, M., Simon, G., Spertus, J., Russo, J., 2002. Depression-related costs in heart failure care. Arch. Intern. Med. 162, 1860–1866.

Sumner, J.A., Kubzansky, L.D., Kabrhel, C., Roberts, A.L., Chen, Q., Winning, A., Gilsanz, P., Rimm, E.B., Glymour, M.M., Koenen, K.C., 2016. Associations of trauma exposure and posttraumatic stress symptoms with venous thromboembolism over 22 years in women. J. Am. Heart Assoc. 5 (5), e003197.

Thombs, B.D., de Jonge, P., Coyne, J.C., Whooley, M.A., Frasure-Smith, N., Mitchell, A.J., Zuidersma, M., Eze-Nliam, C., Lima, B.B., Smith, C.G., Soderlund, K., Ziegelstein, R.C., 2008. Depression screening and patient outcomes in cardiovascular care: a systematic review. J. Am. Med. Assoc. 300 (18), 2161–2171.

Vaccarino, V., Goldberg, J., Rooks, C., Shah, A.J., Veledar, E., Faber, T.L., Votaw, J.R., Forsberg, C.W., Bremner, J.D., 2013. Post-traumatic stress disorder and incidence of coronary heart disease: a twin study. J. Am. Coll. Cardiol. 62 (11), 970–978.

van Melle, J.P., de Jonge, P., Spijkerman, T.A., Tijssen, J.G., Ormel, J., van Veldhuisen, D.J., van den Brink, R.H., van den Berg, M.P., 2004. Prognostic association of depression following myocardial infarction with mortality and cardiovascular events: a meta-analysis. Psychosom. Med. 66, 814–822.

Whiteford, H.A., Harris, M.G., McKeon, G., Baxter, A., Pennell, C., Barendregt, J.J., Wang, J., 2013. Estimating remission from untreated major depression: a systematic review and meta-analysis. Psychol. Med. 43 (8), 1569–1585.

Whittaker, K.S., Krantz, D.S., Rutledge, T., Johnson, B.D., Wawrzyniak, A.J., Bittner, V., Eastwood, J.A., Eteiba, W., Cornell, C.E., Pepine, C.J., Vido, D.A., Handberg, E., Merz, C.N., 2012. Combining psychosocial data to improve prediction of cardiovascular disease risk factors and events: The National Heart, Lung, and Blood Institute—sponsored women's ischemia syndrome evaluation study. Psychosom. Med. 74 (3), 263–270.

Wilbert-Lampen, U., Leistner, D., Greven, S., Pohl, T., Sper, S., Völker, C., Güthlin, D., Plasse, A., Knez, A., Küchenhoff, H., Steinbeck, G., 2008. Cardiovascular events during world cup soccer. N. Engl. J. Med. 358, 475–483.

Yusuf, S., Hawken, S., Ounpuu, S., Dans, T., Avezum, A., Lanas, F., McQueen, M., Budaj, A., Pais, P., Varigos, J., Lisheng, L., INTERHEART investigators, 2004. Effect of potentially modifiable risk factors associated with myocardial infarction in 52 countries (the INTERHEART study): case-control study. Lancet 364, 937–952.

Zapletal, E., LeMaitre, D., Menard, J., Degoulet, P., 1996. The number needed to treat: a clinically useful nomogram in its proper context. Br. Med. J. 312, 426–429.

CHAPTER 3

Translational research from animal models

Angela J. Grippo
Department of Psychology, Northern Illinois University, DeKalb, IL, United States

Contents

Introduction: The translational value of research with animal models

The purpose of this chapter is to discuss the translational value of research with animal models for understanding the interactions of stress, depression, and cardiovascular disease (CVD). Many theories have been postulated to explain the association between mood and CVD, discussed at length throughout this book. Our understanding of specific neurobiological mechanisms, as well as causal and common processes underlying the associations of stress, emotion, and regulation of the cardiovascular system, will be enhanced through interpreting findings from animal research. Some advantages of research with animal model systems include:

(1) the ability to achieve a high degree of experimental control over the environment;

Cardiovascular Implications of Stress and Depression
https://doi.org/10.1016/B978-0-12-815015-3.00003-9

(2) a focus on pathways of communication between the central and peripheral nervous systems; and
(3) an emphasis on integrated neurobiological and biobehavioral processes.

To highlight the value of animal models, the following sections will address the substantive contributions from animal research regarding several mechanisms that may underlie negative mood and CVD. This chapter will first discuss specific stress-induced animal models of depression, including the core behavioral features of the models, operational definitions and measurement of the depressive phenotype, evidence for the validity of the models, and associated cardiovascular outcome measures indicative of dysfunction. The following models are addressed here:

(1) exposure to unpredictable environmental stressors, specifically the chronic mild stress (CMS) model in rodents;
(2) exposure to social stressors, specifically the models of social defeat in rodents, disruption of social bonds and social isolation in rodents, and social subordination stress in non-human primates; and
(3) exposure to physical stressors, specifically the olfactory bulbectomy and coronary artery ligation models in rodents.

This discussion is followed by evidence from research using these models that highlights specific processes and mechanisms underlying the links among stress, depression and CVD, including:

(1) autonomic imbalance;
(2) cardiac instability and arrhythmias;
(3) vascular hyper-reactivity and atherosclerosis;
(4) endocrine and immune dysfunction; and
(5) central nervous system dysregulation.

Given the focus on human research supporting the association of depression and CVD elsewhere in this book (Chapter 1 and Section 1), coupled with a specific discussion of mechanisms underlying this association (Section 2), the following discussion is limited primarily to the translational evidence that is derived from research involving animal models.

Animal models and operational definitions of depression

As presented in Chapter 1, depression is a complex behavioral and physiological condition. Several animal models have been developed for the study of various constructs related to the depressive syndrome, such as the behavioral profile, effectiveness of antidepressant treatments, and associated physiological or neurobiological consequences. Many models have been reviewed in detail between the 1980s and present day [see for an older and newer example: (Slattery and Cryan, 2017; Willner, 1984)]. For the purpose of this chapter, the models that have previously been used to evaluate neurobiological correlates of depression and the association between depression and CVD will be reviewed. Several models have been developed and validated in multiple species, including monkeys, rats, mice, and prairie voles. This section will discuss the manipulations used to induce depressive behaviors in each model, validity characteristics, and specific assays used to operationally define depressive behaviors in animals.

Exposure to unpredictable environmental stressors: Chronic mild stress

Given the influence of stress on many aspects of daily life, models that expose animals to stressors are extremely valuable for improving our understanding of emotion, behavior, and physiology. The CMS model of depression has been used for several decades in rodents for the study of depression and associated conditions. Katz, Roth and Carroll published initial studies describing CMS as a low-aversive, realistic strategy to induce depression-like changes in rodents (Katz et al., 1981; Katz, 1981). The CMS model exposes rodents to a chronic period of unpredictable, mild stressors, involving proximal changes such as disruptions to the animal's cage (e.g., damp bedding, slightly tilted cage, restraint in a smaller cage), as well as distal stressors such as disruptions to the ambient environment (e.g., strobe light, background noise, changes in light/dark cycle). These mild—yet unpredictable and uncontrollable—stressors are considered to realistically mimic daily hassles that are experienced by humans. A combination of several mild stressors is used in a CMS protocol (Table 3.1 provides a sample list), presented in series or in partially-overlapping sessions with the primary objective being to maximize the unpredictable nature of the stressors. Typically, a chronic period of stressors (i.e., 2–5 weeks) is sufficient to induce a variety of behavioral changes that are argued to be relevant to human depression, and these changes can be maintained at a relatively stable level by carrying out the CMS procedure for the desired time period (Muscat and Willner, 1992; Willner et al., 1991). Each individual stressor is of very minor consequence—similar to a human experiencing, for example, one bad day among many average days. However, when multiple stressors are presented to the animal in an unpredictable manner, over a several-week period, it becomes challenging for the animal to adapt to the stressors—similar to a human experiencing a wide variety of unpredictable hassles across many bad days.

Table 3.1 A sample list of stressors used in studies to investigate the association between depressive behaviors and cardiovascular dysregulation with the chronic mild stress (CMS) model.

- Stroboscopic illumination (in a dark room)
- Continuous overnight lighting
- Reversal of light-dark cycle
- Background white noise (either continuous or intermittent)
- Tilted cage
- Damp bedding
- Exposure to an empty water bottle immediately following a brief period of water deprivation
- Exposure to a foreign object in the cage
- Paired or grouped housing
- Restraint

As noted in previous discussions of this paradigm [e.g., Willner, 2005], there is no one specific CMS protocol, and the details, length, and timing of the stressors have varied across laboratories. Each individual stressor is not as important as the presentation of a combination of stressors, over a chronic period of time, in an unpredictable fashion (to put it simply, the whole is certainly greater than the sum of its parts).

The validity and reliability of CMS as a model of depression have been described in detail (Willner, 2005, 2017). For example, this model is centered on the behavioral feature of anhedonia. Anhedonia is a psychological construct, defined in the literature on human psychopathology as a reduced capacity to experience pleasure (American Psychiatric Association, 2013). This sign is central to human affective disorders (American Psychiatric Association, 2013). Katz (1982) reported that anhedonia was a characteristic observed in rats exposed to CMS, and this behavior was reversed by a tricyclic antidepressant. Willner and colleagues continued to characterize the reliability of the hedonic deficit seen in CMS [for instance (Muscat et al., 1992; Willner et al., 1991, 2017)].

In rats, anhedonia is defined as a reduction in responding for a previously demonstrated reinforcing stimulus. This feature can be assessed through motivational and reward-seeking behaviors, such as via the intake of a palatable solution (sucrose or saccharin) or by the animal's willingness to perform a behavior to receive electrical brain stimulation into a region of the brain that is related to dopamine function. The predictable behavioral responses of rodents during these assays of anhedonia have been used to argue for the face validity of CMS as a model of depression (Willner et al., 1991). Rats and mice exposed to CMS show a reduced preference for a sweet solution, compared to unstressed (control) conditions (Willner et al., 1987; Solberg et al., 1999). Similarly, relative to an experimentally established baseline, rats exposed to CMS display reduced responding for rewarding electrical brain stimulation into the ventral tegmental area of the brain (Moreau et al., 1994).

In addition to anhedonia, virtually all other observable and quantifiable features associated with human depression have been observed in the CMS model. For example, rats exposed to CMS display disruptions in spontaneous locomotor activity, exploratory behaviors in a novel environment, sleep patterns, sexual behavior, food intake and body weight (in some studies), cognitive processes, and some comorbidity with anxiety-related behaviors [reviewed recently by (Willner, 2017)]. In fact, Willner has argued that a rat exposed to CMS would meet criteria for a DSM diagnosis of major depressive disorder or major depressive disorder with melancholic features [see recent discussion in (Willner, 2017)]. Indeed, the only features of CMS that have not been observed are subjective experiences (such as suicidal ideations, feelings of guilt or low self-worth) or behaviors that are unique to humans in modern society (such as specific suicide attempts).

Predictive validity has been established in CMS by investigating the effects of several antidepressant and non-antidepressant treatments on the behavioral profile. A detailed discussion of the effects of antidepressant drugs and the value of the CMS model in drug discovery research is reviewed by Willner (2017). To briefly summarize this discussion:

(1) Anhedonia is selectively altered by pharmacological antidepressant treatments over a similar time course as observed in humans, including monoamine oxidase inhibitors, tricyclic antidepressants, serotonin and norepinephrine reuptake inhibitors, serotonin and dopamine receptor agonists, and other atypical antidepressants.

(2) Non-pharmacological therapies, such as electroconvulsive therapy, also can be effective treatments for the consequences of CMS.

(3) It is notable that CMS can positively screen for antidepressant drugs; psychoactive drugs that are not typical treatments for depression in humans (e.g., stimulants, some anxiolytics) do not predictably alter the depressive signs elicited in rats exposed to CMS.

CMS is a particularly useful model of depression because it also possesses a high degree of construct validity. As detailed throughout this book, environmental stress has long been cited to influence both the etiology and maintenance of depression [e.g., Sapolsky, 1996; Willner, 2017]. The mild stressors employed in the CMS procedure represent a realistic analog to the daily stressors experienced by humans (Willner et al., 1992, 2017). Further, the foundation of CMS is focused on a core feature of depression (anhedonia). A study of the diagnostic criteria relating to depression suggests that 95% of depressed patients experience anhedonia (Keller et al., 1995). Also, the anhedonia observed in rats exposed to CMS is a specific hedonic deficit; many studies have demonstrated that it is not predictably affected by changes in body weight or motor activity (Willner, 2005, 2017). Not only is CMS valuable for its link to anhedonia, the stress paradigm also induces several correlated features (as discussed above), enhancing the construct validity of the model.

As discussed, several lines of evidence support the value of CMS as a valid and reliable rodent model of depression. However, following initial studies focused on depression-relevant research questions, additional studies have used the CMS model to investigate outcome measures associated with CVD. For example, CMS produces an elevation in resting heart rate as well as a reduction in resting heart rate variability (Grippo et al., 2002, 2003a, 2006), consistent with the syndromes of both depression and CVD (Carney et al., 1995; Krittayaphong et al., 1997; Pitzalis et al., 2001; Tapanainen et al., 2002; Ferrari et al., 2003; La Rovere et al., 2003; Guzzetti et al., 2005). In addition to basal rate and rhythm disturbances, CMS also is associated with arrhythmias and increased cardiac reactivity to acute stressors (Grippo et al., 2002, 2004, 2006), features that are observed in patients with CVD or its antecedent risk factors, and contribute to mortality (Frasure-Smith et al., 1995; Olshausen et al., 1991; Talman, 1997). Impairments in cardiac output and stroke volume are also observed in the CMS model (Grippo et al., 2006). Finally, vascular disturbances, including endothelial impairments and injury, have likewise been observed in the CMS model (Bouzinova et al., 2012, 2014; Isingrini et al., 2011).

Exposure to social stressors: Social defeat, disrupted social bonds and isolation, and social subordination

In addition to general environmental stressors, experimental investigations involving exposure to specific social stressors are valuable for the study of behavioral and physiological changes in depression and associated cardiovascular outcome measures. Animal models have focused on the consequences of social conflict, such as the social defeat paradigm in rats. Capitalizing on a rodent's natural tendency to defend its territory from an outsider, social defeat involves exposing a rodent to repeated attacks from another (more dominant) animal. The defeated animal displays several behavioral patterns of subordination and stress (Koolhaas et al., 2013).[1] Previous research using this model indicates that subjects of social defeat show behavioral changes associated with human depression including anhedonia, physiological changes such as altered neuroendocrine reactivity, and central dysfunction in brain structures that are relevant to reward, emotion, and stress (Rygula et al., 1998; Becker et al., 1998; Knowland and Lim, 2018). Social defeat coupled with additional social disturbances, such as social isolation following an episode of defeat, is associated with anhedonia (reduced intake of a sucrose solution) as well attenuated body weight gain (Carnevali et al., 2012).

In addition to the social defeat model, depressive behaviors and associated features have been observed in rodent studies focused on social isolation and the disruption of established social bonds. To investigate the mechanisms through which social experiences influence emotion and behavior, studies have focused on a specific social rodent species, the prairie vole. Similar to humans, prairie voles are highly dependent on social interactions for the regulation of behavior, physiology, and brain function. These rodents are among a small percentage of mammals that are considered socially monogamous and cooperative breeders, sharing with humans the capacity to form long-term social bonds, develop extended families, and engage in biparental and alloparental care (Carter et al., 1995). A focus on the causes and consequences of sociality in the prairie vole offers a powerful model for studying processes by which negative and positive social experiences influence behavior, emotion, and physiology.

The disruption of social bonds in the prairie vole model induces behaviors relevant to depression, consistent with behavioral changes observed in the CMS model (discussed above) and those described in human depression (American Psychiatric Association, 2013). Several operational assays have been used to investigate depression-relevant behaviors in prairie voles exposed to social stressors. For example, 4 weeks of social isolation from a same-sex sibling in prairie voles leads to anhedonia, shown via the reduced consumption of a sucrose solution (Grippo et al., 2007a), similar to the anhedonia observed following CMS exposure in rats.

Social isolation from a family member and the disruption of established male-female bonds in prairie voles also produce maladaptive coping responses during a forced swimming stressor (Bosch et al., 2009; McNeal

[1]A video of the social defeat paradigm can be viewed online, associated with this publication: https://www.jove.com/video/4367/the-resident-intruder-paradigm-standardized-test-for-aggression (Koolhaas et al., 2013).

et al., 2014). In this behavioral task, a rodent's behavior is observed for active coping strategies such as swimming, struggling, and actively attempting to escape, versus a maladaptive, immobility response (e.g., floating or treading water just enough so that the head is not submerged). The forced swim test has been used to define helplessness and behavioral despair operationally; and is sensitive to pharmacological antidepressant treatments (Cryan et al., 2005b). Social isolation in prairie voles also has been combined with CMS to produce behavioral deficits relevant to depression in the forced swim test as well as in a related tail-suspension test (McNeal et al., 2017; Grippo et al., 2015),[2] which involves characterizing active coping strategies versus a passive immobility response to being suspended by the tail (Cryan et al., 2005a). Finally, disruption of the social structure of prairie voles produces several additional behavioral features that have relevance to depression, including changes in exploratory behavior, aggression, sexual behavior, social interactions, and social cognition (Sun et al., 2014; Pohl et al., 2019).

In addition to rodent models, non-human primate models of social instability have been developed to investigate depression-relevant research questions. Shively and colleagues have observed behavioral signs of depression and several associated physiological features in female cynomolgus monkeys when they are the subordinate members of a social hierarchy (Shively et al., 2002, 2009). Cynomolgus monkeys organize their behaviors around dominance hierarchies. Females of this species that are socially subordinate receive high levels of aggression from other animals in the social group, and spend a large amount of time time in a fearful, vigilant state, with comparatively less time as the recipients of affiliative behaviors from other monkeys (Shively, 1998). It has been suggested that the environment of female subordinate cynomolgus monkeys mirrors that of low socioeconomic status humans (Shively et al., 2005). These subordinate monkeys display naturalistic behaviors that mimic a depressed state—which can be readily observed—including changes in body

[2]The CMS paradigm used in recent prairie vole studies [e.g., McNeal et al., 2017, Grippo et al., 2015] was based on initial protocols developed in rats (e.g., as discussed in the section *Exposure to Unpredictable Environmental Stressors: Chronic Mild Stress*), however not all of the stressors listed in Table 3.1 have been included in prairie vole protocols. In particular, the paired housing stressor was removed from the two studies listed here, since it was not possible to incorporate this stressor without interfering with the social environmental manipulations in those specific study designs. In place of the paired housing stressor, a foreign object was placed in the cage—a rectangular brick was chosen as the foreign object, given that it takes up a significant amount of space in the cage, is immovable, and is not easily manipulated.

posture and reduced reactivity to environmental stimuli to which other monkeys are actively attending. Subordinate monkeys display several additional characteristics that have been associated with the syndrome of depression, such as impaired endocrine and reproductive function (Shively et al., 2002, 2009).

There is a high degree of validity in animal models focused on social manipulations to induce depression-relevant behaviors. Construct validity can be demonstrated by the strong association of social stressors and negative social experiences with depressive symptoms in humans [e.g., Cacioppo et al., 2010]. For instance, subordinate cynomolgus monkeys receive less social support and higher levels of aggression than monkeys who are considered higher ranking; these animals in turn show several behavioral and physiological indications of being socially stressed (Shively et al., 2009). Similarly, social defeat and social isolation are associated with altered hedonic responsiveness (Becker et al., 1998; Rygula et al., 1998; Grippo et al., 2007b); and the effects of social defeat are dampened when the subordination episodes are followed by a positive social experience (Ruis et al., 1999). Face validity has also been demonstrated in social stress models of depression. For example, in the subordinate cynomolgus monkeys, body posture and behavioral responsiveness to their surroundings mimic a depressed state at face value (Shively and Day, 2015).[3] These models also display predictive validity; for example, both subordinate cynomolgus monkeys and rats exposed to social defeat are insensitive to the negative feedback effects of dexamethasone (Shively et al., 1997; Kaplan et al., 2010; Carnevali et al., 2012), similar to depressed humans (Mokhtari et al., 2013).

These aforementioned models of social stressors focus on behaviors and physiological features that have relevance to human depression. Additionally, social stress models have been used to investigate the interactions of depression and CVD. A depressive phenotype is associated with progressive cardiac disturbances as a function of social stressors in rodents—similar to those observed in rodents following CMS—including reductions in heart rate, increases in heart rate variability, and alterations in circadian rhythmicity of cardiovascular variables (Grippo et al., 2007c; Carnevali et al., 2012). For instance, an episode of social defeat followed by 3 weeks of social isolation led to a disruption of circadian rhythmicity of heart rate in rats, compared to social defeat followed by paired housing (control) conditions (Carnevali

[3]Pictures of subordinate monkeys, compared to dominant monkeys that are actively engaged in the surroundings, can be viewed in Fig. 1 of Shively and Day (2015).

et al., 2012). Additional studies in prairie voles demonstrate that resting and stressor-induced heart rate are elevated, and the recovery of heart rate to resting levels is delayed after a stressor, as a function of 4 weeks of social isolation (relative to paired control conditions) (Grippo et al., 2007c). Subordinate cynomolgus monkeys display similar cardiac characteristics to socially isolated prairie voles, including elevated basal heart rate, exaggerated heart rate reactivity to a stressor, and delayed recovery of heart rate following the stressor (Shively, 1998).

In addition to cardiac rate and rhythm disturbances, social stressors are associated with vascular dysfunction, indicative of an increased risk for CVD. Social subordination stress in female cynomolgus monkeys is associated with the development of atherosclerosis and related vascular dysfunction, such as endothelial injury (Shively et al., 2008; Kaplan et al., 2009; Strawn et al., 1991). Female subordinate cynomolgus monkeys that display signs of depression have four times the coronary artery atherosclerosis compared with those that do not display signs of depression, and the extent of coronary artery atherosclerosis is positively correlated with amount of time animals spend in the depressed posture (Shively et al., 2008). The extent of coronary artery atherosclerosis in this population of monkeys is also positively correlated with 24-h heart rate (Shively et al., 2008). Similarly, prairie voles that display signs of depression as a function of social isolation also exhibit evidence of vascular impairments and injury (Peuler et al., 2012).

Exposure to physical stressors: Coronary artery ligation and olfactory bulbectomy

Aside from the presentation of environmental stressors, some models focus on exposure to physical stressors to investigate depression-relevant research questions. One valuable model, which has been used to study the association of depression and CVD, is the coronary artery ligation model in rodents. Initially designed as a model to investigate CVD-relevant research questions, the coronary artery ligation model includes physically restricting the blood supply to the heart to cause damage. By occluding a large coronary artery, the oxygen supply to the heart becomes insufficient to meet its metabolic demands. This, in turn, leads to myocardial infarction (i.e., a "heart attack") and a progressive decline in cardiac function.

Experimental coronary artery ligation in rodents is characterized by several neurohumoral, electrophysiological and morphological changes that mimic the syndrome of human congestive heart failure (CHF) (Felder et al., 2003; Francis et al., 2001). Dysfunction of the left ventricle due to dilation of the ventricular chamber leads to fluid retention and vasoconstriction, which are mediated by the sympathetic nervous system and the renin-angiotensin-aldosterone system (Francis et al., 2001). Activation of proinflammatory cytokines also is associated with CHF (Francis et al., 2003; Levine et al., 1990). In addition to mediating CHF and cardiovascular outcomes, the regulation of these systems involves central nervous system mechanisms relevant to depression. In particular, forebrain structures such as the paraventricular nucleus, and central mineralocorticoid processes, may mediate the syndrome of CHF (Francis et al., 2003; Felder et al., 2003).

A separate model that has been used to investigate questions related to depression is the olfactory bulbectomy model in rats. The olfactory bulbs constitute a large proportion of the rodent brain, and rodents rely heavily on olfactory information to assess and interact with the environment. The olfactory bulbs have extensive communication with several regions of the forebrain and midbrain (Leonard and Tuite, 1981). Therefore, bilateral removal of the olfactory bulbs (bulbectomy) produces changes in exploratory behaviors, hyperactivity, sexual behavior, cognitive function, and nociception, coupled with several endocrine, immune, and central nervous system alterations that are relevant to depression [see Song and Leonard, 1995; Morales-Medina et al., 2017].

These physical stress models have validity for investigating depression-related research questions. The coronary artery ligation model has been used to investigate behavioral and immunological processes associated with depression, and is valuable for understanding mechanisms that are common to both depression and CVD (Grippo et al., 2003b). For instance, the proinflammatory cytokine tumor necrosis factor-alpha (TNF-α) is associated with heart disease, contributing to cardiomyopathy and left ventricular dysfunction (Kapadia et al., 1998). TNF-α function and related proinflammatory processes also may play an etiological role in the development of depression. Administration of TNF-α or interleukin-2 produce signs associated with depression, including fatigue, lethargy, anorexia, anhedonia, and related sickness behaviors (Anisman et al., 1998; Smith, 1991).

Similarly, the olfactory bulbectomy model has been argued to possess face, predictive, and some limited construct validity (Morales-Medina et al., 2017; Song and Leonard, 2005). As reviewed elsewhere, this model has been used to screen for pharmacological antidepressant treatments—including tricyclic antidepressants and serotonin reuptake inhibitors—and antidepressant drugs alter some of the deficits observed in this model (such as cognitive deficits) (Morales-Medina et al., 2017), making it a useful technique for evaluating the effectiveness of potential novel pharmacological compounds. Further, the disruption of central communication in this model—involving cortical, hippocampal, and amygdala function—has been observed in depression (Song and Leonard, 2005).

Mechanisms, pathways, and physiological correlates

Animal models of depression are valuable experimental approaches for understanding whether exposure to stress exacerbates the process of CVD and the role of depression in this process. Following initial development and discussions of validity of the models discussed above, these models have been used to specifically investigate mechanisms underlying the association of stress, depression, and cardiovascular function. As an introduction to the second section of this book, the following paragraphs highlight specific physiological correlates of depression, and evidence from animal models that provides support for processes underlying the association of depression and CVD.

Autonomic imbalance

Dysfunction of the autonomic nervous system is an important candidate mechanism that links altered mood and cardiovascular disturbances, as presented in Chapter 5 of this book. Research from a variety of animal models of depression supports this hypothesis. For example, many cardiovascular disturbances that are mediated by autonomic mechanisms have been observed in rodent and non-human primate models of depression. As detailed above in the discussion of CMS, the behavioral changes observed in this model are coupled with several cardiovascular disturbances consistent with an increased risk of CVD, including increased resting heart rate, reduced heart rate variability, and exaggerated cardiac reactivity to novel stressors (Grippo et al., 2002, 2003a).

The behavioral changes associated with CMS recover within a few weeks following cessation of the stressors; however, the cardiovascular disruptions do not follow the same time course. For example, at a time point when anhedonia and reduced activity have returned to baseline (pre-stress) levels, animals continue to display higher resting heart rate, lower heart rate variability, and exaggerated cardiac reactivity to a novel stressor (Grippo et al., 2003a). Similarly, administration of fluoxetine to animals exposed to CMS prevents anhedonia—consistent with its antidepressant actions in humans—however, this antidepressant only partially prevents autonomic and cardiac disturbances in CMS-exposed rats (Grippo et al., 2006). These data suggest that remediation of the depressive behaviors is not associated with alleviation of the underlying cardiovascular pathophysiology over the same time course. Similarly, in humans, pharmacotherapy or psychotherapy for depression may partially improve heart rate and heart rate variability, but may not be sufficient to improve cardiovascular status to baseline levels (Carney et al., 2000). This dissociation in both humans and animal models highlights the need for a better understanding of specific neurobiological mechanisms that underlie the link between depression and CVD, to help inform the development of effective therapies.

Similar to the consequences of environmental stressors, social stressors likewise contribute to elevated heart rate, reduced heart rate variability, and exaggerated cardiac reactivity to stressors in both rodents and non-human primates (Grippo et al., 2007c; Carnevali et al., 2012; Shively, 1998). These disturbances suggest that negative social experiences may have a profound influence on the resting state of the cardiovascular system as well as its ability to respond to stressors in an adaptive manner. Considering the association among the social environment, mood, and cardiovascular health (for instance, Chapter 9 of this book), it is noteworthy that these relationships are reciprocal. For example, not only does the disruption of social bonds produce depressive behaviors and autonomic dysfunction, but healthy autonomic function—such as that represented by higher heart rate variability—can predict pair bonding success in prairie voles (Lewis et al., 2017).

The physiological changes observed following CMS and social environmental manipulations implicate autonomic dysregulation as a mechanism through which depression may interact with cardiovascular function. Both increased sympathetic and reduced parasympathetic drive likely mediate the relationship between depression and CVD. Selective blockade of sympathetic inputs to the heart with propranolol hydrochloride indicates that sympathetic drive is elevated in rats exposed to CMS (Grippo et al., 2002), similar to the excess sympathetic drive observed in patients with CVD (e.g., Chapter 5). Similarly, both CMS and olfactory bulbectomy are characterized by changes in sympathoexcitatory reflexes, providing further evidence for sympathetic nervous system dysfunction (Moffitt et al., 2002; Grippo et al., 2008). Further studies using the prairie vole social isolation model have shown that several cardiac changes are mediated by a withdrawal of parasympathetic tone, which may be responsible for cardiovascular morbidity and mortality associated with social isolation. This has been demonstrated by attenuated cardiac responsiveness to the cholinergic receptor antagonist atropine methyl nitrate, along with reduced respiratory sinus arrhythmia, a measure of heart rate variability that is hypothesized to represent parasympathetic regulation of the heart (Grippo et al., 2007c). When integrated together, the research from animal models of depression implicates impaired autonomic regulation of the heart as one mechanism underlying the comorbidity of depression and CVD.

Cardiac instability and arrhythmias

An imbalance of autonomic regulation of the heart, characterized by increased sympathetic and/or reduced parasympathetic tone, increases stress and strain on the heart muscle and favors cardiac instability. Several lines of evidence from animal models highlight cardiac instability as a significant concern associated with depression. CMS is associated with an increased vulnerability to ventricular arrhythmias when the cardiovascular system is challenged with a pro-arrhythmic drug (Grippo et al., 2004). A similar vulnerability has been described in rats following social defeat, with the majority of rhythm disturbances occurring immediately following attacks (Sgoifo et al., 1997). Further, both ventricular and atrial arrhythmias have been observed in socially isolated prairie voles following short-term stressors (social crowding or a forced swimming stressor) (Grippo et al., 2010, 2012a). These results are especially significant when considered in the context of findings from human populations. Patients with comorbid CVD and depression have a higher prevalence of ventricular arrhythmias, compared with patients that are not experiencing depression (Carney et al., 1993). Similarly, following myocardial infarction, patients are at a greater risk of mortality if they have a combination of ventricular arrhythmias and high depressive symptomatology, relative to patients with fewer arrhythmias or those with lower levels of depression (Frasure-Smith et al., 1995).

There are several additional consequences of autonomic imbalance and cardiac instability that can increase the risk of morbidity and mortality in CVD. For instance, these changes can influence the ability of the heart to adequately pump blood, thereby increasing the workload and consequently contributing to further degeneration of cardiac function. In support of this hypothesis, not only does social defeat produce several changes in heart rate and rhythms, but it also produces moderate hypertrophy of the right ventricle (Carnevali et al., 2012). Social isolation in prairie voles may also be associated with cardiac hypertrophy (Grippo et al., 2007c). Related research indicates that the autonomic imbalance in CMS is associated with reduced cardiac output and stroke volume; these changes were partially (but not completely) alleviated with the antidepressant drug fluoxetine (Grippo et al., 2006).

Electrical instability of the heart muscle may also be related to inefficient electrical communication and/or altered neural control of electrical communication. For instance, social isolation in prairie voles reduces levels of the gap junction protein connexin43 in the left ventricle, whereas CMS (but not in combination with social isolation) increases connexin45 protein levels (Grippo et al., 2015). Although the specific interactions of social stress, depression, and gap junction proteins require further study, appropriate connexin43 function is important for maintaining cardiac rhythmicity and facilitating efficient cell-to-cell communication in the heart, whereas increased connexin45 or changes in the connexin43/connexin45 ratio may encourage a pro-arrhythmic state (Bernstein and Morley, 2006; Betsuyaku et al., 2006). The research cited above has focused on left ventricular connexin proteins, but connexin proteins likewise support cellular communication in the brain. The importance of central gap junction protein function in depression has recently been discussed (Wang et al., 2017). Taken together, the research from animal models provides evidence for several aspects of cardiac instability, electrical rhythm disturbances, and associated dysfunction, thereby increasing the likelihood of negative outcomes associated with CVD in depressed patients.

Vascular hyper-reactivity and atherosclerosis

Dysfunction of the cardiac muscle itself is associated with depression. However, the cardiovascular system is a closed system, such that autonomic imbalance not only disrupts cardiac functions, but also vascular functions. Therefore, altered vascular reactivity is another candidate mechanism underlying the comorbidity of depression and CVD (see also Chapter 6). Excess sympathetic drive and/or a reduction in parasympathetic tone can facilitate coagulation, impair blood flow, damage the vascular endothelium, and promote atherosclerotic events.

As detailed above in the discussion of social stress in monkeys, the depressive phenotype is directly associated with evidence of coronary artery atherosclerosis in subordinate monkeys (Shively et al., 2008). Studies with rodents support the data from non-human primates. Both exposure to long-term social isolation in prairie voles and exposure to CMS in mice impairs vascular relaxation, via an impairment of endothelial-mediated relaxation (Peuler et al., 2012; Isingrini et al., 2011). These data support the hypothesis that social stress and depression may be associated with impaired release of anti-inflammatory factors, and/or increased release of contracting factors, from the endothelium. Further evidence suggests that individual differences in vulnerability to anhedonia may predict vascular dysfunction in the CMS model, as both catecholamine uptake pathways and endothelium-dependent relaxation pathways in the vascular wall differ in small arteries as a function of the level of anhedonia in rats (determined by CMS-induced changes in sucrose intake) (Bouzinova et al., 2012, 2014). Additional circulating factors that are relevant to vascular function, such as total and high density lipoprotein cholesterol, are associated with the depressive state in subordinate cynomolgus monkeys (Shively et al., 2008). Taken together, these findings suggest that individual levels of vulnerability and resilience to depressive behaviors (such as anhedonia) may mediate aspects of cardiovascular dysfunction and the development of atherosclerosis. Further discussion of vascular reactivity in the context of depression and CVD, including issues of vulnerability and resilience, is presented by Bouzinova et al. (2015).

Endocrine and immune dysfunction

The endocrine and immune systems are dysfunctional both in mood disorders and in CVD. These systems interact with the autonomic nervous system, the heart muscle, and the vasculature; and are directly and indirectly affected by stressors. Dysfunction of the hypothalamic-pituitary-adrenal (HPA) axis has consistently been observed in animal models of depression (see also Chapter 5). For example, morning cortisol levels are higher in subordinate (vs dominant) cynomolgus monkeys, and cortisol is not suppressed appropriately following either an adrenocorticotropic hormone (ACTH) or a dexamethasone challenge (Shively et al., 1997; Shively, 1998). Indices of neuroendocrine dysregulation, such as altered corticotropin-releasing factor (CRF) receptors, elevated resting or stressor-induced corticosterone and ACTH, lack of sensitivity to a dexamethasone challenge, and adrenal hypertrophy, have been observed as a function of social defeat or CMS in rats (Grippo et al., 2005; Ayensu et al., 1995; Carnevali et al., 2012); exposure to social stressors in prairie voles (Grippo et al., 2007a,b; Bosch et al., 2009; McNeal et al., 2014); and following a combination of social isolation and CMS in prairie voles (McNeal et al., 2017).

A variety of additional endocrine and immune system components, which are related to behavioral, autonomic, and HPA axis dysfunction, have been characterized in several models of depression. Subordinate cynomolgus monkeys display characteristics of ovarian and reproductive disturbances, which are associated with cardiovascular dysfunction (Shively et al.,

2009; Shively and Day, 2015). Plasma levels of oxytocin and vasopressin are elevated in prairie voles following long-term social isolation; these peptides have relevance for understanding reproductive function, stress reactivity, social behavior, and cardiovascular regulation (Grippo et al., 2007a,b; Porges, 2007). Further, rats exposed to CMS exhibit elevated plasma aldosterone and renin activity (Grippo et al., 2005), suggesting also that the renin-angiotensin-aldosterone system may play a role in the association of depression and CVD. This system has been reviewed recently as a possible target for antidepressant therapies (Vian et al., 2017).

The endocrine system interacts with the immune system both in the brain and the peripheral nervous system. As reviewed elsewhere (including in Chapter 8), activation of the immune system is an important component of cardiovascular disorders, and may act on the brain to produce depressive signs and symptoms (Dantzer et al., 2008; Wang et al., 2017; Das, 2000). Neuroimmune processes can further influence the HPA axis response to stressors and impair autonomic regulation of the heart, thereby creating a progression of declining function. This hypothesis is supported by animal research. For instance, CMS elevates circulating pro-inflammatory cytokines including TNF-α and interleukin-1-beta, and increases levels of these cytokines in brain regions associated with HPA regulation, such as the hypothalamus and pituitary (Grippo et al., 2005). Levels of these cytokines are also positively correlated with the amount of anhedonia displayed by rats exposed to CMS (Grippo et al., 2005), mirroring data from human depressed individuals (Suarez et al., 2003).

Further evidence of impaired immune functions in depression comes from observations that social defeat stress is associated with several indices of inflammation, immune reactivity to stress, and altered immune-endocrine communication in the peripheral and central nervous systems (Takahashi et al., 2018). It has also been suggested that differential coping strategies and inflammatory responses during stress may be predictive of vulnerability to depression and CVD. For instance, a passive (i.e., maladaptive) response in rodents exposed to social defeat is associated with increased inflammation in the central nervous system, which may in turn influence the development of depressive behaviors and cardiovascular dysfunction (Carnevali et al., 2018). In addition, impairments in the inability to destroy foreign cells has been observed in socially isolated prairie voles (Scotti et al., 2015) and in rats exposed to CMS (Ayensu et al., 1995). Taken together, these data support the association of depression with inflammatory mechanisms and dysfunction of both innate and adaptive immune processes.

Given the bidirectional nature of the association of mood and cardiovascular function, it is useful to study immune mechanisms underlying the emotional consequences of CVD. Anhedonia is a behavioral feature of the coronary artery ligation model of CHF. For instance, coronary artery ligation in rats is associated with increased circulating levels of TNF-α (Francis et al., 2004), as well as a reduction in responding for rewarding electrical brain stimulation into the lateral hypothalamus, indicative of anhedonia (Grippo et al., 2003b). When plasma TNF-α levels are reduced with etanercept—a soluble TNF receptor that acts as a TNF-α antagonist—the behavioral responding for rewarding electrical brain stimulation is restored to baseline (pre-ligation) levels, indicating a reversal of CHF-induced anhedonia (Grippo et al., 2003b). These results suggest that CHF can induce anhedonia via a physiological mechanism involving inflammation, providing insight into an immune mechanism that may mediate depressive behaviors in individuals with CVD. The precise mechanisms by which peripheral cytokines influence central functions might involve actions at the circumventricular organs, the stimulation of prostaglandins at the vascular-brain interface, or actions on visceral afferent nerves. Given the bidirectional communication between inflammatory processes and central nervous system function, these mechanisms require further investigation in the context of depressive disorders [see for instance (Ma et al., 2017)].

CNS dysregulation

Several central nervous system processes are altered both in mood disorders and CVD, and are affected by behavioral, cardiovascular, endocrine, immune, or other neural inputs. Therefore, central processes and disrupted communication in regions of the brain that support peripheral functions and behavior can mediate the cardiovascular dysfunction associated with depression (also see Chapter 7). For instance, the central serotonin (5-HT) system interacts with endocrine, immune, and autonomic function to influence cardiovascular regulation. The hypothalamic paraventricular nucleus receives serotonergic innervation, thereby affecting HPA axis functions; and this area sends projections to several brain regions that influence sympathetic and parasympathetic outflow to the cardiovascular system (Thayer and Brosschot, 2005). The CMS model has been used to investigate the role of 5-HT in the context of mood and cardiovascular function. Compared to saline, administration of the serotonin reuptake inhibitor fluoxetine prevents the behavioral consequences of CMS (e.g., anhedonia), but only partially prevents the cardiovascular consequences in this model (e.g., changes in heart rate and heart rate variability) (Grippo et al., 2006). These findings suggest that reducing depressive signs and symptoms with traditional therapies may not necessarily reduce the underlying cardiovascular pathophysiology associated with this condition, and also highlight the need for further investigating the role of the 5-HT system in the context of stress, emotion, and cardiovascular regulation [see Nalivaiko and Sgoifo, 2009].

Aside from 5-HT, neuropeptides and neurohormones have been receiving increased attention in the context of emotion and autonomic function. Oxytocin, vasopressin, and CRF play a central role in a complex neuroendocrine network that coordinates social behaviors and neurobiological responses to stressors, including hypothalamic functions that regulate the HPA axis (Porges, 2007; Carter and Altemus, 2005). It is hypothesized that several behavioral and physiological indicators associated with depression are mediated by dysfunction of CRF receptors in the hypothalamic paraventricular nucleus, thereby influencing downstream HPA axis responses, behavioral stress reactivity, and autonomic regulation of peripheral processes. For instance, antagonism of CRF receptors protects against the consequences of social defeat in rats (Wood et al., 2012) and the disruption of social bonds in prairie voles (Bosch et al., 2009). Increased reactivity of the paraventricular nucleus following a behavioral stressor has also been observed in socially isolated prairie voles, and this response is coupled with increased heart rate, decreased heart rate variability, and increased vulnerability to cardiac arrhythmias in the prairie vole model (Grippo et al., 2010). In a related series of studies, social isolation in female (but not male) prairie voles is associated with increased activation of CRF neurons in the paraventricular nucleus following a short-term stressor (Grippo et al., 2007b).

Related to central CRF function, basal levels of oxytocin neurons in the paraventricular nucleus are upregulated, and these neurons are hyper-activated following a short-term behavioral stressor, in socially isolated prairie voles (relative to social pairing) (Grippo et al., 2007a,b). Disrupted oxytocin signaling may play a central role in social behavioral deficits, depressive behaviors, and physiological dysregulation in the prairie vole model (Sun et al., 2014; Pohl et al., 2019; see also Chapter 9). For instance, oxytocin may be released from the hypothalamus to compensate for disrupted neuroendocrine and autonomic communication in prairie voles and humans as a result of disrupted social bonds. To further support this hypothesis, daily administration of oxytocin during a period of 4 weeks of isolation in prairie voles is protective against many of the detrimental behavioral and cardiac effects of isolation, including anhedonia, increased heart rate, reduced parasympathetic tone, and short-term autonomic reactivity to behavioral stressors (Grippo et al., 2009, 2012b). More recent research indicates that the administration of oxytocin to prairie voles prior to a short-term stressor prevents the coupling of paraventricular nucleus activity and downstream HPA processes, and functionally alters the communication between the paraventricular nucleus and cardioregulatory brainstem regions (Yee et al., 2016). Further research focused specifically on pathways of communication between the hypothalamus and brainstem regions that support autonomic outflow will enhance our understanding of central processes involved in autonomic and cardiovascular dysregulation in depressed individuals.

Concluding remarks

The comorbidity of depression and CVD in response to environmental stress is a critical worldwide health concern. In combination with findings from human populations, an emphasis on neurobiological processes using valid, reliable, and relevant animal models is useful for the study of mechanisms underlying this association. Integrative studies that incorporate systems level analyses, such as those described here, will continue to enhance our understanding of the bidirectional association of emotion and cardiovascular regulation in response to stress by providing insight into causal and common pathways of communication. Along with continued mechanistic research, additional benefits will stem from strategies that encourage an open dialog among researchers from multiple scientific fields, including clinical, social and experimental psychology, cardiology, immunology, endocrinology, and neuroscience—a view which is clearly supported by the interdisciplinary discussions in the current book. Comparative studies conducted by multi-disciplinary teams of researchers—such as those focused on multiple species or parallel human/animal study designs—will provide additional support for the utility of animal research in the context of mood disorders and cardiovascular dysfunction. These combined efforts can lead to enhanced treatment strategies, such as a greater emphasis on patient-focused therapies and personalized treatment protocols, improving the quality of life for individuals with comorbid psychological and cardiovascular conditions.

References

American Psychiatric Association, 2013. Diagnostic and Statistical Manual of Mental Disorders, fifth ed. American Psychiatric Association, Arlington, VA.

Anisman, H., Kokkinidis, L., Borowski, T., Merali, Z., 1998. Differential effects of interleukin (IL)-1β, IL-2, and IL-6 on responding for rewarding lateral hypothalamic stimulation. Brain Res. 779, 177–187.

Ayensu, W.K., Pucilowski, O., Mason, G.A., Overstreet, D.H., Rezvani, A.H., Janowsky, D.S., 1995. Effects of chronic mild stress on serum complement activity, saccharin preference, and corticosterone levels in Flinders lines of rats. Physiol. Behav. 57, 165–169.

Becker, C., Zeau, B., Rivat, C., Blugeot, A., Hamon, M., Benoliel, J.-J., 1998. Repeated social defeat-induced depression-like behavioral and biological alterations in rats: involvement of cholecystokinin. Mol. Psychiatry 13, 1079–1092.

Bernstein, S.A., Morley, G.E., 2006. Gap junctions and propagation of the cardiac action potential. Adv. Cardiol. 42, 71–85.

Betsuyaku, T., Nnebe, N.S., Sundset, R., Patibandla, S., Krueger, C.M., Yamada, K.A., 2006. Overexpression of connexin45 increases susceptibility to ventricular tachyarrhythmias in vivo. Am. J. Physiol. Heart Circ. Physiol. 290, H163–H171.

Bosch, O.J., Nair, H.P., Ahern, T.H., Neumann, I.D., Young, L.J., 2009. The CRF system mediates increased passive stress-coping behavior following the loss of a bonded partner in a monogamous rodent. Neuropsychopharmacology 34, 1406–1415.

Bouzinova, E.V., Moller-Nielsen, N., Boedtkjer, D.B., Broegger, T., Wiborg, O., Aalkjaer, C., Matchkov, V.V., 2012. Chronic mild stress-induced depression-like symptoms in rats and abnormalities in catecholamine uptake in small arteries. Psychosom. Med. 74, 278–287.

Bouzinova, E.V., Norregaard, R., Boedtkjer, D.M., Razgovorova, I.A., Moeller, A.M., Kudryavtseva, O., Wilborg, O., Aalkjaer, C., Matchkov, V.V., 2014. Association between endothelial dysfunction and depression-like symptoms in chronic mild stress model of depression. Psychosom. Med. 76, 268–276.

Bouzinova, E.V., Wiborg, O., Aalkjaer, C., Matchkov, V.V., 2015. Role of peripheral vascular resistance for the association between major depression and cardiovascular disease. J. Cardiovasc. Pharmacol. 65, 299–307.

Cacioppo, J.T., Hawkley, L.C., Thisted, R.A., 2010. Perceived social isolation makes me sad: 5-year cross-lagged analyses of loneliness and depressive symptomatology in the Chicago health, aging, and social relations study. Psychol. Aging 25, 453–463.

Carnevali, L., Mastorci, F., Graiani, G., Razzoli, M., Trombini, M., Pico-Alfonso, M.A., Arban, R., Grippo, A.J., Quaini, F., Sgoifo, A., 2012. Social defeat and isolation induce clear signs of a depression-like state, but modest cardiac alterations in wild-type rats. Physiol. Behav. 106, 143–150.

Carnevali, L., Statello, R., Sgoifo, A., 2018. The utility of rodent models of stress for disentangling individual vulnerability to depression and cardiovascular comorbidity. Curr. Cardiol. Rep. 20, 111.

Carney, R.M., Freedland, K.E., Rich, M.W., Smith, L.J., Jaffe, A.S., 1993. Ventricular tachycardia and psychiatric depression in patients with coronary artery disease. Am. J. Med. 95, 23–28.

Carney, R.M., Saunders, R.D., Freedland, K.E., Stein, P., Rich, M.W., Jaffe, A.S., 1995. Association of depression with reduced heart rate variability in coronary artery disease. Am. J. Cardiol. 76, 562–564.

Carney, R.M., Freedland, K.E., Stein, P.K., Skala, J.A., Hoffman, P., Jaffe, A.S., 2000. Change in heart rate and heart rate variability during treatment for depression in patients with coronary heart disease. Psychosom. Med. 62, 639–647.

Carter, C.S., Altemus, M., 2005. Oxytocin, vasopressin and depression. In: Den Boer, J.A., George, M.S., Ter Horst, G.J. (Eds.), Current and Future Developments in Psychopharmacology. Benecke N.I, Amsterdam.

Carter, C.S., DeVries, A.C., Getz, L.L., 1995. Physiological substrates of mammalian monogamy: the prairie vole model. Neurosci. Biobehav. Rev. 19, 303–314.
Cryan, J.F., Mombereau, C., Vassout, A., 2005a. The tail suspension test as a model for assessing antidepressant activity: review of pharmacological and genetic studies in mice. Neurosci. Biobehav. Rev. 29, 571–625.
Cryan, J.F., Page, M.E., Lucki, I., 2005b. Differential behavioral effects of the antidepressants reboxetine, fluoxetine, and moclobemide in a modified forced swim test following chronic treatment. Psychopharmacology (Berl) 182, 335–344.
Dantzer, R., O'Connor, J.C., Freund, G.G., Johnson, R.W., Kelley, K.W., 2008. From inflammation to sickness and depression: when the immune system subjugates the brain. Nat. Rev. Neurosci. 9, 46–57.
Das, U.N., 2000. Free radicals, cytokines and nitric oxide in cardiac failure and myocardial infarction. Mol. Cell. Biochem. 215, 145–152.
Felder, R.B., Francis, J., Zhang, Z.H., Wei, S.G., Weiss, R.M., Johnson, A.K., 2003. Heart failure and the brain: new perspectives. Am. J. Physiol. Regul. Integr. Comp. Physiol. 284, R259–R276.
Ferrari, R., Censi, S., Mastrorilli, F., Boraso, A., 2003. Prognostic benefits of heart rate reduction in cardiovascular disease. Eur. Heart J. Suppl. 5 (Suppl. G), G10–G14.
Francis, J., Weiss, R.M., Wei, S.G., Johnson, A.K., Felder, R.B., 2001. Progression of heart failure after myocardial infarction in the rat. Am. J. Physiol. Regul. Integr. Comp. Physiol. 281, R1734–R1745.
Francis, J., Weiss, R.M., Johnson, A.K., Felder, R.B., 2003. Central mineralocorticoid receptor blockade decreases plasma TNF-α after coronary artery ligation in rats. Am. J. Physiol. Regul. Integr. Comp. Physiol. 284, R328–R335.
Francis, J., Zhang, Z.H., Weiss, R.M., Felder, R.B., 2004. Neural regulation of the proinflammatory cytokine response to acute myocardial infarction. Am. J. Physiol. Heart Circ. Physiol. 287, H791–H797.
Frasure-Smith, N., Lespérance, F., Talajic, M., 1995. Depression and 18-month prognosis after myocardial infarction. Circulation 91, 999–1005.
Grippo, A.J., Moffitt, J.A., Johnson, A.K., 2002. Cardiovascular alterations and autonomic imbalance in an experimental model of depression. Am. J. Physiol. Regul. Integr. Comp. Physiol. 282, R1333–R1341.
Grippo, A.J., Beltz, T.G., Johnson, A.K., 2003a. Behavioral and cardiovascular changes in the chronic mild stress model of depression. Physiol. Behav. 78, 703–710.
Grippo, A.J., Francis, J., Weiss, R.M., Felder, R.B., Johnson, A.K., 2003b. Cytokine mediation of experimental heart failure-induced anhedonia. Am. J. Physiol. Regul. Integr. Comp. Physiol. 284, R666–R673.
Grippo, A.J., Santos, C.M., Johnson, R.F., Beltz, T.G., Martins, J.B., Felder, R.B., Johnson, A.K., 2004. Increased susceptibility to ventricular arrhythmias in a rodent model of experimental depression. Am. J. Physiol. Heart Circ. Physiol. 286, H619–H626.
Grippo, A.J., Francis, J., Beltz, T.G., Felder, R.B., Johnson, A.K., 2005. Neuroendocrine and cytokine profile of chronic mild stress-induced anhedonia. Physiol. Behav. 84, 697–706.
Grippo, A.J., Beltz, T.G., Weiss, R.M., Johnson, A.K., 2006. The effects of chronic fluoxetine treatment on chronic mild stress-induced cardiovascular changes and anhedonia. Biol. Psychiatry 59, 309–316.
Grippo, A.J., Cushing, B.S., Carter, C.S., 2007a. Depression-like behavior and stressor-induced neuroendocrine activation in female prairie voles exposed to chronic social isolation. Psychosom. Med. 69, 149–157.
Grippo, A.J., Gerena, D., Huang, J., Kumar, N., Shah, M., Ughreja, R., Carter, C.S., 2007b. Social isolation induces behavioral and neuroendocrine disturbances relevant to depression in female and male prairie voles. Psychoneuroendocrinology 32, 966–980.

Grippo, A.J., Lamb, D.G., Carter, C.S., Porges, S.W., 2007c. Social isolation disrupts autonomic regulation of the heart and influences negative affective behaviors. Biol. Psychiatry 62, 1162–1170.

Grippo, A.J., Moffitt, J.A., Johnson, A.K., 2008. Evaluation of baroreceptor reflex function in the chronic mild stress rodent model of depression. Psychosom. Med. 70, 435–443.

Grippo, A.J., Trahanas, D.M., Zimmerman II, R.R., Porges, S.W., Carter, C.S., 2009. Oxytocin protects against negative behavioral and autonomic consequences of long-term social isolation. Psychoneuroendocrinology 34, 1542–1553.

Grippo, A.J., Sgoifo, A., Mastorci, F., McNeal, N., Trahanas, D.M., 2010. Cardiac dysfunction and hypothalamic activation during a social crowding stressor in prairie voles. Auton. Neurosci. 156, 44–50.

Grippo, A.J., Moffitt, J.A., Sgoifo, A., Jepson, A.J., Bates, S.L., Chandler, D.L., McNeal, N., Preihs, K., 2012a. The integration of depressive behaviors and cardiac dysfunction during an operational measure of depression: investigating the role of negative social experiences in an animal model. Psychosom. Med. 74, 612–619.

Grippo, A.J., Pournajafi-Nazarloo, H., Sanzenbacher, L., Trahanas, D.M., McNeal, N., Clarke, D.A., Porges, S.W., Carter, C.S., 2012b. Peripheral oxytocin administration buffers autononomic but not behavioral responses to environmental stressors in isolated prairie voles. Stress 15, 149–161.

Grippo, A.J., Moffitt, J.A., Henry, M.K., Firkins, R., Senkler, J., McNeal, N., Wardwell, J., Scotti, M.A., Dotson, A., Schultz, R., 2015. Altered connexin 43 and connexin 45 protein expression in the heart as a function of social and environmental stress in the prairie vole. Stress 18, 107–114.

Guzzetti, S., La Rovere, M.T., Pinna, G.D., Maestri, R., Borroni, E., Porta, A., Mortara, A., Malliani, A., 2005. Different spectral components of 24 h heart rate variability are related to different modes of death in chronic heart failure. Eur. Heart J. 26, 357–362.

Isingrini, E., Surget, A., Belzung, C., Freslon, J.-L., Frisbee, J., O'Donnell, J., Camus, V., d'Audiffret, A., 2011. Altered aortic vascular reactivity in the unpredictable chronic mild stress model of depression in mice: UCMS causes relaxation impairment to ACh. Physiol. Behav. 103, 540–546.

Kapadia, S., Dibbs, Z., Kurrelmeyer, K., Karla, D., Seta, Y., Wang, F., Bozkurt, B., Oral, H., Sivasubramanian, N., Mann, D.L., 1998. The role of cytokines in the failing human heart. Cardiol. Clin. 16, 645–656.

Kaplan, J.R., Chen, H., Manuck, S.B., 2009. The relationship between social status and atherosclerosis in male and female monkeys as revealed by meta-analysis. Am. J. Primatol. 71, 732–741.

Kaplan, J.R., Chen, H., Appt, S.E., Lees, C.J., Franke, A.A., Berga, S.L., Wilson, M.E., Manuck, S.B., Clarkson, T.B., 2010. Impairment of ovarian function and associated health-related abnormalities are attributable to low social status in premenopausal monkeys and not mitigated by a high-isoflavone soy diet. Hum. Reprod. 25, 3083–3094.

Katz, R.J., 1981. Animal model of depression: effects of electroconvulsive shock therapy. Neurosci. Biobehav. Rev. 5, 273–277.

Katz, R., 1982. Animal model of depression: pharmacological sensitivity of a hedonic deficit. Pharmacol. Biochem. Behav. 16, 965–968.

Katz, R.J., Roth, K.A., Carroll, B.J., 1981. Acute and chronic stress effects on open field activity in the rat: implications for a model of depression. Neurosci. Biobehav. Rev. 5, 247–251.

Keller, M.B., Klein, D.N., Hirschfeld, R.M., Kocsis, J.H., McCullough, J.P., Miller, I., First, M.B., Holzer, C.P., Keitner, G.I., Marin, D.B., 1995. Results of the DSM-IV mood disorders field trial. Am. J. Psychiatry 152, 843–849.

Knowland, D., Lim, B.K., 2018. Circuit-based frameworks of depressive behaviors: the role of reward circuitry and beyond. Pharmacol. Biochem. Behav. 174, 42–52.

Koolhaas, J.M., Coppens, C.M., de Boer, S.F., Buwalda, B., Meerlo, P., Timmermans, P.J., 2013. The resident-intruder paradigm: a standardized test for aggression, violence and social stress. J. Vis. Exp. 77, e4367. https://doi.org/10.3791/4367.

Krittayaphong, R., Cascio, W.E., Light, K.C., Sheffield, D., Golden, R.N., Finkel, J.B., Glekas, G., Koch, G.G., Sheps, D.S., 1997. Heart rate variability in patients with coronary artery disease: differences in patients with higher and lower depression scores. Psychosom. Med. 59, 231–235.

La Rovere, M.T., Pinna, G.D., Maestri, R., Mortara, A., Capomolla, S., Febo, O., Ferrari, R., Franchini, M., Gnemmi, M., Opasich, C., Riccardi, P.G., Traversi, E., Cobelli, F., 2003. Short-term heart rate variability strongly predicts sudden cardiac death in chronic heart failure patients. Circulation 107, 565–570.

Leonard, B.E., Tuite, M., 1981. Anatomical, physiological, and behavioural aspects of olfactory bulbectomy in the rat. Int. Rev. Neurobiol. 22, 251–286.

Levine, B., Kalman, J., Mayer, L., Fillit, H.M., Packer, M., 1990. Elevated circulating levels of tumor necrosis factor in severe chronic heart failure. N. Engl. J. Med. 323, 236–241.

Lewis, R., Wilkins, B., Benjamin, B., Curtis, J.T., 2017. Cardiovascular control is associated with pair-bond success in male prairie voles. Auton. Neurosci. 208, 93–102.

Ma, L., Demin, K.A., Kolesnikova, T.O., Kharsko, S.L., Zhu, X., Yuan, X., Song, C., Meshalkina, D.A., Leonard, B.E., Tian, L., Kaleuff, A.V., 2017. Animal inflammation-based models of depression and their application to drug discovery. Expert Opin. Drug Discovery 12, 995–1009.

McNeal, N., Scotti, M.L., Wardwell, J., Chandler, D.L., Bates, S.L., LaRocca, M.A., Trahanas, D.M., Grippo, A.J., 2014. Disruption of social bonds induces behavioral and physiological dysregulation in male and female prairie voles. Auton. Neurosci. 180, 9–16.

McNeal, N., Appleton, K.M., Johnson, A.K., Scotti, M.L., Wardwell, J., Murphy, R., Bishop, C., Knecht, A., Grippo, A.J., 2017. The protective effects of social bonding on behavioral and pituitary-adrenal axis reactivity to chronic mild stress in prairie voles. Stress 20, 175–182.

Moffitt, J.A., Grippo, A.J., Holmes, P.V., Johnson, A.K., 2002. Olfactory bulbectomy attenuates cardiovascular sympathoexcitatory reflexes in rats. Am. J. Physiol. Heart Circ. Physiol. 283, H2575–H2583.

Mokhtari, M., Arfken, C., Boutros, N., 2013. The DEX/CRH test for major depression: a potentially useful diagnostic test. Psychiatry Res. 208, 131–139.

Morales-Medina, J.C., Iannitti, T., Freeman, A., Caldewell, H.K., 2017. The olfactory bulbectomized rat as a model of depression: the hippocampal pathway. Behav. Brain Res. 317, 562–575.

Moreau, J.-L., Bourson, A., Jenck, F., Martin, J.R., Mortas, P., 1994. Curative effects of the atypical antidepressant mianserin in the chronic mild stress-induced anhedonia model of depression. J. Psychiatry Neurosci. 19, 51–56.

Muscat, R., Willner, P., 1992. Suppression of sucrose drinking by chronic mild unpredictable stress: a methodological analysis. Neurosci. Biobehav. Rev. 16, 507–517.

Muscat, R., Papp, M., Willner, P., 1992. Reversal of stress-induced anhedonia by the atypical antidepressants, fluoxetine and maprotiline. Psychopharmacology (Berl) 109, 433–438.

Nalivaiko, E., Sgoifo, A., 2009. Central 5-HT receptors in cardiovascular control during stress. Neurosci. Biobehav. Rev. 33, 95–106.

Olshausen, K.V., Witt, T., Pop, T., Treese, N., Bethge, K.P., Meyer, J., 1991. Sudden cardiac death while wearing a Holter monitor. Am. J. Cardiol. 67, 381–386.

Peuler, J.D., Scotti, M.L., Phelps, L.E., McNeal, N., Grippo, A.J., 2012. Chronic social isolation in the prairie vole induces endothelial dysfunction: implications for depression and cardiovascular disease. Physiol. Behav. 106, 476–484.

Pitzalis, M.V., Iacoviello, M., Todarello, O., Fioretti, A., Guida, P., Massari, F., Mastropasqua, F., Russo, G.D., Rizzon, P., 2001. Depression but not anxiety influences the autonomic control of heart rate after myocardial infarction. Am. Heart J. 141, 765–771.
Pohl, T.T., Young, L.J., Bosch, O.J., 2019. Lost connections: oxytocin and the neural, physiological, and behavioral consequences of disrupted relationships. Int. J. Psychophysiol. 136, 54–63.
Porges, S.W., 2007. The polyvagal perspective. Biol. Psychol. 74, 116–143.
Ruis, M.A.W., te Brake, J.H.A., Buwalda, B., De Boer, S.F., Meerlo, P., Korte, S.M., Blokhuis, H.J., Koolhaas, J.M., 1999. Housing familiar male wildtype rats together reduces the long-term adverse behavioural and physiological effects of social defeat. Psychoneuroendocrinology 24, 285–300.
Rygula, R., Abumaria, N., Havemann-Reinecke, U., Rüther, E., Hiemke, C., Zernig, G., Fuchs, E., Flügge, G., 1998. Pharmacological validation of a chronic social stress model of depression in rats: effects of reboxetine, haloperidol and diazepam. Behav. Pharmacol. 19, 183–196.
Sapolsky, R.M., 1996. Why stress is bad for your brain. Science 273, 749–750.
Scotti, M.L., Carlton, E.D., Demas, G.E., Grippo, A.J., 2015. Social isolation disrupts innate immune responses in both male and female prairie voles and enhances agonistic behavior in female prairie voles (*Microtus ochrogaster*). Horm. Behav. 70, 7–13.
Sgoifo, A., De Boer, S.F., Westenbroek, C., Maes, F.W., Beldhuis, H., Suzuki, T., Koolhaas, J.M., 1997. Incidence of arrhythmias and heart rate variability in wild-type rats exposed to social stress. Am. J. Physiol. Heart Circ. Physiol. 273, H1754–H1760.
Shively, C.A., 1998. Social subordination, stress, behavior, and central monoaminergic function in female cynomolgus monkeys. Biol. Psychiatry 44, 882–891.
Shively, C.A., Day, S.M., 2015. Social inequalities in health in nonhuman primates. Neurobiol. Stress 1, 156–163.
Shively, C.A., Laber-Laird, K., Anton, R.F., 1997. Behavior and physiology of social stress and depression in female cynomolgus monkeys. Biol. Psychiatry 41, 871–882.
Shively, C.A., Williams, J.K., Laber-Laird, K., Anton, R.F., 2002. Depression and coronary artery atherosclerosis and reactivity in female cynomolgus monkeys. Psychosom. Med. 64, 699–706.
Shively, C.A., Register, T.C., Friedman, D.P., Morgan, T.M., Thompson, J., Lanier, T., 2005. Social stress-associated depression in adult female cynomolgus monkeys (*Macaca fascicularis*). Biol. Psychol. 69, 67–84.
Shively, C.A., Register, T.C., Adams, M.R., Golden, D.L., Willard, S.L., Clarkson, T.B., 2008. Depressive behavior and coronary artery atherogenesis in adult female cynomolgus monkeys. Psychosom. Med. 70, 637–645.
Shively, C.A., Musselman, D.L., Willard, S.L., 2009. Stress, depression, and coronary artery disease: modeling comorbidity in female primates. Neurosci. Biobehav. Rev. 33, 133–144.
Slattery, D.A., Cryan, J.F., 2017. Modelling depression in animals: at the interface of reward and stress pathways. Psychopharmacology (Berl) 234, 1451–1465.
Smith, R.S., 1991. The macrophage theory of depression. Med. Hypotheses 35, 298–306.
Solberg, L.C., Horton, T.H., Turek, F.W., 1999. Circadian rhythms and depression: effects of exercise in an animal model. Am. J. Physiol. Regul. Integr. Comp. Physiol. 276, R152–R161.
Song, C., Leonard, B.E., 1995. The effect of olfactory bulbectomy in the rat, alone or in combination with antidepressants and endogenous factors, on immune function. Hum. Psychopharmacol. 10, 7–18.
Song, C., Leonard, B.E., 2005. The olfactory bulbectomised rat as a model of depression. Neurosci. Biobehav. Rev. 29, 627–647.
Strawn, W.B., Bondjers, G., Kaplan, J.R., Manuck, S.B., Schwenke, D.C., Hansson, C.A., Clarkson, S., Clarkson, T.B., 1991. Endothelial dysfunction in response to psychosocial stress in monkeys. Circ. Res. 68, 1270–1279.

Suarez, E.C., Krishnan, R.R., Lewis, J.G., 2003. The relation of severity of depressive symptoms to monocyte-associated proinflammatory cytokines and chemokines in apparently healthy men. Psychosom. Med. 65, 362–368.
Sun, P., Smith, A.S., Lei, K., Liu, Y., Wang, Z., 2014. Breaking bonds in male prairie vole: long-term effects on emotional and social behavior, physiology, and neurochemistry. Behav. Brain Res. 265, 22–31.
Takahashi, A., Flanigan, M.E., McEwen, B.S., Russo, S.J., 2018. Aggression, social stress, and the immune system in humans and animal models. Front. Behav. Neurosci. 12, 56.
Talman, W.T., 1997. The central nervous system and cardiovascular control in health and disease. In: LOW, P.A. (Ed.), Clinical Autonomic Disorders, second ed. Lippincott-Raven Publishers, Philadelphia.
Tapanainen, J.M., Thomsem, P.E.B., Køber, L., Torp-Pedersen, C., Mäkikallio, T.H., Still, A.-M., Lindgren, K.S., Huikuri, H.V., 2002. Fractal analysis of heart rate variability and mortality after an acute myocardial infarction. Am. J. Cardiol. 90, 347–352.
Thayer, J.F., Brosschot, J.F., 2005. Psychosomatics and psychopathology: looking up and down from the brain. Psychoneuroendocrinology 30, 1050–1058.
Vian, J., Pereira, C., Chavarria, V., Köhler, C., Stubbs, B., Quevedo, J., Kim, S.-W., Carvalho, A.F., Berk, M., Fernandes, B.S., 2017. The renin-angiotensin system: a possible new target for depression. BMC Med. 15 (144), 1–13.
Wang, Q., Jie, W., Liu, J.-H., Yang, J.-M., Gao, T.-M., 2017. An astroglial basis of major depressive disorder? An overview. Glia 65, 1227–1250.
Willner, P., 1984. The validity of animal models of depression. Psychopharmacology (Berl) 83, 1–16.
Willner, P., 2005. Chronic mild stress (CMS) revisited: consistency and behavioural-neurobiological concordance in the effects of CMS. Neuropsychobiology 52, 90–110.
Willner, P., 2017. The chronic mild stress (CMS) model of depression: history, evaluation and usage. Neurobiol. Stress 6, 78–93.
Willner, P., Towell, A., Sampson, D., Sophokleous, S., Muscat, R., 1987. Reduction of sucrose preference by chronic unpredictable mild stress, and its restoration by a tricyclic antidepressant. Psychopharmacology (Berl) 93, 358–364.
Willner, P., Sampson, D., Papp, M., Phillips, G., Muscat, R., 1991. Animal models of anhedonia. In: Soubrié, P. (Ed.), Anxiety, Depression, and Mania. Animal Models of Psychiatric Disorders. Karger, Basel.
Willner, P., Muscat, R., Papp, M., 1992. Chronic mild stress-induced anhedonia: a realistic animal model of depression. Neurosci. Biobehav. Rev. 16, 525–534.
Wood, S.K., McFadden, K.V., Grigoriadis, D., Bhatnagar, S., Valentino, R.J., 2012. Depressive and cardiovascular disease comorbidity in a rat model of social stress: a putative role for corticotropin-releasing factor. Psychopharmacology (Berl) 222, 325–336.
Yee, J.R., Kenkel, W.M., Frijling, J.L., Dodhia, S., Onishi, K.G., Tovar, S., Saber, M.J., Lewis, G.F., Liu, W., Porges, S.W., Carter, C.S., 2016. Oxytocin promotes functional coupling between paraventricular nucleus and both sympathetic and parasympathetic cardioregulatory nuclei. Horm. Behav. 80, 82–91.

CHAPTER 4

Intervention research on therapies that aim to treat depression and cardiovascular disease

Casey E. Cavanagh[a,b], Allison E. Gaffey[a,b], Lindsey Rosman[a,b], Matthew M. Burg[a,b,c]
[a]Section of Cardiovascular Medicine, Department of Internal Medicine, Yale School of Medicine, New Haven, CT, United States
[b]VA Connecticut Healthcare System, West Haven, CT, United States
[c]Department of Anesthesiology, Yale School of Medicine, New Haven, CT, United States

Contents

Large, prospective, epidemiological studies of initially healthy individuals provide clear evidence that a history of major depressive disorder (MDD) increases risk of incident heart disease, with meta-analyses showing depression to carry a relative risk of 1.64 for cardiac incidence (Rugulies, 2002). Approximately 15–20% of patients with heart disease—e.g., stable coronary artery disease (CAD), unstable angina, or acute coronary syndrome (ACS) event—meet criteria for major depression (Carney and Freedland, 2008), while approximately 40% evidence clinically meaningful levels of depressive symptoms (Celano and Huffman, 2011), each of which increases risk for cardiac recurrence and mortality (Nicholson et al., 2006). Among ACS

Cardiovascular Implications of Stress and Depression
https://doi.org/10.1016/B978-0-12-815015-3.00004-0

patients with depression during hospital admission, more than half have depressive symptoms prior to their cardiac event. In addition to the increased risk for cardiac recurrence and early mortality, depression in these patients is associated with increased rates of disability, and health-care expenditures (Lichtman et al., 2008).

Given the adverse impact of comorbid depression on clinical outcomes and quality of life in patients with heart disease, a large clinical trial effort has been mounted over the past 3 decades to determine whether treating depression in these patients improves psychosocial and medical outcomes. These depression clinical trials have particularly been focused on immediate and recent post-ACS patients, and more recently, heart failure (HF) patients. This effort has specifically examined the effectiveness of cognitive-behavioral, pharmacological, and non-traditional interventions. The selection of interventions for these trials has been guided by the depression treatment evidence base (Archer et al., 2012; Donker et al., 2009; Hare et al., 2014; Linde et al., 2015). The aims of these trials have included both the mitigation of depression symptoms and improvement in medical outcomes. While some trials have shown improvements in depressive symptoms or quality of life, the effect of improved depression on cardiovascular disease (CVD) related outcomes has been disappointing at best. Furthermore, questions concerning the cost-effectiveness, feasibility, and resource implications of depression treatment in these patients remain to be addressed. In this chapter we provide a focused review of the intervention literature concerning depression in patients with CVD. We first review the literature on psychosocial/behavioral and related interventions, followed by reviews of pharmacotherapy interventions, and then a review of combined psychosocial/pharmacologic interventions.

Psychosocial and behavioral interventions

Psychosocial and behavioral interventions tested in clinical trials for depression in cardiac patients have included cognitive-behavioral therapy (CBT), collaborative care (CC), cardiac rehabilitation (CR), and stress management. In this section we review this literature and highlight both the successes and critical evidence gaps.

Cognitive-behavioral therapy

CBT is among the most efficacious treatments for depression in otherwise healthy individuals (Butler et al., 2006). Because of this, the potential benefits of this therapeutic approach for cardiac populations have been tested in multiple studies. CBT is a psychotherapy based on the principle that mental health problems and psychological distress such as depression, in large part result from, and are maintained by cognitive processes. In CBT, patients are taught to identify common cognitive distortions in their thinking and to challenge or restructure (cognitive restructuring) these thoughts. In addition to the building of these cognitive skills—identifying cognitive distortions in the context of daily experiences and testing these out in a Socratic process—patients are also engaged in behavioral activation—e.g., to "jump-start" behaviors that previously engendered positive effects, but that the person has stopped in the context of depression, the development of active problem-solving skills, relaxation training, and coping skills training to better respond to stressful and thus depressogenic life demands. Specific target populations have included patients with depression post-ACS, after sudden cardiac arrest, after coronary artery bypass (CABG) surgery, and in those with HF and/or implantable cardioverter defibrillators (ICDs) (c.f., Dekker, 2011; Salmoirago-Blotcher and Ockene, 2009; Habibovic et al., 2013).

Clinical trials evaluating CBT for the treatment of depression in patients with heart disease are characterized by this standard course of therapy, usually delivered during individual weekly sessions by a range of health care professionals, including psychologists, psychiatrists, social workers, and nurses for a prescribed period of time and/or until a threshold reduction of depression symptom severity is achieved (c.f., Freedland et al., 2009, 2015; Doering et al., 2013). In two studies, the effects of a brief CBT intervention were tested (Cully et al., 2017; Dekker et al., 2012), while in others, the effects of combined CBT + cardiac rehabilitation were tested (Frizelle et al., 2004). Often, therapy materials are adapted for the cardiac population—e.g., homework pages—to address concerns that are common to patients with heart disease. These adaptations can include avoidance of reminders of an acute cardiac event, the importance of adherence to medical treatments and recommendations, attention to health risk behavior change, and shock anxiety in patients with ICDs (c.f., Kohn et al., 2000).

Aggregate findings from a number of relatively small clinical trials suggests that CBT has efficacy for reducing depression severity among patients with heart disease (Salmoirago-Blotcher and Ockene, 2009; Dekker, 2011; Habibovic et al., 2013) with some data also demonstrating good depression remissions rates (Freedland et al., 2015). Yet, while depression symptoms show improvement in most clinical trials, there is limited evidence that this effect carries over to improved cardiovascular (CV) outcomes—e.g., decreasing cardiac recurrence/mortality or risk of arrhythmia requiring shock for termination in ICD patients (Dekker, 2011; Salmoirago-Blotcher and Ockene, 2009). One small study showed improved exercise capacity in depressed ICD patients (Frizelle et al., 2004), others reported longer cardiac event-free survival among depressed patients who received CBT (Cowan et al., 2001; Dekker et al., 2012), and two others reported that CBT was associated with fewer hospitalizations (Dekker et al., 2012; Freedland et al., 2015). Despite these findings, drawing conclusions has been limited by small sample sizes (Dekker, 2011; Salmoirago-Blotcher and Ockene, 2009) and under representation by women and ethnic minorities (Dekker, 2011).

Collaborative care

CC programs are defined by a multidisciplinary approach to care that includes a structured depression care management plan and frequent follow-up visits to assess the effects of the plan and make adjustments as necessary to reduce depression symptom severity and promote depression remission (Katon et al., 2010). This multidisciplinary approach is usually delivered in the context of primary care, with primary care providers or general practitioners supported by cardiologists, palliative care providers, psychiatrists, psychologists, social workers, and nurses (Huffman et al., 2011a). The multidisciplinary team together reviews cases and develops treatment recommendations focused on assessment of psychological distress, psychoeducation, and goal setting for symptom management. Some CC programs may also include elements of psychotherapy, such as active problem solving and/or behavioral activation (Bekelman et al., 2018). Although CC may also involve pharmacotherapy, trials combining these two therapeutic approaches are reviewed later in the chapter.

Despite the increasing use of CC programs for patients with chronic illness such as chronic pain and diabetes, few trials have tested the effectiveness of this approach for treating depression in patients with heart disease (Huffman et al., 2011a; Morgan et al., 2013). In one trial, CC for depressed cardiac in-patients was associated with improved cardiac related symptoms at 6-month follow-up (Huffman et al., 2011b), while in another, cardiac outpatients randomized to CC exercised significantly more than patients in the control group and experienced lower 10-year CVD risk (Morgan et al., 2013). Yet, in another trial for patients with chronic HF, mortality in the CC group was comparable to that in the usual care (UC) group (Bekelman et al., 2018). The Montreal Heart Attack Readjustment

Trial (M-HART) (Frasure-Smith et al., 2002) tested the effectiveness of a psychosocial intervention similar to CC in post-ACS patients. In this trial, patients randomized to the intervention received CC nursing home visits when they evidenced elevated symptoms of distress, with these visits focusing on problem-solving and support. Results of this trial revealed that distressed men in the intervention group had reduced cardiac mortality at follow-up, while in contrast, distressed women had higher cardiac morality. Of note, post-hoc analyses of trial data revealed that for trial participants in the intervention group, those who demonstrated reduced distress in association with the nursing visits had lower cardiac mortality at follow-up, while those who evidenced no change—or an increase in distress—after the nursing visits, had increased mortality (Cossette et al., 2001). This post-hoc finding indicates that in the testing of any intervention in a clinical trial for depression, it is essential to demonstrate that the intervention improves the most proximal dependent measure, in this case, distress.

Cardiac rehabilitation and exercise

Evidence suggests that exercise is moderately effective in reducing depression symptom severity and promoting depression remission (Cooney et al., 2013). Given that exercise-based CR programs reduce CV mortality, there has been extensive research testing the effects of exercise and CR on depression in CVD (Anderson et al., 2016). These trials have included post-ACS, post-CABG, and post-percutaneous coronary intervention patients, and have been conducted internationally (Zheng et al., 2019). CR programs often include a structured exercise program conducted in clinic or at home, complemented by relaxation training stress management, patient education, and nutritional counseling, which are usually delivered by nurses, physicians, and/or physiotherapists (Pesah et al., 2017). Unlike CBT and CC programs, there is strong evidence supporting the efficacy of CR programs in treating depression and reducing CV related mortality. A recent systematic review and meta-analysis of six studies including a combined 703 patients found that CR programs were associated with significant reductions in depression scores (Zheng et al., 2019). An earlier meta-analysis also found these programs to be efficacious in treating depression and associated with fewer coronary heart disease (CHD) events and lower mortality (Rutledge et al., 2013). Other research demonstrates improved CHD biomarkers, lipids, and exercise capacity for patients participating in CR programs (Blumenthal et al., 2016).

Other exercise programs such as t'ai chi, have also been tested to evaluate the effects on depression in patients with heart disease. In one pilot study, Redwine and colleagues assigned 28 HF patients to t'ai chi training or UC (Redwine et al., 2012). T'ai chi training was provided by a certified holistic health practitioner with experience working with chronically ill older adults. Patients in the t'ai chi group received two trainings per week for 12 weeks and were instructed to practice daily on the five non-instructor led training days. Compared to the UC group, the t'ai chi group showed significant reductions in depression symptom severity. The promising finding from this small trial suggests the need for larger trials to test patient acceptance, and both depression and medical outcomes associated with t'ai chi for depressed patients with various heart disease presentations.

Stress management/relaxation training

Multiple studies have shown that stress management and/or relaxation training can be effective for improving event-free survival among patients with heart disease (c.f., Black et al., 1998; Cully et al., 2017; Frizelle et al., 2004; Lewin et al., 2009; Cowan et al., 2001; Pesah et al., 2017; Blumenthal et al., 2016). In one notable large clinical trial conducted in Sweden (Gulliksson et al., 2011), patients randomized to the CBT condition demonstrated a 41% lower rate of combined fatal/non-fatal cardiac events, and a 45% lower rate of recurrent ACS vs UC. Yet, few studies have tested stress management for depression in patients with heart disease. Freedland and colleagues compared the efficacy of CBT, supportive stress management, and UC in treating depression in 123 post-CABG patients (Freedland et al., 2009). Supportive stress management was defined by coping skills training and progressive relaxation training, controlled breathing, and relaxing imagery. Patients in both the CBT and supportive stress management groups showed significantly lower depression scores compared to the UC group. Furthermore at 3 months, depression scores were significantly lower in the CBT group vs the supportive stress management group, suggesting that CBT may be effective for treating depression in this patient group.

Caregiver interventions

Research suggests that partners and caregivers of patients with CVD often experience psychological distress, including depression (Chung et al., 2010, 2016). Dougherty and colleagues found that during the first 12 months after ICD implant, partners of patients with ICDs demonstrated depression scores that were significantly higher than those of the patients (Dougherty et al., 2016, 2019; Dougherty and Thompson, 2009). In a study of patients with HF, 27.5% of caregivers reported depressive symptoms, which was associated with poorer functional status and higher caregiver burden (Chung et al., 2010). In addition, caregivers of patients with HF and depression experience more caregiver burden and a lower mental quality of life (Chung et al., 2016). Evidence also indicates that caregiver psychological distress is associated with patient HF symptoms and that there is a significant relationship between higher caregiver psychological distress and clinical event risk in HF patients (Bidwell et al., 2017). These data suggest that providing support to caregivers and involving caregivers in interventions may influence patient depression and potentially CV outcomes.

Several studies have tested caregiver interventions in the context of HF. A review of eight such caregiver interventions found that psychoeducation and support groups are the most common, often supplemented with follow-up in-person sessions, home visits, telephone calls, and telemonitoring (Evangelista et al., 2016). Only four of the eight studies measured depression and in most there were no significant effects (Lofvenmark et al., 2013; Agren et al., 2012; McMillan et al., 2013; Piette et al., 2011). Yet, one study showed a significant reduction in depression for caregivers who received mobile health (mHealth) based feedback about the HF patient (Piette et al., 2015); few studies measured patient depression or patient CV outcomes (McMillan et al., 2013; Agren et al., 2012). Therefore, it is difficult to determine if interventions focused on the caregiver have a significant impact on patient outcomes in this population.

Caregiver interventions have also been examined in partners of patients with ICDs. In one study of 301 patients with ICDs and their partners, Dougherty and colleagues tested the effects of a patient plus partner (P+P) intervention as compared to a patient-only (P-only) intervention (Dougherty et al., 2019). The P-only intervention included education about patient experiences, 10 phone contacts focused on patient progress facilitated by a cardiovascular nurse, access to a research nurse 24/7 for questions or concerns, and an informational video about the ICD. The P+P intervention included these same components and seven additional elements: managing physical symptoms, returning to usual physical activities, reviewing safe exercise practices, coping with depression and anxiety, coping with and managing ICD shocks, maintaining surveillance, and learning how to use the emergency medical system. Patient depression scores in the P+P intervention decreased significantly vs the P-only intervention. In contrast, there were no significant reductions or significant group differences in partner depression scores. Additionally, there were no group differences in patient or partner hospitalizations. The number of outpatient ICD visits was lower in the P+P group. These promising findings suggest that further testing of caregiver interventions is warranted.

Nutritional interventions

Recent research suggests that some nutritional supplements, such as omega-3 fatty acids (FAs) may be efficacious in treating depression, particularly as adjunctive therapy with selective serotonin reuptake inhibitors (SSRIs) (Rechenberg, 2016). More specifically, research suggests that eicosapentaenoic acid (EPA)-predominant formulations of omega-3 highly unsaturated fatty acids may be effective for depression (Hallahan et al., 2016). Similarly, there is evidence that consumption of fish or fish oil is associated with reduction in CHD mortality, fatal myocardial infarction (MI), and sudden cardiac death, and that omega-3 polyunsaturated fatty acids are associated with reductions in total mortality (Mozaffarian and Wu, 2011; Leon et al., 2008; Manik and Varon, 2009). Although the American Heart Association's current guidelines (Siscovick et al., 2017) recommend the use of omega-3 FAs in patients with prior CHD, HF, or reduced left ventricular ejection fraction (LVEF), a recent meta-analysis found that omega-3 FAs were not associated with fatal or nonfatal CHD or major vascular events (Aung et al., 2018).

In the Omega-3 Supplementation for Co-morbid Depression and Heart Failure Treatment (OCEAN), 108 patients with HF and MDD were randomized to one of three conditions: (1) 4 capsules of 400/200 EPA/DHA 500 mg per capsule (2:1 EPA/DHA); (2) 4 capsules of EPA 500 mg per capsule (high EPA); (3) 4 capsules of placebo (corn oil) daily for 12 weeks (Jiang et al., 2018). While analyses suggested that omega-3s improved depressive symptoms, psychosocial functioning, and increased endogenous omega-3 levels, treatment with either dose of omega-3s was not associated with significant improvements in depression.

Pharmacotherapy

Preliminary assessments of pharmacologic agents in cardiac patients included very small groups of patients with and without depression and were designed as observational studies rather than placebo-controlled randomized controlled trials (RCTs). Results provided initial evidence that antidepressants could benefit both psychological and cardiovascular outcomes (Roose et al., 1998a,b; Strik et al., 1998; Shapiro et al., 1999). The safety of potential pharmacological treatments was of central interest to investigators, particularly given the known cardiotoxic effects of older agents and the emergence of the new SSRIs class of medication. Although studies found fewer cardiotoxic effects of SSRIs in contrast with the older class of tricyclic antidepressants, there was evidence that SSRIs would elevate cardiovascular risks including altered heart rate, decreased heart rate variability, and increased systolic blood pressure (Roose et al. 1998a,b; Strik et al., 1998; Shapiro et al., 1999). The Sertraline Antidepressant Heart Attack Randomized Trial (SADHEART) was an early trial designed to test the safety of one particular SSRI for patients with acute MI or unstable angina (Glassman et al., 2002). This relatively small trial found sertraline to be safe, with no untoward effects vs placebo on LVEF, cardiac arrhythmias, or various electrocardiography (ECG/EKG) parameters. There was modest improvement in depression symptom severity and overall depression remission, and a lower incidence of major adverse cardiac events at the end of the 24-week treatment period, though this latter finding was not statistically significant. A more recent randomized trial testing the safety of this SSRI for depression in HF patients also had demonstrated safety. Furthermore, while treatment was not associated with a greater reduction in depression or improvement in cardiovascular status vs placebo, remission of depression *was* associated with better HF related outcomes (Jiang et al., 2011).

The Myocardial Infarction and Depression-Intervention Trial (MIND-IT) advanced this area of research by introducing a stepped protocol design (De Jonge et al., 2007). The primary goal of MIND-IT was to determine if treatment with an antidepressant (mirtazapine) as compared to UC could reduce the incidence of cardiac events in patients who had a recent ACS event and screened positive for depression (De Jonge et al., 2006a,b). Patients who refused mirtazapine or were nonresponsive to the drug after 8 weeks were offered citalopram as a second treatment option. Of those randomized to treatment, 45 received no treatment (due to refusal); 94 were randomized to mirtazapine vs placebo and of these 46 were switched to citalopram because they remained depressed while receiving the initial medication; 40 received non-pharmacologic depression treatment; and 45 received no treatment. Receiving an antidepressant did not improve depression or reduce cardiac events (Van Melle et al., 2007). In secondary analyses, patients categorized as responders or nonresponders to treatment (based on demonstrating at least a 50% decrease in depression symptoms at 24 weeks) had a lower cardiac event rate than nonresponders (De Jonge et al., 2007).

Like the SADHART study, the Korean, Escitalopram for Depression in Acute Coronary Syndrome (EsDEPACS) investigation tested 24-weeks of escitalopram on cardiovascular indices in 300 post-ACS patients (Kim et al., 2014). Unique to this study, treatment effects on quality of life and occupational functioning were also evaluated. Escitalopram was superior to placebo for depression, occupational functioning, and quality of life (Kang et al., 2015).

In a related but distinct group of studies, the efficacy of SSRIs for depression and functional recovery following an acute cerebrovascular event has been tested. Fluoxetine for Motor Recovery after Acute Ischemic Stroke (FLAME) was a French multi-center clinical trial (Chollet et al., 2011) that randomized patients with moderate to severe motor deficit ($n=231$) to 3-months of fluoxetine or placebo. Fluoxetine led to better recovery and lower symptoms of depression. Another large investigation of SSRI treatment in ischemic stroke, the Korean Efficacy of Early Administration of Escitalopram on Depressive and Emotional Symptoms (EMOTION) (Kim et al., 2017), randomized patients ($n=405$) to 3-months of escitalopram vs placebo. Escitalopram was found to be successful for alleviating mild (but not moderate or severe) depression, and there was no difference in adverse events between escitalopram and placebo. Although these studies in stroke lacked sufficient power, it appears that SSRIs may be safe in this patient group and may offer a promising approach to improving functional, cognitive, or depressive symptoms.

Other pharmacotherapy treatment

Further research examining depression and CV outcomes has tested the merit of administering omega-3 FAs to enhance the effects of SSRIs or as a stand-alone treatment. As previously discussed, a robust literature suggests that omega-3s are effective when treating psychiatric patients (e.g., Grosso et al., 2014). Carney et al. (2009) conducted an RCT of patients with CHD and depression ($n=122$), who were treated with sertraline and then randomized to receive omega-3 or a corn oil placebo for 10 weeks. Both groups improved over time and patients who received omega-3s did not show significantly improved depression symptoms compared to those who received placebo. In another study, the effect of augmentation with 12 weeks of omega-3 FAs on depressive symptoms in patients with CAD participating in CR ($n=92$) was assessed. Notably, the sample consisted of those with and without depression (Mazereeuw et al., 2016a,b). Compared to placebo, treatment with omega-3s did not significantly reduce depression or enhance general cognitive performance, though most patients reported mild or no symptoms of depression at baseline.

Underwhelming results from RCTs examining omega-3 treatments may relate to the small sample sizes and warrant replication with larger groups. Dose-dependent relations between treatment with omega-3s and SSRIs, longer treatment, or treatment with solely omega-3s could all influence results (Carney et al., 2009; Grosso et al., 2014). Given the varied literature on the efficaciousness of SSRIs alone for cardiac patients, determining the most effective delivery for antidepressant medication is a requisite first step before complementing the medication with omega-3 FAs or other supplements.

Combined behavioral and psychopharmacology interventions

As reviewed earlier, behavioral treatments and psychologically supportive interventions, such as CBT, CC, CR, and stress management, provide a powerful toolkit for addressing depression and improving CV outcomes in patients. Findings from the several reviews provided especially compelling evidence as to the value of CBT interventions in reducing depressive symptoms (Salmoirago-Blotcher and Ockene, 2009; Dekker, 2011; Habibovic et al., 2013) with some additional evidence that CBT interventions may improve CV outcomes (Dekker, 2011; Salmoirago-Blotcher and Ockene, 2009). Given the variability in RCT outcomes, as well as the differing cost and resources required for treatment, it is understandable that combining behavioral and pharmacological treatments as well as the comparative efficacy of behavioral compared to pharmacological interventions are inquiries of great interest to researchers and clinicians alike.

The Enhancing Recovery in CHD Patients (ENRICHD) Trial was the first multi-center randomized clinical trial that was large enough ($n=2481$) to test the effects of depression treatment on a primary endpoint of recurrent MI and mortality in patients with major depression after MI (Berkman et al., 2003). For this trial, up to 6-months of individual cognitive therapy (CT) for depression, augmented by pharmacologic treatment with sertraline for patients with severe depression, or with depression not sufficiently responsive to CT, was chosen for the intervention. This individual therapy was further augmented by group-based psychotherapy that included both depression remission strategies and stress reduction strategies, though for logistical reasons, only about one third of patients randomized to treatment participated in group sessions. Six-month improvements in depression symptom severity were modest vs UC, with no differential effects on medical outcomes, and group differences on depression at 12-month follow-up were largely gone. Lastly, there was a troubling sign that minority women assigned to the treatment group had worse medical outcomes, which was similar to the earlier finding in the 'case-management' for distress, M-HART trial, again highlighting the importance of delivering an intervention suited to the target patient population.

Among the questions raised by the M-HART and ENRICHD trials was the timing of depression intervention for post-ACS patients, the method with which to choose appropriate treatment, and risks of the intervention when assessing CV morbidities and cardiac mortality (Burg et al., 2008). For example, in ENRICHD, the improvement in depression symptom severity observed among patients randomized to UC—"spontaneous remission"—may have been a consequence of assessing depression over too short a time period post-ACS, with many patients who "appeared" eligible merely having an expected adjustment to having experienced a major life event. With this reasoning, many patients who did not have a true depression syndrome may have been consented and randomized, thereby reducing the power of the study to find an effect of treatment. Similarly, the trend toward harm among women in both trials may have been a consequence of a study intervention not well suited to the patient's needs.

These issues informed the design of two small depression trials for post-ACS patients, the multi-center Coronary Patients Evaluation Study (COPES; $n = 209$) (Davidson et al., 2010) and the subsequent, Comparison of Depression Interventions after Acute Coronary Syndrome (CODIACS; $n = 150$) Vanguard Trial (Davidson et al., 2013). The largely identical study protocols were based upon the design of the earlier Improving Mood-Promoting Access to Collaborative Treatment (IMPACT) Trial, which was based in a CC model and entailed a patient preference, stepped care approach to depression treatment (Unützer et al., 2002). For both COPES and CODIACS, post-ACS patients were screened for elevated depression symptoms at the time of ACS, and then again at 1- and 3-months post-event. Patients with persistent depression elevation were then randomized to intervention or UC. Intervention lasted for up to 6-months and consisted of a patient choice between brief problem-solving therapy (PST) or an SSRI; for CODIACS, PST was delivered centrally by phone rather than local therapists trained in the protocol. Evaluations were conducted at 2-month intervals. If the patient was on a sufficient improvement trajectory, treatment as chosen remained in effect. If the patient was not improving sufficiently, treatment was intensified, or switched, with the patient participating in the decision. Comparable groups of patients preferred counseling compared to medication. Patients randomized to intervention in both the COPES and CODIACS trials reported high satisfaction with depression care. Both trials also found the patient preference, stepped-care approach to produce significant improvement in depression symptom severity vs UC. Furthermore, in COPES, randomization to the treatment condition was associated with a lower risk of cardiac recurrence or death at the end of the 6-month treatment, an effect that did not persist after treatment ended.

Another CC based clinical trial, the Bypassing the Blues (BtB) Trial ($n=259$) tested this type of intervention on patient self-reported mental and physical health-related quality of life after CABG (Rollman and Belnap, 2011). Patients who endorsed at least moderate depression symptom severity were randomized to either an 8-month course of nurse-delivered telephone-based CC supervised by a psychiatrist and primary care expert, or to their physicians' UC. Patients who screened negative for depression served as an additional control. Patients randomized to CC received regular nurse care manager phone contacts to, (1) provide basic education on depression and treatment options, (2) assess for depression treatment preferences, (3) offer a depression self-management workbook and review lessons, if applicable, (4) promote adherence/adjust depression pharmacotherapy in concert with the patient's PCP and monitor for medication side effects, and (5) monitor the patient's treatment response. Referral to a mental health provider was also part of the CC condition, as needed according to the patient's treatment choice and depression severity. When declined any active depression treatment, the nurse care manager continued to contact the patient monthly to reassess mood and encourage treatment initiation. Patients randomized to CC exhibited greater improvement in depression and quality of life vs UC, with men showing greater improvement than women. In secondary analyses, patients who were consistently taking antidepressant medication at baseline and follow-up showed the least improvement in depression and quality of life (Meyer et al., 2014).

Using a distinct psychotherapy approach, the Canadian Cardiac Randomized Evaluation of Antidepressant and Psychotherapy Efficacy (CREATE) Trial, randomized patients with CAD and MDD ($n=284$) to

12 weeks of citalopram or a placebo, concurrent with 12 weeks of interpersonal psychotherapy (IPT) (Lespérance et al., 2007) vs UC. Clinical management was also offered to all patients. Compared to placebo, citalopram produced a better treatment response, reducing 12-week depression symptoms with a moderate effect size. Patients who received the antidepressant also demonstrated significantly lower readmission rates. IPT did not prove more advantageous than clinical management for reducing depression severity. When joining these findings with those of other clinical trials of SSRIs alone, adding clinical management to psychotropic treatment appears to be a reasonable first step for treating CAD patients with a complementary dual approach.

A multi-component approach has also been tested to prevent minor or major depression after stroke (Robinson et al., 2008). In this trial, non-depressed acute stroke patients were randomized to a 12-month course of treatment with escitalopram or placebo, in addition to PST. For 6 months after treatment, individuals who received escitalopram and PST exhibited fewer depression symptoms and were less likely to experience new onset major depression. In additional analyses, the benefits of escitalopram appeared significantly greater than the placebo although PST had comparable effects to placebo. A primary study limitation was the absence of a sufficient control group. It is also notable that the behavioral treatment groups were not blinded, which might have resulted in bias. In a follow-up study, compared to the problem-solving or placebo groups, discontinuing escitalopram was associated with greater depressive symptoms in the 6 months after ending treatment (Mikami et al., 2011). These results serve as an important warning; if patients are prescribed an antidepressant following a stroke or another acute event, it may be critical to titrate the dosage over time rather than abruptly, and to monitor patient symptoms in the months after terminating treatment.

Finally, although psychosocial interventions are the most common behavioral techniques combined with and compared to pharmacotherapy, investigators have also evaluated other behavioral methods, in particular, aerobic exercise. This is based in part on the benefits of exercise more broadly for patients with heart disease, and findings from the depression literature concerning the effects of aerobic exercise (Cooney et al., 2013). The Understanding the Prognostic Benefits of Exercise and Antidepressant Therapy (UPBEAT) study, was designed to determine the efficacy of exercise versus an SSRI (sertraline) for alleviating depression and enhancing CV biomarkers for patients with CHD and comorbid depression (Blumenthal et al., 2007). Patients were randomized to 90 min of weekly aerobic exercise, sertraline, or placebo for 16 weeks. After this 16-week period the entire sample showed improvement in depression symptom severity, with exercise and sertraline being superior to placebo but comparable in their effects (Blumenthal et al., 2012). Furthermore, compared to placebo, exercise and sertraline both improved heart rate variability, although exercise was found to have a greater effect on this variable than sertraline. There was no effect for either intervention on additional biomarkers—e.g., endothelial function, inflammation, platelet function—though this may have been a consequence of insufficient power, given the size of effect observed. As noted by the investigators, the effectiveness of exercise for depression in this cardiac population may also relate to the degree of engagement required of participants to complete the study (Blumenthal et al., 2012). Findings from this clinical trial suggest that exercise may be an alternative to antidepressants for treatment of depression and may be particularly attractive for treating those who cannot tolerate the risks of SSRIs.

Limitations

The research concerning depression treatment in CVD is notable for several limitations. For example, studies are generally characterized by an underrepresentation of women and ethnic minorities and small sample sizes, which limits generalizability and the ability to detect effects (Dekker, 2011). The issue of sample size is of particular note when the desire is to determine whether treating depression can improve medical outcomes, since the 1-year event rate after an acute cardiac event is now quite low. Any clinical trial effort would therefore require an exceedingly large sample size with commensurate costs, a major limitation. For example, a post hoc power calculation by the SADHART team determined that an RCT with a sample size of 4000 patients (rather than the 469 participants in SADHART) was necessary to firmly conclude that treatment with sertraline lowered the risk of adverse events by 20% (Glassman et al., 2002), and event rates have improved further over the intervening 16 years.

Another issue is that most studies do not provide information about treatment fidelity, or patient acceptance of, and adherence to treatment (Salmoirago-Blotcher and Ockene, 2009). Low treatment acceptance can lead to low treatment fidelity/treatment adherence, thereby reducing the overall impact of the intervention condition and producing the observed small effect sizes and mixed findings. An example of this can be seen in the secondary analysis from

M-HART (Cossette et al., 2001), and in the "post-mortem" from ENRICHD (Burg and Czajkowski, 2011). An additional issue concerning pharmacotherapy interventions specifically is that the medications and doses tested may be inappropriate for, or poorly tolerated by cardiac patients. This too can lead to low treatment acceptance and low fidelity/non-adherence.

There is also the issue of when best to target depression treatment. For example, in both ENRICHD and SADHART there were high rates of spontaneous remission (Berkman et al., 2003; Glassman et al., 2002). It is also important to note that depression is a syndrome defined by a constellation of symptoms, some of which are necessary—depressed mood, anhedonia—and some of which vary from person to person. Furthermore, depression is a recurring/remitting condition with first onset usually in early adulthood, associated with a major life event. Yet, among depressed cardiac patients, some have the expected lifetime history, while others have what is called "incident depression"—e.g., first onset in later adulthood, in association with the cardiac event. The cardiotoxicity of different symptom clusters—cognitive, affective, somatic—has been a focus of the depression/heart disease research, as has the importance of lifetime course vs incident depression in this patient group (Bush et al., 2001; De Jonge et al., 2006a,b; Van Melle et al., 2007). If the aim of a clinical trial is to determine whether treating depression in patients with CVD improves medical outcomes, these issues become critical.

Many of clinical trials included only patients meeting diagnostic criteria for major depression. Yet, the research linking depression to cardiac recurrence and early mortality in heart disease demonstrates that a relatively low threshold of depression symptom severity—a threshold well below that necessary to meet diagnostic criteria—carries risk (Nicholson et al., 2006). Again, if the goal is to improve event-free survival after an acute cardiac event, failing to include the full range of depression associated risk among the inclusion criteria limits the impact of any clinical trial. A final limitation to note is the failure in most trials to follow patients for a sufficient period after depression treatment (Kronish et al., 2012; Rollman and Belnap, 2011; Dekker, 2011). As noted earlier, depression is a remitting recurring disorder, and thus failure to account for recurrence, which is evident in essentially all depression CVD clinical trials, severely limits the impact of any findings (Ye et al., 2014).

Summary of findings and future considerations

The evidence linking a lifetime history of depression to incident CVD, and also linking depression in the setting of a cardiac event to cardiac recurrence and early mortality, is strong. Notwithstanding the conclusions of a recent meta-analysis (Richards et al., 2018), the evidence demonstrating efficacy of depression treatment—regardless of modality—for patients with heart disease, or for any effect on medical outcomes among those with depression is however, mixed at best. The strength of evidence varies considerably across cardiac populations, with notable support for the use of specific psychological interventions to treat depression in CVD.

The clinical trial evidence is strongest with regard to CBT, CC, and CR/exercise programs for improving depression symptom severity and promoting remission of major depression in cardiac populations (Berkman et al., 2003; Salmoirago-Blotcher and Ockene, 2009; Dekker, 2011; Habibovic et al., 2013; Zheng et al., 2019; Blumenthal et al., 2012, 2016), especially when these modalities are combined with pharmacotherapy (Davidson et al., 2010, 2013; Lespérance et al., 2007; Robinson et al., 2008; Berkman et al., 2003; Rollman and Belnap, 2011). Other interventions focused on caregiver and nutrition have demonstrated inconsistent findings. Studies investigating omega-3 FAs found no significant improvements in depressive symptoms (Grosso et al., 2014; Mazereeuw et al., 2016b; Carney et al., 2009; Jiang et al., 2018). In addition, research suggests that caregiver interventions, for example in patients with HF, do not significantly improve depression in the patient being cared for (Lofvenmark et al., 2013; Agren et al., 2012; McMillan et al., 2013; Piette et al., 2011), though findings from one trial demonstrate improvement in ICD patients' depression severity associated with a psychosocial intervention targeting both the patient and their partner (Dougherty et al., 2019).

While support for psychosocial interventions to treat depression in CVD is relatively modest, evidence that treating depression can improve medical outcomes in patients with heart disease—e.g., prevent cardiac recurrence and/or improve event-free survival—is very limited. A notable exception is the two small multi-center clinical trials by Davidson and colleagues—COPES and CODIACS, each of which addressed many of the limitations described above. These two trials are noteworthy for the use of a patient preference, stepped approach to care previously found acceptable to medical patients with depression. In each of these trials there was also a 3-month period of watchful waiting after a cardiac event, and only those patients with persistent depression—both at the time of their event and 3-months later—were randomized. In addition, a diagnosis of MDD was not necessary to meet eligibility. Rather a threshold score on the Beck Depression Inventory ($\geq$10) that was previously associated with cardiac recurrence and mortality was used. In both trials, ACS patients assigned to the intervention condition demonstrated significant reduction in depression symptoms vs UC, and patients expressed high satisfaction with the intervention. In the COPES trial, treatment was also associated with a reduced risk of death or hospitalization for MI/unstable angina at the end of the 6-month treatment period. The effect overall did not persist after treatment ceased, however. The results of these two trials offer the most promise moving forward, as they provide a model for testing of an intervention "program" that is, (1) acceptable to the patient population of interest—those with CVD, (2) utilizes a screening/recruitment approach that maximizes the identification of patients at highest risk for depression associated cardiac recurrence/mortality, and (3) delivers a relatively low cost intervention that is easily disseminable in the context of busy clinical settings (Unützer et al., 2002). While COPES and CODIACS provide a promising pathway forward for future depression/CVD clinical trials, they also offer a cautionary note in that when treatment ended, depression eventually returned. This raises the specter of depression as a chronic aspect of CVD that requires an ongoing surveillance and treatment effort, much in the way hypertension, diabetes, and hypercholesterolemia do. Trial designers should consider how best to address this fact—e.g., how to build in ongoing surveillance after symptoms remit, and re-engagement with booster sessions when symptoms re-emerge. Psychotherapy approaches that include self-surveillance, perhaps utilizing electronic (eHealth) technologies that allow reporting of symptom elevations to the healthcare provider—should be included in future such trials.

We see two additional issues of critical importance moving forward. One question concerns the appropriate final target of intervention: improvement in depression symptom severity/remission, or this improvement in the service of a reduction in cardiac recurrence/mortality. Addressing this issue may require a clearer understanding of the pathways by which depression contributes to CVD risk—whether through autonomic disruption, platelet activity, inflammation, or health risk behavior/adherence. Indeed, it may be that the different depression presentations and constellations contribute to CVD outcomes by different pathways, each of which might need consideration in future clinical trial design—e.g., through the selection of which treatment approach to test.

The final issue concerns whether attention to depression in CVD patients is at worst benign. This is the assumption made by the leadership of the major cardiologic, nursing, and psychiatric/psychologic professional organizations in their published recommendations that screening for depression in

CVD patients, and referral for care be universally implemented (Litchman et al., 2008). This recommendation was made in the absence of any evidence base of support and thus, in the absence of any evidence that this approach would not lead to patient harm. This too is a critical question needing attention in any future, large scale efforts directed toward depression in patients with CVD.

References

Agren, S., Evangelista, L.S., Hjelm, C., Stromberg, A., 2012. Dyads affected by chronic heart failure: a randomized study evaluating effects of education and psychosocial support to patients with heart failure and their partners. J. Card. Fail. 18, 359–366.

Anderson, L., Thompson, D.R., Oldridge, N., Zwisler, A.D., Rees, K., Martin, N., Taylor, R.S., 2016. Exercise-based cardiac rehabilitation for coronary heart disease. Cochrane Database Syst. Rev. 1, CD001800.

Archer, J., Bower, P., Gilbody, S., Lovell, K., Richards, D., Gask, L., Dickens, C., Coventry, P., 2012. Collaborative care for depression and anxiety problems. Cochrane Database Syst. Rev. 10, CD006525.

Aung, T., Halsey, J., Kromhout, D., Gerstein, H.C., Marchioli, R., Tavazzi, L., Geleijnse, J.M., Rauch, B., Ness, A., Galan, P., Chew, E.Y., Bosch, J., Collins, R., Lewington, S., Armitage, J., Clarke, R., 2018. Omega-3 Treatment Trialists' Collaboration, Associations of omega-3 fatty acid supplement use with cardiovascular disease risks: meta-analysis of 10 trials involving 77917 individuals. JAMA Cardiol. 3, 225–234.

Bekelman, D.B., Allen, L.A., Mcbryde, C.F., Hattler, B., Fairclough, D.L., Havranek, E.P., Turvey, C., Meek, P.M., 2018. Effect of a collaborative care intervention vs usual care on health status of patients with chronic heart failure: the CASA randomized clinical trial. JAMA Intern. Med. 178, 511–519.

Berkman, L.F., Blumenthal, J., Burg, M., Carney, R.M., Catellier, D., Cowan, M.J., Czajkowski, S.M., Debusk, R., Hosking, J., Jaffee, A., 2003. Effects of treating depression and low perceived social support on clinical events after myocardial infarction: the enhancing recovery in coronary heart disease patients (ENRICHD) randomized trial. JAMA 289, 3106–3116.

Bidwell, J.T., Lyons, K.S., Lee, C.S., 2017. Caregiver well-being and patient outcomes in heart failure: a meta-analysis. J. Cardiovasc. Nurs. 32, 372–382.

Black, J.L., Allison, T.G., Williams, D.E., Rummans, T.A., Gau, G.T., 1998. Effect of intervention for psychological distress on rehospitalization rates in cardiac rehabilitation patients. Psychosomatics 39, 134–143.

Blumenthal, J.A., Sherwood, A., Rogers, S.D., Babyak, M.A., Murali Doraiswamy, P., Watkins, L., Hoffman, B.M., O'connell, C., Johnson, J.J., Patidar, S.M., 2007. Understanding prognostic benefits of exercise and antidepressant therapy for persons with depression and heart disease: the UPBEAT study—rationale, design, and methodological issues. Clin. Trials 4, 548–559.

Blumenthal, J.A., Sherwood, A., Babyak, M.A., Watkins, L.L., Smith, P.J., Hoffman, B.M., O'hayer, C.V.F., Mabe, S., Johnson, J., Doraiswamy, P.M., 2012. Exercise and pharmacological treatment of depressive symptoms in patients with coronary heart disease: results from the UPBEAT (Understanding the prognostic benefits of exercise and antidepressant therapy) study. J. Am. Coll. Cardiol. 60, 1053–1063.

Blumenthal, J.A., Sherwood, A., Smith, P.J., Watkins, L., Mabe, S., Kraus, W.E., Ingle, K., Miller, P., Hinderliter, A., 2016. Enhancing cardiac rehabilitation with stress management training: a randomized, clinical efficacy trial. Circulation 133, 1341–1350.

Burg, M.M., Czajkowski, S., 2011. The ENRICHD clinical trial: lessons learned. In: Allan, R., Scheidt, S. (Eds.), Heart and Mind: The Practice of Cardiac Psychology, second ed. American Psychological Association, Washington, DC.
Burg, M.M., Lespérance, F., Rieckmann, N., Clemow, L., Skotzko, C., Davidson, K.W., 2008. Treating persistent depressive symptoms in post-ACS patients: the project COPES phase-I randomized controlled trial. Contemp. Clin. Trials 29, 231–240.
Bush, D.E., Ziegelstein, R.C., Tayback, M., Richter, D., Stevens, S., Zahalsky, H., Fauerbach, J.A., 2001. Even minimal symptoms of depression increase mortality risk after acute myocardial infarction. Am. J. Cardiol. 88, 337–341.
Butler, A.C., Chapman, J.E., Forman, E.M., Beck, A.T., 2006. The empirical status of cognitive-behavioral therapy: a review of meta-analyses. Clin. Psychol. Rev. 26, 17–31.
Carney, R.M., Freedland, K.E., 2008. Depression in patients with coronary heart disease. Am. J. Med. 121, S20–S27.
Carney, R.M., Freedland, K.E., Rubin, E.H., Rich, M.W., Steinmeyer, B.C., Harris, W.S., 2009. Omega-3 augmentation of sertraline in treatment of depression in patients with coronary heart disease: a randomized controlled trial. JAMA 302, 1651–1657.
Celano, C.M., Huffman, J.C., 2011. Depression and cardiac disease: a review. Cardiol. Rev. 19, 130–142.
Chollet, F., Tardy, J., Albucher, J.-F., Thalamas, C., Berard, E., Lamy, C., Bejot, Y., Deltour, S., Jaillard, A., Niclot, P., 2011. Fluoxetine for motor recovery after acute ischaemic stroke (FLAME): a randomised placebo-controlled trial. Lancet Neurol. 10, 123–130.
Chung, M.L., Pressler, S.J., Dunbar, S.B., Lennie, T.A., Moser, D.K., 2010. Predictors of depressive symptoms in caregivers of patients with heart failure. J. Cardiovasc. Nurs. 25, 411–419.
Chung, M.L., Lennie, T.A., Mudd-Martin, G., Dunbar, S.B., Pressler, S.J., Moser, D.K., 2016. Depressive symptoms in patients with heart failure negatively affect family caregiver outcomes and quality of life. Eur. J. Cardiovasc. Nurs. 15, 30–38.
Cooney, G.M., Dwan, K., Greig, C.A., Lawlor, D.A., Rimer, J., Waugh, F.R., McMurdo, M., Mead, G.E., 2013. Exercise for depression. Cochrane Database Syst. Rev. 9, CD004366.
Cossette, S., Frasure-Smith, N., Lespérance, F., 2001. Clinical implications of a reduction in psychological distress on cardiac prognosis in patients participating in a psychosocial intervention program. Psychosom. Med. 63, 257–266.
Cowan, M.J., Pike, K.C., Budzynski, H.K., 2001. Psychosocial nursing therapy following sudden cardiac arrest: impact on two-year survival. Nurs. Res. 50, 68–76.
Cully, J.A., Stanley, M.A., Petersen, N.J., Hundt, N.E., Kauth, M.R., Naik, A.D., Sorocco, K., Sansgiry, S., Zeno, D., Kunik, M.E., 2017. Delivery of brief cognitive behavioral therapy for medically ill patients in primary care: a pragmatic randomized clinical trial. J. Gen. Intern. Med. 32, 1014–1024.
Davidson, K.W., Rieckmann, N., Clemow, L., Schwartz, J.E., Shimbo, D., Medina, V., Albanese, G., Kronish, I., Hegel, M., Burg, M.M., 2010. Enhanced depression care for patients with acute coronary syndrome and persistent depressive symptoms: coronary psychosocial evaluation studies randomized controlled trial. Arch. Intern. Med. 170, 600–608.
Davidson, K.W., Bigger, J.T., Burg, M.M., Carney, R.M., Chaplin, W.F., Czajkowski, S., Dornelas, E., Duer-Hefele, J., Frasure-Smith, N., Freedland, K.E., 2013. Centralized, stepped, patient preference–based treatment for patients with post–acute coronary syndrome depression: CODIACS vanguard randomized controlled trial. JAMA Intern. Med. 173, 997–1004.
De Jonge, P., Ormel, J., Van Den Brink, R.H., Van Melle, J.P., Spijkerman, T.A., Kuijper, A., Van Veldhuisen, D.J., Van Den Berg, M.P., Honig, A., Crijns, H.J., 2006a. Symptom dimensions of depression following myocardial infarction and their relationship with somatic health status and cardiovascular prognosis. Am. J. Psychiatry 163, 138–144.
De Jonge, P., Van Den Brink, R.H., Spijkerman, T.A., Ormel, J., 2006b. Only incident depressive episodes after myocardial infarction are associated with new cardiovascular events. J. Am. Coll. Cardiol. 48, 2204–2208.

De Jonge, P., Mangano, D., Whooley, M.A., 2007. Differential association of cognitive and somatic depressive symptoms with heart rate variability in patients with stable coronary heart disease: findings from the heart and soul study. Psychosom. Med. 69, 735–739.
Dekker, R.L., 2011. Cognitive therapy for depression in patients with heart failure: a critical review. Heart Fail. Clin. 7, 127–141.
Dekker, R.L., Moser, D.K., Peden, A.R., Lennie, T.A., 2012. Cognitive therapy improves three-month outcomes in hospitalized patients with heart failure. J. Card. Fail. 18, 10–20.
Doering, L.V., Chen, B., Cross Bodan, R., Magsarili, M.C., Nyamathi, A., Irwin, M.R., 2013. Early cognitive behavioral therapy for depression after cardiac surgery. J. Cardiovasc. Nurs. 28, 370–379.
Donker, T., Griffiths, K.M., Cuijpers, P., Christensen, H., 2009. Psychoeducation for depression, anxiety and psychological distress: a meta-analysis. BMC Med. 7, 79.
Dougherty, C.M., Thompson, E.A., 2009. Intimate partner physical and mental health after sudden cardiac arrest and receipt of an implantable cardioverter defibrillator. Res. Nurs. Health 32, 432–442.
Dougherty, C.M., Fairbanks, A.M., Eaton, L.H., Morrison, M.L., Kim, M.S., Thompson, E.A., 2016. Comparison of patient and partner quality of life and health outcomes in the first year after an implantable cardioverter defibrillator (ICD). J. Behav. Med. 39, 94–106.
Dougherty, C.M., Thompson, E.A., Kudenchuk, P.J., 2019. Patient plus partner trial: a randomized controlled trial of 2 interventions to improve outcomes after an initial implantable cardioverter-defibrillator. Heart Rhythm 16, 453–459.
Evangelista, L.S., Stromberg, A., Dionne-Odom, J.N., 2016. An integrated review of interventions to improve psychological outcomes in caregivers of patients with heart failure. Curr. Opin. Support. Palliat. Care 10, 24–31.
Frasure-Smith, N., Lespérance, F., Gravel, G., Masson, A., Juneau, M., Bourassa, M.G., 2002. Long-term survival differences among low-anxious, high-anxious and repressive copers enrolled in the montreal heart attack readjustment trial. Psychosom. Med. 64, 571–579.
Freedland, K.E., Skala, J.A., Carney, R.M., Rubin, E.H., Lustman, P.J., Davila-Roman, V.G., Steinmeyer, B.C., Hogue Jr., C.W., 2009. Treatment of depression after coronary artery bypass surgery: a randomized controlled trial. Arch. Gen. Psychiatry 66, 387–396.
Freedland, K.E., Carney, R.M., Rich, M.W., Steinmeyer, B.C., Rubin, E.H., 2015. Cognitive behavior therapy for depression and self-care in heart failure patients: a randomized clinical trial. JAMA Intern. Med. 175, 1773–1782.
Frizelle, D.J., Lewin, R.J.P., Kaye, G., Hargreaves, C., Hasney, K., Beaumont, N., Moniz-Cook, E., 2004. Cognitive-behavioural rehabilitation programme for patients with an implanted cardioverter defibrillator: a pilot study. Br. J. Health Psychol. 9, 381–392.
Glassman, A.H., O'connor, C.M., Califf, R.M., Swedberg, K., Schwartz, P., Bigger Jr., J.T., Krishnan, K.R.R., Van Zyl, L.T., Swenson, J.R., Finkel, M.S., 2002. Sertraline treatment of major depression in patients with acute MI or unstable angina. JAMA 288, 701–709.
Grosso, G., Pajak, A., Marventano, S., Castellano, S., Galvano, F., Bucolo, C., Drago, F., Caraci, F., 2014. Role of omega-3 fatty acids in the treatment of depressive disorders: a comprehensive meta-analysis of randomized clinical trials. PLoS One 9, e96905.
Gulliksson, M., Burrell, G., Vessby, B., Lundin, L., Toss, H., Svärdsudd, K., 2011. Randomized controlled trial of cognitive behavioral therapy vs standard treatment to prevent recurrent cardiovascular events in patients with coronary heart disease: secondary prevention in uppsala primary health care project (SUPRIM). Arch. Intern. Med. 171, 134–140.
Habibovic, M., Burg, M.M., Pedersen, S.S., 2013. Behavioral interventions in patients with an implantable cardioverter defibrillator: lessons learned and where to go from here? Pacing Clin. Electrophysiol. 36, 578–590.
Hallahan, B., Ryan, T., Hibbeln, J.R., Murray, I.T., Glynn, S., Ramsden, C.E., Sangiovanni, J.P., Davis, J.M., 2016. Efficacy of omega-3 highly unsaturated fatty acids in the treatment of depression. Br. J. Psychiatry 209, 192–201.

Hare, D.L., Toukhsati, S.R., Johansson, P., Jaarsma, T., 2014. Depression and cardiovascular disease: a clinical review. Eur. Heart J. 35, 1365–1372.
Huffman, J.C., Mastromauro, C.A., Sowden, G., Fricchione, G.L., Healy, B.C., Januzzi, J.L., 2011a. Impact of a depression care management program for hospitalized cardiac patients. Circ. Cardiovasc. Qual. Outcomes 4, 198–205.
Huffman, J.C., Mastromauro, C.A., Sowden, G.L., Wittmann, C., Rodman, R., Januzzi, J.L., 2011b. A collaborative care depression management program for cardiac inpatients: depression characteristics and in-hospital outcomes. Psychosomatics 52, 26–33.
Jiang, W., Krishnan, R., Kuchibhatla, M., Cuffe, M.S., Martsberger, C., Arias, R.M., O'connor, C.M., Investigators, S.-C., 2011. Characteristics of depression remission and its relation with cardiovascular outcome among patients with chronic heart failure (from the SADHART-CHF Study). Am. J. Cardiol. 107, 545–551.
Jiang, W., Whellan, D.J., Adams, K.F., Babyak, M.A., Boyle, S.H., Wilson, J.L., Patel, C.B., Rogers, J.G., Harris, W.S., O'connor, C.M., 2018. Long-chain omega-3 fatty acid supplements in depressed heart failure patients: results of the OCEAN trial. JACC Heart Fail. 6, 833–843.
Kang, H.-J., Stewart, R., Bae, K.-Y., Kim, S.-W., Shin, I.-S., Hong, Y.J., Ahn, Y., Jeong, M.H., Yoon, J.-S., Kim, J.-M., 2015. Predictors of depressive disorder following acute coronary syndrome: results from K-Depacs and EsDEPACS. J. Affect. Disord. 181, 1–8.
Katon, W.J., Lin, E.H.B., Von Korff, M., Ciechanowski, P., Ludman, E.J., Young, B., Peterson, D., Rutter, C.M., McGregor, M., McCulloch, D., 2010. Collaborative care for patients with depression and chronic illness. N. Engl. J. Med. 363, 2611–2620.
Kim, J.-M., Bae, K.-Y., Kang, H.-J., Kim, S.-W., Shin, I.-S., Hong, Y.-J., Kim, J.-H., Shin, H.-Y., Ahn, Y., Kim, J.-K., 2014. Design and methodology for the Korean observational and escitalopram treatment studies of depression in acute coronary syndrome: K-DEPACS and EsDEPACS. Psychiatry Investig. 11, 89–94.
Kim, J.S., Lee, E.-J., Chang, D.-I., Park, J.-H., Ahn, S.H., Cha, J.-K., Heo, J.H., Sohn, S.-I., Lee, B.-C., Kim, D.-E., 2017. Efficacy of early administration of escitalopram on depressive and emotional symptoms and neurological dysfunction after stroke: a multicentre, double-blind, randomised, placebo-controlled study. Lancet Psychiatry 4, 33–41.
Kohn, C.S., Petrucci, R.J., Baessler, C., Soto, D.M., Movsowitz, C., 2000. The effect of psychological intervention on patients' long-term adjustment to the ICD:a prospective study. Pacing Clin. Electrophysiol. 23, 450–456.
Kronish, I.M., Carson, A.P., Davidson, K.W., Muntner, P., Safford, M.M., 2012. Depressive symptoms and cardiovascular health by the American heart association's definition in the reasons for geographic and racial differences in stroke (REGARDS) study. PLoS One 7, e52771.
Leon, H., Shibata, M.C., Sivakumaran, S., Dorgan, M., Chatterley, T., Tsuyuki, R.T., 2008. Effect of fish oil on arrhythmias and mortality: systematic review. BMJ 337, a2931.
Lespérance, F., Frasure-Smith, N., Koszycki, D., Laliberté, M.-A., Van Zyl, L.T., Baker, B., Swenson, J.R., Ghatavi, K., Abramson, B.L., Dorian, P., 2007. Effects of citalopram and interpersonal psychotherapy on depression in patients with coronary artery disease: the Canadian cardiac randomized evaluation of antidepressant and psychotherapy efficacy (CREATE) trial. JAMA 297, 367–379.
Lewin, R.J., Coulton, S., Friezelle, D.J., Kaye, G., Cox, H., 2009. A brief cognitive behavioural preimplantation and rehabilitation programme for patients receiving an implantable cardioverter-defibrillator improves physical health and reduces psychological morbidity and unplanned readmissions. Heart 95, 63–69.
Lichtman, J.H., Bigger, J.T., Blumenthal, J.A., Frasure-Smith, N., Kaufmann, P.G., Lespérance, F., Mark, D.B., Sheps, D.S., Taylor, C.B., Froelicher, E.S., 2008. Depression and coronary heart disease. Circulation 118, 1768–1775.
Linde, K., Sigterman, K., Kriston, L., Rucker, G., Jamil, S., Meissner, K., Schneider, A., 2015. Effectiveness of psychological treatments for depressive disorders in primary care: systematic review and meta-analysis. Ann. Fam. Med. 13, 56–68.

Litchman, J.H., Bigger Jr., J.T., Blumenthal, J.A., Frasure-Smith, N., Kaufmann, P.G., Lespérance, F., Mark, D.B., Sheps, D.S., Taylor, C.B., Froelicher, E.S., 2008. Depression and coronary heart disease. Recommendations for screening, referral, and treatment. A science advisory from the American heart association prevention committee of the council on cardiovascular nursing, council on clinical cardiology, council on epidemiology and prevention, and interdisiciplinary council on quality of care and outcomes research. Circulation 118, 1768–1775.

Lofvenmark, C., Saboonchi, F., Edner, M., Billing, E., Mattiasson, A.C., 2013. Evaluation of an educational programme for family members of patients living with heart failure: a randomised controlled trial. J. Clin. Nurs. 22, 115–126.

Manik, P., Varon, J., 2009. Omega-3 dietary supplements and the risk of cardiovascular events: a systematic review. Clin. Cardiol. 32, 365–372.

Mazereeuw, G., Herrmann, N., Ma, D.W., Hillyer, L.M., Oh, P.I., Lanctôt, K.L., 2016a. Omega-3/omega-6 fatty acid ratios in different phospholipid classes and depressive symptoms in coronary artery disease patients. Brain Behav. Immun. 53, 54–58.

Mazereeuw, G., Herrmann, N., Oh, P.I., Ma, D.W., Wang, C.T., Kiss, A., Lanctôt, K.L., 2016b. Omega-3 fatty acids, depressive symptoms, and cognitive performance in patients with coronary artery disease: analyses from a randomized, double-blind, placebo-controlled trial. J. Clin. Psychopharmacol. 36, 436–444.

McMillan, S.C., Small, B.J., Haley, W.E., Zambroski, C., Buck, H.G., 2013. The COPE intervention for caregivers of patients with heart failure: an adapted intervention. J. Hosp. Palliat. Nurs. 15.

Meyer, T., Belnap, B.H., Herrmann-Lingen, C., He, F., Mazumdar, S., Rollman, B.L., 2014. Benefits of collaborative care for post-CABG depression are not related to adjustments in antidepressant pharmacotherapy. J. Psychosom. Res. 76, 28–33.

Mikami, K., Jorge, R.E., Moser, D.J., Arndt, S., Jang, M., Solodkin, A., Small, S.L., Fonzetti, P., Hegel, M.T., Robinson, R.G., 2011. Increased frequency of first-episode poststroke depression after discontinuation of escitalopram. Stroke 42, 3281–3283.

Morgan, M.A., Coates, M.J., Dunbar, J.A., Reddy, P., Schlicht, K., Fuller, J., 2013. The TrueBlue model of collaborative care using practice nurses as case managers for depression alongside diabetes or heart disease: a randomised trial. BMJ Open 3, e002171

Mozaffarian, D., Wu, J.H., 2011. Omega-3 fatty acids and cardiovascular disease: effects on risk factors, molecular pathways, and clinical events. J. Am. Coll. Cardiol. 58, 2047–2067.

Nicholson, A., Kuper, H., Hemingway, H., 2006. Depression as an aetiologic and prognostic factor in coronary heart disease: a meta-analysis of 6362 events among 146 538 participants in 54 observational studies. Eur. Heart J. 27, 2763–2774.

Pesah, E., Supervia, M., Turk-Adawi, K., Grace, S.L., 2017. A review of cardiac rehabilitation delivery around the world. Prog. Cardiovasc. Dis. 60, 267–280.

Piette, J.D., Holtz, B., Beard, A.J., Blaum, C., Greenstone, C.L., Krein, S.L., Tremblay, A., Forman, J., Kerr, E.A., Ann Arbor, P.S.C., 2011. Improving chronic illness care for veterans within the framework of the patient-centered medical home: experiences from the Ann Arbor patient-aligned care team laboratory. Transl. Behav. Med. 1, 615–623.

Piette, J.D., Striplin, D., Marinec, N., Chen, J., Aikens, J.E., 2015. A randomized trial of mobile health support for heart failure patients and their informal caregivers: impacts on caregiver-reported outcomes. Med. Care 53, 692–699.

Rechenberg, K., 2016. Nutritional interventions in clinical depression. Clin. Psychol. Sci. 4, 144–162.

Redwine, L.S., Tsuang, M., Rusiewicz, A., Pandzic, I., Cammarata, S., Rutledge, T., Hong, S., Linke, S., Mills, P.J., 2012. A pilot study exploring the effects of a 12-week t'ai chi intervention on somatic symptoms of depression in patients with heart failure. J. Altern. Complement. Med. 18, 744–748.

Richards, S.H., Anderson, L., Jenkinson, C.E., Whalley, B., Rees, K., Davies, P., Bennett, P., Liu, Z., West, R., Thompson, D.R., Taylor, R.S., 2018. Psychological interventions for coronary heart disease: cochrane systematic review and meta-analysis. Eur. J. Prev. Cardiol. 25, 247–259.

Robinson, R.G., Jorge, R.E., Moser, D.J., Acion, L., Solodkin, A., Small, S.L., Fonzetti, P., Hegel, M., Arndt, S., 2008. Escitalopram and problem-solving therapy for prevention of poststroke depression: a randomized controlled trial. JAMA 299, 2391–2400.

Rollman, B.L., Belnap, B.H., 2011. The bypassing the blues trial: collaborative care for post-CABG depression and implications for future research. Cleve. Clin. J. Med. 78, S4–S12.

Roose, S.P., Glassman, A.H., Attia, E., Woodring, S., Giardina, E.-G.V., Bigger Jr., J.T., 1998a. Cardiovascular effects of fluoxetine in depressed patients with heart disease. Am. J. Psychiatry 155, 660–665.

Roose, S.P., Laghrissi-Thode, F., Kennedy, J.S., Nelson, J.C., Bigger Jr., J.T., Pollock, B.G., Gaffney, A., Narayan, M., Finkel, M.S., Mccafferty, J., 1998b. Comparison of paroxetine and nortriptyline in depressed patients with ischemic heart disease. JAMA 279, 287–291.

Rugulies, R., 2002. Depression as a predictor for coronary heart disease: a review and meta-analysis. Am. J. Prev. Med. 23, 51–61.

Rutledge, T., Redwine, L.S., Linke, S.E., Mills, P.J., 2013. A meta-analysis of mental health treatments and cardiac rehabilitation for improving clinical outcomes and depression among patients with coronary heart disease. Psychosom. Med. 75, 335–349.

Salmoirago-Blotcher, E., Ockene, I.S., 2009. Methodological limitations of psychosocial interventions in patients with an implantable cardioverter-defibrillator (ICD) A systematic review. BMC Cardiovasc. Disord. 9, 56.

Shapiro, P.A., Lespérance, F., Frasure-Smith, N., O'connor, C.M., Baker, B., Jiang, J.W., Dorian, P., Harrison, W., Glassman, A.H., 1999. An open-label preliminary trial of sertraline for treatment of major depression after acute myocardial infarction (the SADHAT Trial). Am. Heart J. 137, 1100–1106.

Siscovick, D.S., Barringer, T.A., Fretts, A.M., Wu, J.H., Lichtenstein, A.H., Costello, R.B., Kris-Etherton, P.M., Jacobson, T.A., Engler, M.B., Alger, H.M., Appel, L.J., Mozaffarian, D., American Heart Association Nutrition Committee Of The Council On, L., Cardiometabolic, H, Council On, E., Prevention, Council On Cardiovascular Disease In The, Y, Council On, C., Stroke, N, Council On Clinical, C, 2017. Omega-3 polyunsaturated fatty acid (fish oil) supplementation and the prevention of clinical cardiovascular disease: a science advisory from the American heart association. Circulation 135, e867–e884.

Strik, J., Honig, A., Lousberg, R., Cheriex, E., Van, H.P., 1998. Cardiac side-effects of two selective serotonin reuptake inhibitors in middle-aged and elderly depressed patients. Int. Clin. Psychopharmacol. 13, 263–267.

Unützer, J., Katon, W., Callahan, C.M., Williams Jr., J.W., Hunkeler, E., Harpole, L., Hoffing, M., Della Penna, R.D., Noël, P.H., Lin, E.H., 2002. Collaborative care management of late-life depression in the primary care setting: a randomized controlled trial. JAMA 288, 2836–2845.

Van Melle, J.P., De Jonge, P., Honig, A., Schene, A.H., Kuyper, A.M., Crijns, H.J., Schins, A., Tulner, D., Van Den Berg, M.P., Ormel, J., 2007. Effects of antidepressant treatment following myocardial infarction. Br. J. Psychiatry 190, 460–466.

Ye, S., Shaffer, J.A., Rieckmann, N., Schwartz, J.E., Kronish, I.M., Ladapo, J.A., Whang, W., Burg, M.M., Davidson, K.W., 2014. Long-term outcomes of enhanced depression treatment in patients with acute coronary syndromes. Am. J. Med. 127, 1012–1016.

Zheng, X., Zheng, Y., Ma, J., Zhang, M., Zhang, Y., Liu, X., Chen, L., Yang, Q., Sun, Y., Wu, J., Yu, B., 2019. Effect of exercise-based cardiac rehabilitation on anxiety and depression in patients with myocardial infarction: a systematic review and meta-analysis. Heart Lung 48, 1–7.

SECTION 2

Mechanisms linking stress and depression with cardiovascular disease

CHAPTER 5

Autonomic and neuroendocrine response to stress

Kevin T. Larkin, Leah A. Brown, Alaina G. Tiani
Department of Psychology, West Virginia University, Morgantown, WV, United States

Contents

The central nervous system communicates with all other vital organ systems in the body, including the heart and blood vessels of the circulatory system, in two ways: (a) through the direct transmission of neural signals via two branches of the autonomic nervous system (ANS; i.e., the sympathetic and parasympathetic nervous systems), and (b) through the neuroendocrine (NE) system via releasing hormones into the circulation that influence responsive organs downstream. Because of these recognized pathways connecting the neural and cardiovascular systems, it is not surprising that early investigations aimed at uncovering the mechanisms through which psychological factors like stress and depression increase risk for CVD highlighted the roles of the ANS and NE systems (e.g., Carney et al., 1995).

Cardiovascular Implications of Stress and Depression
https://doi.org/10.1016/B978-0-12-815015-3.00005-2

The autonomic nervous system

Evolution equipped mammals with a sophisticated sympathetic nervous system (SNS) that enabled them to respond quickly to a multitude of threats encountered in their immediate environmental surroundings, including exposure to predators as well as dangers associated with natural threats and disasters. In brief, the SNS response, commonly referred to as the "fight-or-flight" response, is characterized by a re-allocation of the body's energy resources (e.g., oxygen, glucose) to the musculature and other organ systems that need to be activated to support the increased energy demand required for fighting or fleeing, including the heart and vasculature (Andreassi, 2007). Because functioning of other organ systems has a lower priority when confronting threat or danger, energy resources are drawn from the digestive and reproductive systems during SNS activation. The SNS is paired with an active parasympathetic nervous system (PNS) that is designed to reduce the physiological activation of the SNS and return the body's energy to its resting state where blood flow resumes to the digestive and reproductive systems. Neuroanatomically, SNS activity is linked to the limbic system and brain stem regions of the brain and extends to peripheral destinations directly though neurons comprising an extensive network of cells called the sympathetic ganglia located bilateral to the spinal cord between the neck and tailbone. Two neural networks that, for the most part, function using norepinephrine as a neurotransmitter comprise the SNS: alpha-adrenergic fibers and beta-adrenergic fibers. In the cardiovascular system, both alpha- and beta-adrenergic networks innervate the vasculature and function to constrict and dilate blood vessels respectively. In contrast, beta-adrenergic neurons alone innervate the heart itself and trigger increases in heart rate (HR) and cardiac contractility. The PNS also has direct neural connections to most of the same organ systems innervated by the SNS but largely employs acetylcholine as a neurotransmitter. PNS activity is not routed through the sympathetic ganglia, but instead is connected directly to peripheral organs through several cranial nerves, including the vagal nerve, and the pelvic splanchnic nerves in the sacrum. In the cardiovascular system, the vagal nerve transmits neural signals to the heart, leading to reductions in HR and cardiac contractility, but has no direct influence on arterial vasoconstriction or dilation. In this regard, the SNS and PNS represent distinctive networks of peripheral neurons that alter physiological functioning of the body, much like the accelerator and brake both alter the speed of a motor vehicle (Andreassi, 2007).

Although the direct measurement of ANS activity is challenging to obtain, numerous methods of measuring target organ system activity using surface electrodes that reflect underlying ANS functioning (e.g., HR, skin conductance, skin temperature, pupillometry) are available. For purposes of understanding the mechanisms involved in the association between exposure to stress and subsequent onset of CVD, the most popular indices of measurement are HR, blood pressure (BP), and heart rate variability (HRV). Because both HR and BP are significantly influenced by cardiac activity that is jointly innervated by both the SNS and PNS, it is impossible to determine whether increases in either of these parameters result from increased SNS activity, decreased PNS activity, or some combination of the two. There are certain measures of HRV, however, that are not influenced by SNS activity, and primarily reflect PNS activity (e.g., respiratory sinus arrhythmia, high-frequency HRV detected via spectral analysis). For these measures, elevated HRV reflects more PNS activity.

The neuroendocrine system

The central nervous system also communicates with other organs and systems in the body using a less direct route, the blood stream itself. Under the control of the pituitary gland, conveniently located next to the hypothalamus in the brain, releasing factors are secreted into the blood that stimulate hormone production and release from other endocrine glands in the body (e.g., thyroid, adrenal, gonad glands). Using the same method, the pituitary gland can direct reduced hormone production and/or release. Because the NE system uses the blood stream as its method of communicating with other bodily organs and systems instead of direct neural impulses, the NE response occurs more slowly than the autonomic responses of the SNS and PNS.

The primary hormones that impact cardiovascular functioning include cortisol released by the adrenal cortex and epinephrine secreted by the adrenal medulla (Lovallo, 2005). The pathway involved in releasing cortisol into the blood stream begins in the hypothalamus through the release of two peptides, corticotropin releasing factor (CRF) and arginine vasopressin that jointly stimulate the release of adrenocorticotropic hormone (ACTH) from the anterior pituitary where it travels through the blood stream to the adrenal cortex and directs the release of cortisol (or corticosterone in non-human mammals). Epinephrine, on the other hand, is released into the bloodstream from the adrenal medulla following direct SNS transmission from the brain stem (Lovallo, 2005). Both cortisol and epinephrine influence sensitivity of beta-adrenergic receptors in the SNS and are needed to ensure proper ANS functioning and survival; cortisol also enhances the sensitivity of alpha-adrenergic receptors in the SNS, catecholamine synthesis, and glucose production (Lovallo, 2005). To ensure the proper balance of these stress responsive hormones, negative feedback loops exist that adjust epinephrine and cortisol levels in the blood through ongoing monitoring by the central nervous system.

In contrast to measures of ANS activity, measures of NE activity require the chemical analysis of bodily fluids or tissues, most often urine, blood, or saliva. Urinalysis is not helpful for measuring NE responses to acute environmental stress due to the lengthy duration of time required for fluids to pass through the body and be excreted, but can provide information about exposure to chronic environmental stressors. Assays of cortisol are also complicated by its natural diurnal variation that requires measurements to be made at comparable times during the day for all individuals participating in studies examining NE responses to stress.

Pathological patterns of autonomic nervous and neuroendocrine system functioning

Based on the work of both Cannon (1915) and Selye (1955), it is well known that the ANS and NE systems of mammals respond to acute stress in predictable ways. The normal physiological response when exposed to threat detected in the surrounding environment involves activation of the SNS (i.e., increased HR, BP, respiration rate, and skin conductance) and NE systems (i.e., increased secretion of epinephrine and cortisol) and slowing of the PNS (i.e., reduced HRV and gastric motility). This response pattern provides the organism with the energy resources needed to ward off threat or escape from it. Once the source of the threat dissipates, the response subsides and energy returns to organs and systems responsible for sustaining life, like digestion. By responding in this manner, the body attempts to stabilize physiological functioning by modifying energy distribution to the body's vital organs, a process termed allostasis (McEwen, 2002). Because allostasis aims to provide organisms with the optimal amount of energy required to eliminate or evade threatening stimuli, it is predictable that the magnitude and duration of the body's physiological response varies based upon the characteristics of the threatening stimuli as well as the health status of the organism. One would expect, for example, larger and more prolonged SNS responses to occur while attempting to outrun a tiger than a tarantula. However, this is not always the case and frequently ANS and NE activation is not proportional to the energy requirements of the body while confronting the source of the threat or after the threat has passed. Furthermore, presence of diseases like atherosclerosis is associated with exaggerated blood pressure reactions to acute stress (Kamarck et al., 1997) indicating that allostasis is influenced by one's cardiovascular health status. When the body's physiological response struggles to match the energy demands needed to confront the environmental stressor, allostatic load is increased and ultimately the challenge to sustain an optimal response may become more taxing than the environmental stress itself (McEwen, 2002), particularly as exposure to environmental stressors persists. In this regard, failure to maintain allostasis in response to acute environmental stress becomes an important agent that leads to patterns of ANS or NE dysfunction during exposure to chronic stress that may differ from patterns detected in response to acute stress presentations (e.g., hyper-arousal to acute stress may lead to hypo-arousal in response to chronic stress).

Several types of allostatic load have been identified (see McEwen, 2002), but for purposes of this chapter, we will highlight three. The first type is characterized by an overactive ANS and NE response that provides considerably more energy to the body than the organism needs to confront the source of the stress. Consider when a musician's HR increases to comparable levels detected when outrunning a bear in the woods and performing in front of an audience. Although the magnitude of the increased HR may be appropriate for evading the bear, the metabolic demands to ensure survival during a musical performance are likely to be far lower. This type of allostatic load could lead to increased risk for CVD among persons who display exaggerated or frequent ANS or NE responses to environmental stimuli in comparison to their low-reactive counterparts or those who confront stressors that evoke physiological responses less frequently in their lives. To assess this pattern of allostatic load, change in BP, HR, or cortisol is calculated from periods of rest to levels detected during acute stress presentations, adjusting for any pre-task differences in physiologic tone or by calculating residualized change scores (Manuck et al., 1989).

A second type of allostatic load occurs when the ANS and NE responses fail to provide the energy required for the organism to confront the source of the threat successfully. This blunted physiological response then renders organisms helpless in confronting challenges required to adapt to ever-changing environments, and consequently fails to equip them with a cardiovascular system capable of responding to these changes. Like the measurement of exaggerated reactivity, blunted reactivity is assessed by calculating change in BP, HR, or cortisol from periods of rest to levels of those parameters observed during acute stress presentations.

The third type of allostatic load mentioned here is prolonged recovery of the physiological responses to stress. In this case, it is neither the magnitude nor the frequency of the HR of BP reactions to stress that matters, but instead, the failure of the body's response system to subside once the source of the threat has been eliminated or evaded that contributes to allostatic load. In this regard, persons who exhibit normal HR and BP reactions to stress may be at greater risk for CVD because of the prolonged duration of SNS activity in comparison to persons whose ANS or NE responses return to resting levels rapidly once the source of threat has been eliminated or evaded. A number of different strategies have been employed to assess recovery rate, including time to recover to pre-task resting levels, analyses of recovery slopes, and calculating area under the recovery curve.

The remaining sections of this chapter are organized around three questions pertinent to understanding the mechanisms responsible for the association between exposure to stress and atherosclerotic disease processes: (1) do organisms exposed to environmental stress acquire maladaptive ANS and NE response profiles that contribute to allostatic load?, (2) do the various patterns of allostatic load exhibited by the ANS and NE systems predict subsequent CVD and other related CVD outcomes?, and (3) what role do the various patterns of allostatic load play in explaining the association between depression and risk for CVD?

Exposure to stress and allostatic load

Our contemporary understanding of allostatic load is that it creates a cumulative burden on normal bodily functioning that ultimately leads to increased risk for numerous diseases, including CVD (McEwen, 2002). In this regard, exposure to chronically stressful environments would be expected to be associated with increased prevalence of the various types of allostatic load. There are several bodies of literature that have examined whether ANS and NE responses to stress are associated with various types of life stress, including residing in environments characterized as being low in socioeconomic status, exposure to stressful work environments, being exposed to adverse environments during childhood, and exposure to traumatic events.

Stress and exaggerated physiological response

There is ample evidence that exposure to life-threatening events can be accompanied by exaggerated autonomic responses to recollections of these traumatic events, particularly among patients diagnosed with post-traumatic stress disorder (PTSD) (Pole, 2007). Regarding cardiovascular parameters, larger HR and BP responses and greater reductions in HRV in response to stress are commonly observed among PTSD patients when compared to controls during exposures to standardized trauma stimuli as well as stimuli specifically associated with the patient's reported trauma (e.g., Nachar et al., 2014; Pole, 2007; Sack et al., 2004). In fact, the association between increased trauma-related arousal and SNS reactivity and PTSD is so robust that it serves as one of the diagnostic criteria for PTSD (American Psychiatric Association, 2013) and has been proposed as a tool that could prove useful in diagnosing PTSD (Castro-Chapman et al., 2018). This exaggerated cardiovascular activation is detected in response to non-trauma-related stimuli as well; for example, in Pole's (2007) meta-analysis of studies examining autonomic arousal during presentations of startling audio stimuli, the increased HR response among PTSD patients stood out as the most distinctive feature of their physiological response profile. Furthermore, the exaggerated HR response to startle probes has been shown to dissipate following successful treatment of PTSD (Griffin et al., 2012), lending further credence to the association between PTSD and exaggerated cardiovascular responses to stress.

Although inclusive of exposure to traumatic events, the constellation of negative early life experiences termed adverse childhood experiences extends the trauma literature by including instances of physical and emotional neglect, parental divorce, and observing family members misusing substances, experiencing mental health problems, and/or being incarcerated in its definition. It is not surprising then that some research has documented increased stress responses among those with significant histories of exposure to adverse childhood experiences. Heim et al. (2000), for example, reported elevated ACTH and cortisol responses to standard laboratory challenges among women exposed to adverse negative childhood experiences, particularly among those who experienced negative life events in adulthood and/or were seeking treatment for depression. Interestingly, women who experienced adverse childhood experiences, but who were not depressed, did not display the exaggerated NE responses to stress. In related studies, high quality early relationships with one's parents have been shown to be associated with decreased cortisol reactivity while completing stressful tasks in adulthood (e.g., Engert et al., 2010; Pruessner et al., 2004) and stable characteristics of one's family-of-origin have been associated with reduced BP responses to interpersonal laboratory stressors (e.g., Larkin et al., 2011).

There is evidence for an association between increased ANS reactivity and exposure to other types of stress. For example, Lepore et al. (1997) revealed that HR and BP responses to acute mental arithmetic and speech challenges were elevated among participants who reported higher levels of chronic stress in their lives. In a review of the literature examining the association between life stress and cardiovascular reactivity, Gump and Matthews (1999) found that a majority of studies showed a positive association between life stress and the magnitude of HR and BP reactivity to acute laboratory stressors. Similar associations have been found between increased HR and BP reactivity to stress and several specific life stressors, including residing in crowded living conditions (Fleming et al., 1987), being exposed to discrimination (Richman et al., 2007), serving as a caregiver for patients with dementia (Uchino et al., 1992), and being employed in stressful occupations (Lumley et al., 2014). In a longitudinal study, BP reactivity to laboratory stressors was shown to increase over time among adolescents who experienced increased negative life events, and that these alterations were associated with increased intima-media thickness, a recognized measure of subclinical atherosclerosis (Low et al., 2009). Similar relations between measures of environmental stress and increased NE responses to stress have been reported (Wirtz et al., 2013).

Finally, there are a significant number of animal studies supporting the linkage between exposure to environmental stress and increased cardiovascular responding to acute stress presentations. For example, in rodents exposed to uncontrollable mild stress, increased HR and BP responses to air jet stress have been reported (Grippo et al., 2002) and prairie voles housed in stressful social isolation living conditions exhibit increased HR and HRV reactions to a forced-swim test in comparison to voles housed together (Normann et al., 2018). Roman et al. (2004) confirmed these findings by showing that spontaneously hypertensive rats exhibit greater ACTH and cortisol responses to the stress of being handled, but only when exposed to uncontrollable stressful living conditions.

Stress and blunted physiological response

Although ample evidence supports an association between exposure to stress and exaggerated reactions to acute stressors, there are several studies that demonstrate reduced or blunted SNS or neuroendocrine reactions have also been associated with life stress. In fact, Gump and Matthews (1999) reported that a substantial minority of studies examining the influence of life stress on cardiovascular reactivity to laboratory stressors demonstrated significantly reduced HR and/or BP reactions among individuals reporting high levels of life stress. Several studies conducted since that time have confirmed the association between life stress and blunted cardiovascular response profiles (e.g., Chida and Hamer, 2008; Ginty and Conklin, 2011; Musante et al., 2000; Phillips et al., 2005). Similarly, although studies have shown that workers with high levels of occupational stress exhibit exaggerated cardiovascular responses to mental stressors, several studies have reported an association between job stress and blunted cardiovascular responses to stress (Hamer et al., 2006; Siegrist and Klein, 1990).

Numerous investigations of children and adolescents who were mistreated early in life have often revealed blunted cardiovascular responses to laboratory stressors. For example, Evans et al. (2007) reported that poverty experienced during childhood was associated with blunted BP responses to acute stressors measured during adolescence. More recently, Voellmin et al. (2015) reported that young healthy women with histories of adverse childhood experiences exhibited blunted endocrine and cardiovascular responses to stress. A substantial number of studies examining the quality of early life experiences and stress responses have detected an association with blunted NE response profiles, most frequently cortisol (e.g., Carpenter et al., 2007, 2011; Lovallo, 2013; Luecken et al., 2009; Taylor et al., 2004). Blunted ACTH and cortisol responses to stress have also been reported in patients diagnosed with PTSD (e.g., Santa Ana et al., 2006).

Animal research has also demonstrated blunted neuroendocrine activity following exposure to stressful living conditions. Based upon this literature, Pitman et al. (2012) reviewed studies that supported both (a) increased ANS and NE arousal during a single prolonged stress exposure consisting of restraint or forced swim tests, and (b) blunted NE activity a week later that was attributed to enhanced negative feedback from the hypothalamic-pituitary-adrenal (HPA) axis.

Because measures of reactivity are calculated in reference to pre-stress rest periods in both humans and animals, blunted reactivity could result from either (a) dampened activation during acute stress presentations or (b) increased ANS and/or NE system tone that exists during pre-stress rest periods. It is possible, for example, that stressful life circumstances are associated with blunted reactions to acute stress because they lead to increased ANS or NE tone present during pre-stress baseline periods; consequently, the attenuated change in reactivity observed with this response pattern likely reflects elevated levels during pre-task rest periods more than reduced arousal during stress. Indeed, there is some evidence to suggest that exposure to traumatic stress as well as other stressful life events leads to increased tonic levels of ANS or NE activity (e.g., Evans, 2003; Fleming et al., 1987; Pole, 2007).

Stress and delayed recovery

As methods for calculating recovery from stress became more sophisticated, researchers turned to examining relations between exposure to life stress and prolonged recovery from acute stress presentations. In an early review of this literature, Pieper and Brosschot (2005) concluded that exposure to various forms of life stress (e.g., job stress, catastrophic events, exposure to racism) resulted in delays in HR and BP recovery from stress, even after controlling for health behaviors likely to be influenced by stress (e.g., smoking, alcohol use). Chida and Hamer (2008) confirmed this relation in their meta-analytic review of literature examining measures of life stress and HR and BP reactivity to and recovery from exposure to acute stress. More recently, Boylan et al. (2018) demonstrated a comparable association between prolonged cardiovascular recovery from stress and low socio-economic status (SES) but failed to demonstrate an association between measures of cardiovascular reactivity to stress and SES. Although prolonged HR and BP recovery from acute stress has been consistently observed among samples reporting high levels of life stress, comparable associations between life stress and delayed recovery of NE parameters are not commonly reported (Chida and Hamer, 2008).

Allostatic load and risk for cardiovascular disease

As noted in the previous section, substantial evidence exists to show that exposure to numerous types of environmental stress are associated with all three types of allostatic load: exaggerated physiologic responses to stress, blunted physiologic responses to stress, and prolonged recovery rates of physiologic activation elicited by stress. However, in order to demonstrate a causal linkage between the various types of allostatic load and disease endpoints, it is important to examine the associations between patterns of allostatic load and underlying disease processes, in this case atherosclerosis and/or clinical phenomena (e.g., stroke, myocardial infarction) through which atherosclerosis presents itself in health care settings. Although assessments of ANS and/or NE reactivity to physical, emotional, or mental challenges have been conducted on cardiac patients and shown to be greater than reactivity to stress observed in non-disease controls (e.g., Corse et al., 1982), these findings do little to support a causal model linking stress with disease endpoints through allostatic load. Once the cardiovascular system is damaged, it could easily be argued that the exaggerated reactivity resulted from the damaged physiology rather than causing it. In this regard, studies that hold the most promise are prospective in nature, assessing patterns of allostatic load long before CVD onset, presumably at a time when the circulatory system was free of disease. Because experimental animal studies permit greater degrees of control over exposure to stress and other factors that contribute to onset of CVD (e.g., dietary fat content), they also can be useful in linking patterns of allostatic load to onset of CVD.

Exaggerated physiological response to stress and risk for CVD

Krantz and Manuck (1984) conducted the initial review of the literature examining physiologic response to stress with CVD, specifically coronary heart disease and essential hypertension. Only one prospective study was available at that time (Keys et al., 1971) and it yielded evidence that DBP response to a cold pressor task predicted subsequent CVD over two decades later. They also reviewed the literature examining the association between ANS and NE reactivity and Type A behavior pattern and concluded that increased stress responses were more commonly detected among individuals with Type A behavior pattern in contrast to Type B counterparts.

Krantz and Manuck (1984) also reported findings from studies conducted on cynomolgus macaques that demonstrated a link between exposure to social environmental stress and atherosclerosis. For example, Kaplan et al. (1982) showed that dominant macaques exhibited greater atherosclerosis than subordinate macaques, but only when exposed to environmental stress. In a related study, Manuck et al. (1983) assessed HR reactivity to the threat of capture among these macaques and showed that high HR reactive animals had significantly greater atherosclerosis than their lower HR reactive counterparts. Animal research using spontaneously hypertensive rats (SHRs) showed that the exaggerated HR, BP, and catecholamine response to stress observed in these rodents preceded hypertension onset (Hallbäck and Folkow, 1974). Casto and Printz (1990) replicated these findings by demonstrating that SHRs exhibited exaggerated BP and HR in contrast to non-hypertensive Wistar Kyoto control rats that exhibited a bradycardic response (decrease in HR) in response to stress. Based on this pattern of findings, it was concluded that the exaggerated responses to stress observed among SHRs were associated with an impaired PNS response to stress.

A significant amount of prospective research was spawned by what has become known as the reactivity hypothesis, which was then reviewed by Treiber et al. (2003) years later. Studies included in their review each measured cardiovascular reactivity to a range of cognitive or behavioral stressors and then followed participants for several years to assess relations with elevated blood pressure, change in ventricular mass, or carotid atherosclerosis. Findings from the studies with the longest follow-up periods convincingly demonstrated greater incidence of hypertension among high reactive participants, but studies with shorter follow-up periods showed less consistent results. Studies conducted on samples consisting of patients already diagnosed with some evidence of CVD revealed stronger effects than those on healthy samples, lending further credence to the fact that one's health status moderated these effects. This was particularly true for studies of clinical disease endpoints like new cardiac events. Data from studies conducted on healthy samples often did not show associations between reactivity and new cardiac events, but those on at-risk samples often showed robust associations between HR and BP reactions to stress and CVD endpoints.

In 2010, Chida and Steptoe conducted a meta-analysis of prospective cohort studies examining whether greater cardiovascular reactivity to acute laboratory mental stressors predicted future cardiovascular risk status. By this time, 36 studies had been conducted to examine this relation prospectively. Results revealed that increased cardiovascular reactivity was again associated with subsequent cardiovascular endpoints, including increased blood pressure or onset of essential hypertension and increased prevalence of carotid atherosclerosis.

Although most research examining the reactivity hypothesis with respect to CVD outcomes has examined HR and BP responses to stress, Krantz and Manuck (1984) also identified circulating catecholamines and cortisol as parameters of interest with respect to the reactivity hypothesis. Although it is well known that exposure to stress elicits responses in the NE system, the number of studies examining the relation between NE reactivity to stress and CVD outcomes is relatively sparse. One example is the work of Hamer et al. (2012) who assessed cortisol responses of healthy men and women exposed to cognitive stressors and followed for 3 years. At follow-up, results revealed that increased cortisol responses to stress predicted increased coronary artery calcification, a recognized risk factor for CVD.

Blunted physiological response to stress and risk for CVD

In contrast to the literature examining exaggerated cardiovascular reactivity to stress and CVD, evidence linking diminished or blunted reactivity and risk for CVD is sparse. Although many observations between blunted reactivity profiles and poor health outcomes have been reported, they typically only relate peripherally to CVD. In a review on health-related outcomes associated with blunted reactivity profiles, Lovallo (2011) presented evidence that low levels of stress reactivity (measured via BP, HR, and cortisol levels) were just as problematic for maintaining health as exaggerated reactivity profiles, but appeared to originate in different parts of the cortex. Specifically, if the limbic system (primarily amygdala) did not function by attaching any emotional significance to a stressful situation and did not command a behavioral response, the peripheral cardiovascular responses would be diminished. Evidence shows that blunted reactivity has been shown to be associated with poorer immune function, high levels of obesity that result from excessive food intake, eating disorders, depression, and a range of addictive behaviors including smoking, gambling, and alcoholism, but not CVD (see Lovallo, 2011).

Although there is no direct evidence linking blunted reactivity to risk for CVD, it could be argued that the behavioral sequelae of blunted cardiovascular responding may indeed increase risk for CVD. Evidence for this hypothesis can be obtained from the work of Ginty et al. (2016) who showed that diminished HR reactivity to acute stress was associated with increased carotid intima-media thickness, and that this association was mediated by increased body mass index and smoking behavior. This suggests that these important health behaviors linked to CVD risk may indeed be associated with blunted cardiovascular response to stress.

Prolonged recovery from stress and risk for CVD

Like studies of allostatic load and life stress, the majority of investigations have focused on the magnitude of reactivity in predicting subsequent CVD rather than measuring length of recovery following exposure to stress. However, there have now been several prospective studies that have examined the role of this type of allostatic load and risk for CVD. In Pieper and Brosschot's (2005) review, initial evidence for linking prolonged cardiovascular recovery from stress to CVD outcomes was noted but reported to be modest. By the time of Chida and Steptoe's (2010) meta-analysis, a sufficient number of studies had been conducted to appraise the nature of this relation. Results indicated that prolonged HR and BP recovery from stress was associated with worse cardiovascular health outcomes, including increased SBP and DBP as well as increased intima-media thickness in the carotid arteries.

In 2015, Panaite et al. examined 37 prospective studies comprising the literature on cardiovascular recovery from stress as a predictor of future CVD outcomes. With a mean follow-up period across studies of 7.7 years, the analysis revealed that delayed physiologic recovery from stress predicted subsequent adverse CVD outcomes. Findings were significant for both participants at-risk for CVD during the initial assessment as well as healthy participants, although stronger associations were detected among samples of healthy individuals. Time from initial assessment to follow-up revealed that longer durations were associated with smaller effect sizes, and exposure to physical challenges predicted CVD outcomes better than exposure to psychological challenges.

Depression and allostatic load

Although exposure to stress is known to increase allostatic load through numerous physiologic pathways, evidence only exists for linking exaggerated cardiovascular reactivity to stress and prolonged recovery from stress with CVD outcomes directly. Blunted physiologic reactivity to stress may be associated with CVD outcomes, but likely through its linkage with numerous health behaviors known to be related to CVD (e.g., smoking, poor dietary factors and/or obesity). To examine the pathophysiologic mechanisms that potentially link stress with CVD through depression or the experience of depressive mood, it is important to examine the literature exploring various patterns of physiologic responding to stress and depression.

Depression and exaggerated physiological response to stress

Given the known association between exaggerated physiologic reactivity to stress and CVD outcomes and the prospective evidence identifying depression as a psychosocial risk factor for CVD, it was only natural that scientists began to examine relations between HR and BP reactivity to stress and depression. In an early review of this literature, however, Kibler and Ma (2004) failed to uncover any convincing evidence of a significant relation between measures of HR or BP reactivity and depression, although the overall effect size supported a positive association between these two variables. Among these types of studies, several have supported the hypothesis that depression was associated with an exaggerated ANS response to stress (e.g., Light et al., 1998; Matthews et al., 2005). Additionally, there was some evidence that an exaggerated endocrine response to stress was associated with depression (Mausbach et al., 2005); in this study, caregivers of Alzheimer's patients with more depressive symptoms exhibited greater levels of plasma norepinephrine in response to a psychological stress task than those with lower levels of depression.

Based on the literature showing that depression is associated with reductions in HRV among non-cardiac populations (Kemp et al., 2010), research examining the relation between cardiovascular reactivity to stress and depression has also examined measures of HRV reactivity to stress. Hughes and Stoney (2000) found an association between depressive symptoms and decreased high-frequency HRV (i.e., less PNS activation) during stress tasks in a group of college students, but Liang et al. (2015) reported an increasing trend in high frequency HRV in response to a mental arithmetic task among males diagnosed with major depressive disorder in comparison to a non-depressed control group.

Depression and blunted physiological response to stress

Although some studies have found a positive association between depression and the magnitude of physiologic responding to stress, there is also considerable evidence showing that depression is associated with a blunted physiological response to stress. For example, Salomon et al. (2009) showed that individuals diagnosed with a major depressive disorder exhibited significantly less HR, SBP, and cardiac output reactivity to stress than non-depressed controls. Similar associations between blunted cortisol reactivity and depression have also been reported (de Rooij et al., 2010). Finally, prospective evidence of the relation between blunted HR response profiles and subsequent worsening of depressive symptoms has also been reported (Phillips et al., 2011), providing some evidence that the blunted response pattern is not simply a consequence of experiencing depression.

With the growing interest in measurement of HRV as a physiologic marker for subsequent disease, 26 studies examining research on HRV reactivity to acute stress and depression in adults, adolescents, and children were reviewed by Hamilton and Alloy (2016). The majority of studies on depressed adults found blunted RSA reactivity (attenuated PNS activation) to a range of stimuli, including speech tasks, cognitive stressors, and emotion-inducing stressors. Findings from studies on adults who were successfully treated for depression revealed no significant differences in HRV response profiles between treated patients and healthy controls, possibly indicating that treatment of depression eradicates this dysfunctional physiologic response profile. Findings from studies that measured depressive symptoms in healthy adults and/or children were mixed, with some finding depressed affect to be associated with blunted response HRV reactions to stress, some with exaggerated HRV reactivity, and some finding no association between HRV reactivity and depression.

Depression and prolonged recovery from stress

Congruent with findings linking depression with both exaggerated and blunted physiological response patterns, there is some evidence that depression is associated with prolonged physiological recovery from stress. For example, patients diagnosed with major depressive disorders have been shown to exhibit prolonged HR recovery following speech and mirror tracing tasks compared to non-depressed controls (Salomon et al., 2009). Similarly, significant associations between self-reported depressive symptoms and slower HR recovery following a treadmill stress test have been reported (Hughes et al., 2006), a finding attributed to differences in exercise capacity. In a review of studies examining plasma or cortisol responses to stress, Burke et al. (2005) reported that individuals diagnosed with depressive disorders have higher cortisol levels during recovery periods following stress than individuals who are not clinically depressed.

Summary and conclusions

Based on the literature reviewed in this chapter, it is safe to conclude that there is strong evidence supporting the role of the ANS and HPA axis in explaining how exposure to environmental stress is translated into the pathophysiology of atherosclerosis and its clinical manifestations of CVD. All three patterns of allostatic load considered here, exaggerated reactivity to stress, blunted reactivity to stress, and prolonged recovery from stress, reflect important physiologic pathways involved in translating environmental stimuli to bodily disease responses. First, research examining various types of environmental stress, including exposure to trauma or adverse childhood experiences and other types of stressful living or occupational settings, has shown a relation between exposure to environmental stress and either exaggerated or blunted physiologic responses to acute stress as well as delayed recovery from stress. Second, exaggerated physiologic reactivity to stress as well as delayed recovery from stress have been shown to predict onset of CVD and subclinical cardiovascular complications, including essential hypertension, increased left ventricular mass, and increased intima media in carotid arteries. Although prospective studies have not shown an association between blunted physiologic responses to stress and subsequent CVD, evidence shows that these blunted response profiles are predictive of a range of health behaviors (e.g., smoking, obesity) that are already established behavioral risk factors for CVD. Finally, research on depressed patients or persons exhibiting depressive symptoms has yielded evidence linking this mood state with all three dysfunctional patterns of allostatic load.

Although studies comprising this literature suggest that dysfunctional ANS and NE systems are involved in explaining how stress and depression are associated with CVD, their involvement is far more complex than initially thought. There is indeed evidence to support the reactivity hypothesis initially proposed by Krantz and Manuck (1984) that exposure to stress elicits a cascade of physiological events that include exaggerated ANS and NE responses to stress that promote atherosclerosis. Findings that these exaggerated reactions have been observed among patients diagnosed with depressive disorders (Kibler and Ma, 2004; Light et al., 1998; Matthews et al., 2005) further corroborate this hypothesis and could be used to explain how an emotional state like depression is associated with increased risk for CVD. However, based on research examining the other two forms of allostatic load, it is clear that such an explanation is far from complete.

One of the most perplexing questions arising from the pattern of findings pertaining to ANS and NE response to stress is how both exaggerated and blunted reactivities can be associated with exposure to stress and/or experiencing depressed mood. Some divergent response patterns are expected and not of concern here (e.g., exaggerated HR reactivity accompanied by blunted HRV). However, most of the divergent response patterns described in this chapter involved the same physiological parameter or two parameters expected to react similarly (e.g., increased HR and increased cortisol reactions). There are several reasons that might account for these divergent response profiles. First, as noted by Gump and Matthews (1999) exposure to stress could evoke either increased or decreased cardiovascular responses to acute stress presentations and findings that emerge from a given study will vary based upon participant characteristics and/or contextual stress factors that are responsible for these divergent findings. Methods for testing nonlinear associations between chronic stress exposure and patterns of reactivity have confirmed than quadratic models showing that both exaggerated and dampened HR and BP responses to stress are associated with chronic stress fit the data better than linear models (Chatkoff et al., 2010). We have yet to identify any participant or contextual factors that lead toward increased or decreased cardiovascular reactivity to stress, but future empirical work should elucidate these explanatory factors. Second, it is possible that blunted physiologic reactions occur due to elevated autonomic tone during pre-task rest periods so that reactivity appears to be attenuated when change scores are calculated. As noted earlier, there is evidence that exposure to stress can result in increased resting blood pressures or cortisol levels (Evans, 2003; Fleming et al., 1987; Pole, 2007); should a study participant arrive in the laboratory already exhibiting elevated SNS arousal, the magnitude of reactivity to a laboratory stressor may appear to be dampened. Finally, it is also possible that the relation between stress and patterns of allostatic load change over time. As initially proposed by Selye (1955), the acute stress reaction differs dramatically from the reaction to stress that persists (e.g., chronic stress). Thus, it is possible that while exaggerated stress reactions may be associated with the earlier phases of responding to life stress, blunted reactions may occur after the body acclimates to the level of environmental stress. Some empirical work already demonstrates that these developmental changes are apparent in assessing ANS and NE response to stress (Pitman et al., 2012). These hypotheses warrant additional attention as we grapple with understanding how stress can be linked with both exaggerated and blunted cardiovascular response profiles.

Krantz and Manuck (1984) noted in their initial appraisal of the reactivity hypothesis that the nature of the stress task employed in such investigations may moderate study findings. Across studies included in this review, a full range of tasks were employed, including physical motor, cognitive, interpersonal, and emotionally-evocative challenges. It is well known that the magnitude and patterning of physiologic responding varies based on important characteristics of the tasks employed to elicit the physiological response (Manuck et al., 1989). Because such a broad range of tasks have been employed in studies comprising this literature, it seems important to determine whether specific tasks possess greater validity in predicting CVD or the relation between depression and CVD than others. Relatedly, the lack of normative information regarding responses to each task precludes ascertaining whether a given participant's response is above average (i.e., exaggerated) or below average (i.e., blunted). To serve as a useful clinical tool, normative data regarding physiologic response profiles need to be developed so that reliable categorizations of exaggerated or blunted response profiles can be made for individual patients. Without such reference points, the clinical utility of stress reactivity testing in clinical settings is limited.

Moderation of patterns of allostatic responding is known to involve several individual difference characteristics, including participant sex, race, and disease status. For example, research has shown that men exhibit higher SBP and epinephrine responses to acute stressors than women, but women exhibit greater HR responses to stress than men (Stoney et al., 1987). Interestingly, cardiovascular response profiles of post-menopausal women differ from pre-menopausal women in that they more closely resemble the acute physiological response profiles of men (Saab et al., 1989). However, treatment with hormone replacement returns women's response patterns to their pre-menopausal reactivity patterns (Matthews et al., 2001; McCubbin et al., 2002), suggesting that estrogen plays an important role in explaining sex differences in responding to stress.

Regarding race, in contrast to Caucasians, African-Americans have been reported to exhibit elevated BP responses to stress, lower HR responses to stress (Anderson et al., 1992), and delayed BP recovery from stress (Schuler and O'Brien, 1997). Additionally, African-Americans secrete more endothelin-1, a vasoconstrictive peptide released by the blood vessel wall, than Caucasians (Treiber et al., 2002) as well as displaying lower sodium excretion following sodium loading (Luft et al., 1985), suggesting that these racial differences in cardiovascular reactivity to stress involves vasoactive agents that impact renal functioning.

Finally, in addition to the presence of CVD, both diabetes and obesity have been shown to impact ANS and NE response profiles. Diabetic patients reportedly exhibit elevated blood pressure responses to stress in contrast to non-diabetic controls (Shapiro et al., 1963) and obese women have been shown to display increased HR reactions to mental arithmetic tasks than women of normal weight (Putnam and Rennert, 1984). Given the documented differences in cardiovascular response profiles associated with participant sex, race, and disease status, it will be important for future work exploring relations among stress, depression, and CVD to examine the potential moderating influences of these individual difference characteristics.

Although ANS and NE reactivity to stress play an important role in explaining how stress and depression lead to CVD, they clearly are not the only physiologic mechanisms involved. Certainly, ANS and NE reactivities interact with many of the other physiologic mechanisms described in the other chapters in this volume. Future empirical work aimed at uncovering these important connections and identifying their causal relations will bring us closer to understanding how stress and depression lead to CVD. By understanding these associations, health care providers will possess a greater comprehension of how external stimuli cause atherosclerotic blockage and consequently be more inclined to assess these important psychophysiological risk factors for CVD.

References

American Psychiatric Association, 2013. Diagnostic and Statistical Manual of Mental Disorders—(DSM-5), fifth ed. American Psychiatric Association Publishing, Washington, DC.

Anderson, N.B., McNeilly, M., Myers, H.F., 1992. A contextual model for research of race differences in autonomic reactivity. In: Johnson, E.H., Gentry, W.D., Julius, S. (Eds.), Personality, Elevated Blood Pressure, and Essential Hypertension. Hemisphere Publishing Corporation, Washington, DC, pp. 197–216.

Andreassi, J.L., 2007. Psychophysiology: Human Behavior & Physiological Response, 5th ed. Lawrence Erlbaum Associates, Mahwah, NJ.

Boylan, J.M., Cundiff, J.M., Matthews, K.A., 2018. Socioeconomic status and cardiovascular responses to standardized stressors: a systematic review and meta-analysis. Psychosom. Med. 80, 278–293.

Burke, H.M., Davis, M.C., Otte, C., Mohr, D.C., 2005. Depression and cortisol responses to psychological stress: a meta-analysis. Psychoneuroendocrinology 30 (9), 846–856.

Cannon, W.B., 1915. Bodily Changes in Pain, Hunger, Fear and Rage: An Account of Recent Researches into the Function of Emotional Excitement. Appleton.

Carney, R.M., Freedland, K.E., Rich, M.W., Jaffe, A.S., 1995. Depression as a risk factor for cardiac events in established coronary heart disease: a review of possible mechanisms. Ann. Behav. Med. 17 (2), 142–149.

Carpenter, L.L., Carvalho, J.P., Tyrka, A.R., Wier, L.M., Mello, A.F., Mello, M.F., Anderson, G.M., Wilkinson, C.W., Price, L.H., 2007. Decreased adrenocorticotropic hormone and cortisol responses to stress in healthy adults reporting significant childhood maltreatment. Biol. Psychiatry 62 (10), 1080–1087.

Carpenter, L.L., Shattuck, T.T., Tyrka, A.R., Geracioti, T.D., Price, L.H., 2011. Effect of childhood physical abuse on cortisol stress response. Psychopharmacology 214 (1), 367–375.

Casto, R., Printz, M.P., 1990. Exaggerated response to alerting stimuli in spontaneously hypertensive rats. Hypertension 16 (3), 290–300.

Castro-Chapman, P.W., Orr, S.P., Berg, J., Pineles, S.L., Yanson, J., Salomon, K., 2018. Heart rate reactivity to trauma-related imagery as a measure of PTSD symptom severity: examining a new cohort of Veterans. Psychiatry Res. 261, 574–580.

Chatkoff, D.K., Maier, K.J., Klein, C., 2010. Nonlinear associations between chronic stress and cardiovascular reactivity and recovery. Int. J. Psychophysiol. 77, 150–156.

Chida, Y., Hamer, M., 2008. Chronic psychosocial factors and acute physiological responses to laboratory-induced stress in healthy populations: a quantitative review of 30 years of investigations. Psychol. Bull. 134 (6), 829–885.

Chida, Y., Steptoe, A., 2010. Greater cardiovascular responses to laboratory mental stress are associated with poor subsequent cardiovascular risk status: a meta-analysis of prospective evidence. Hypertension 55 (4), 1026–1032.

Corse, C.D., Manuck, S.B., Cantwell, J.D., Giordani, B., Matthews, K.A., 1982. Coronary-prone behavior pattern and cardiovascular response in persons with and without coronary heart disease. Psychosom. Med. 44, 449–459.

de Rooij, S.R., Schene, A.H., Phillips, D.I., Roseboom, T.J., 2010. Depression and anxiety: associations with biological and perceived stress reactivity to a psychological stress protocol in a middle-aged population. Psychneuroendocrinology 35 (6), 866–877.

Engert, V., Buss, C., Khalili-Mahani, N., Wadiwalla, M., Dedovic, K., Pruessner, J.C., 2010. Investigating the association between early life parental care and stress responsivity in adulthood. Dev. Neuropsychol. 35 (5), 570–581.
Evans, G.W., 2003. A multimethodological analysis of cumulative risk and allostatic load among rural children. Dev. Psychol. 39 (5), 924–933.
Evans, G.W., Kim, P., Ting, A.H., Tesher, H.B., Shannis, D., 2007. Cumulative risk, maternal responsiveness, and allostatic load among young adolescents. Dev. Psychol. 43 (2), 341–351.
Fleming, I., Baum, A., Davidson, L.M., Rectanus, E., McArdle, S., 1987. Chronic stress as a factor in physiologic response to a challenge. Health Psychol. 6 (3), 221–237.
Ginty, A.T., Conklin, S.M., 2011. High perceived stress in relation to life events is associated with blunted cardiac reactivity. Biol. Psychol. 86, 383–385.
Ginty, A.T., Williams, S.E., Jones, A., Roseboom, T.J., Phillips, A.C., Painter, R.C., Carroll, D., de Rooij, S.R., 2016. Diminished heart rate reactivity to acute psychological stress is associated with enhanced carotid intima-media thickness through adverse health behaviors. Psychophysiology 53 (6), 769–775.
Griffin, M.G., Resick, P.A., Galovski, T.E., 2012. Does physiologic response to loud tones change following cognitive-behavioral treatment for posttraumatic stress disorder? J. Trauma. Stress. 25, 25–32.
Grippo, A.J., Moffitt, J.A., Johnson, A.K., 2002. Cardiovascular alterations and autonomic imbalance in an experimental model of depression. Am. J. Physiol. Regul. Integr. Comp. Physiol. 282, R1333–R1341.
Gump, B.B., Matthews, K.A., 1999. Do background stressors influence reactivity to and recovery from acute stressors. J. Appl. Soc. Psychol. 29 (3), 469–494.
Hallbäck, M., Folkow, B., 1974. Cardiovascular responses to acute mental 'stress' in spontaneously hypertensive rats. Acta Physiol. Scand. 90 (4), 684–698.
Hamer, M., Williams, E., Vuonovirta, R., Giacobazzi, P., Gibson, E.L., Steptoe, A., 2006. The effects of effort-reward imbalance on inflammatory and cardiovascular responses to mental stress. Psychosom. Med. 68, 408–413.
Hamer, M., Endrighi, R., Venuraju, S.M., Lahiri, A., Steptoe, A., 2012. Cortisol responses to mental stress and the progression of coronary artery calcification in healthy men and women. PLoS One 7 (2), e31356.
Hamilton, J., Alloy, L., 2016. Atypical reactivity of heart rate variability to stress and depression across development: systematic review of the literature and directions for future research. Clin. Psychol. Rev. 50, 67–79.
Heim, C., Newport, D.J., Heit, S., Graham, Y.P., Wilcox, M., Bonsall, R., Miller, A.H., Nemeroff, C.B., 2000. Pituitary-adrenal and autonomic responses to stress in women after sexual and physical abuse in childhood. J. Am. Med. Assoc. 284 (5), 592–597.
Hughes, J., Stoney, C., 2000. Depressed mood is related to high-frequency heart rate variability during stressors. Psychosom. Med. 62 (6), 796–803.
Hughes, J., Casey, E., Luyster, F., Doe, V., Waechter, D., Rosneck, J., Josephson, R., 2006. Depression symptoms predict heart rate recovery after treadmill stress testing. Am. Heart J. 151 (5), 1122e1–1122e6.
Kamarck, T.W., Susan, A., Everson, S.A., Kaplan, G.A., Jennings, J.R., Salonen, R., Salonen, J.T., 1997. Exaggerated blood pressure responses during mental stress are associated with enhanced carotid atherosclerosis in middle-aged Finnish men. Circulation 96, 3842–3848.
Kaplan, J.R., Manuck, S.B., Clarkson, T.B., Lusso, F.M., Taub, D.M., 1982. Social status, environment, and atherosclerosis in cynomolgus monkeys. Arterioscler. Thromb. Vasc. Biol. 2 (5), 359–368.
Kemp, A.H., Quintana, D.S., Gray, M.A., Felmingham, K.L., Brown, K., Gatt, J.M., 2010. Impact of depression and antidepressant treatment on heart rate variability: a review and meta-analysis. Biol. Psychiatry 67, 1067–1074.

Keys, A., Taylor, H.L., Blackburn, H., Brozek, J., Anderson, J.T., Somonson, E., 1971. Mortality and coronary heart disease among men studied for 23 years. Arch. Intern. Med. 128, 201–214.
Kibler, J.L., Ma, M., 2004. Depressive symptoms and cardiovascular reactivity to laboratory behavioral stress. Int. J. Behav. Med. 11 (2), 81–87.
Krantz, D.S., Manuck, S.B., 1984. Acute psychophysiologic reactivity and risk of cardiovascular disease: a review and methodologic critique. Psychol. Bull. 96 (3), 435–464.
Larkin, K.T., Frazer, N.L., Wheat, A.L., 2011. Responses to interpersonal conflict among young adults: influence of family of origin. Pers. Relat. 18 (4), 657–667.
Lepore, S.J., Miles, H.J., Levy, J.S., 1997. Relation of chronic and episodic stressors to psychological distress, reactivity, and health problems. Int. J. Behav. Med. 4 (1), 39–59.
Liang, C., Lee, J., Chen, C., Chang, Y., 2015. Reactive heart rate variability in male patients with first-episode major depressive disorder. Prog. Neuro-Psychopharmacol. Biol. Psychiatry 56, 52–57.
Light, K.C., Kothandapani, R.V., Allen, M.T., 1998. Enhanced cardiovascular and catecholamine responses in women with depressive symptoms. Int. J. Psychophysiol. 28 (2), 157–166.
Lovallo, W.R., 2005. Stress & Health: Biological and Psychological Interactions, second ed. Sage, Thousand Oaks, CA.
Lovallo, W.R., 2011. Do low levels of stress reactivity signal poor states of health? Biol. Psychol. 86 (2), 121–128.
Lovallo, W.R., 2013. Early life adversity reduces stress reactivity and enhances impulsive behavior: implications for health behaviors. Int. J. Psychophysiol. 90, 8–16.
Low, C.A., Salomon, K., Matthews, K.A., 2009. Chronic life stress, cardiovascular reactivity, and subclinical cardiovascular disease in adolescents. Psychosom. Med. 71, 927–931.
Luecken, L.J., Kraft, A., Hagan, M.J., 2009. Negative relationships in the family-of-origin predict attenuated cortisol in emerging adults. Horm. Behav. 55 (3), 412–417.
Luft, F., Grim, C., Weinberger, M., 1985. Electrolyte and volume homeostasis in blacks. In: Hall, W., Saunders, E., Shulman, N. (Eds.), Hypertension in Blacks: Epidemiology, Pathophysiology, and Treatment. Yearbook Medical, Chicago, pp. 115–131.
Lumley, M.A., Chi, W., Wiholm, C., Slatcher, R.B., Sandmark, H.S., Wang, S., Hytter, A., Arnetz, A.B., 2014. The relationship of chronic and momentary work stress to cardiac reactivity in female managers: feasibility of a smart phone-assisted assessment system. Psychosom. Med. 76, 512–518.
Manuck, S.B., Kaplan, J.R., Clarkson, T.B., 1983. Behaviorally induced heart rate reactivity and atherosclerosis in cynomolgus monkeys. Psychosom. Med. 45, 95–108.
Manuck, S.B., Kasprowicz, A.L., Monroe, S.M., Larkin, K.T., Kaplan, J.R., 1989. Psychophysiologic reactivity as a dimension of individual differences. In: Handbook of Research Methods in Cardiovascular Behavioral Medicine. Springer, Boston, MA, pp. 365–382.
Matthews, K.A., Flory, J.D., Owens, J.F., Harris, K.F., Berga, S.L., 2001. Influence of estrogen replacement therapy on cardiovascular responses to stress of healthy postmenopausal women. Psychophysiology 38, 391–398.
Matthews, S.C., Nelesen, R.A., Dimsdale, J.E., 2005. Depressive symptoms are associated with increased systemic vascular resistance to stress. Psychosom. Med. 67 (4), 509–513.
Mausbach, B., Dimsdale, J., Ziegler, M., Mills, P., Ancoli-Israel, S., Patterson, T., Grant, I., 2005. Depressive symptoms predict norepinephrine response to a psychological stressor task in Alzheimer's caregivers. Psychosom. Med. 67 (4), 638–642.
McCubbin, J.A., Helfer, S.G., Switzer, F.S., Price, T.M., 2002. Blood pressure control and hormone replacement therapy in postmenopausal women at risk for coronary heart disease. Am. Heart J. 143, 711–717.
McEwen, B., 2002. The End of Stress as We Know It. Joseph Henry Press, Washington, DC.

Musante, L., Treiber, F.A., Kapuku, G., Moore, D., Davis, H., Strong, W.B., 2000. The effects of life events on cardiovascular reactivity to behavioral stressors as a function of socio-economic status, ethnicity, and sex. Psychosom. Med. 62 (6), 760–767.
Nachar, N., Lavoie, M.E., Marchand, A., O'Conner, K.P., Guay, S., 2014. The effect of talking about psychological trauma with a significant other on heart rate reactivity in individuals with posttraumatic stress disorder. Psychiatry Res. 219, 171–176.
Normann, M.C., McNeal, N., Dagner, A., Ihm, E., Woodbury, M., Grippo, A.J., 2018. The influence of environmental enrichment on cardiovascular and behavioral responses to social stress. Psychosom. Med. 80, 271–277.
Panaite, V., Salomon, K., Jin, A., Rottenberg, J., 2015. Cardiovascular recovery from psychological and physiological challenge and risk for adverse cardiovascular outcomes and all-cause mortality. Psychosom. Med. 77 (3), 215–226.
Phillips, A.C., Carroll, D., Ring, C., Sweeting, H., West, P., 2005. Life events and acute cardiovascular reactions to mental stress: a cohort study. Psychosom. Med. 67 (3), 384–392.
Phillips, A.C., Hunt, K., Der, G., Carroll, D., 2011. Blunted cardiac reactions to acute psychological stress predict symptoms of depression five years later: evidence from a large community study. Psychophysiology 48, 142–148.
Pieper, S., Brosschot, J.F., 2005. Prolonged stress-related cardiovascular activation: Is there any? Ann. Behav. Med. 30 (2), 91–103.
Pitman, R.K., Rasmusson, A.M., Koenen, K.C., Shin, L.M., Orr, S.P., Gilbertson, M.W., Milad, M.R., Liberzon, I., 2012. Biological studies of post-traumatic stress disorder. Nat. Rev. Neurosci. 13, 769–787.
Pole, N., 2007. The psychophysiology of posttraumatic stress disorder: a meta-analysis. Psychol. Bull. 133 (5), 725–746.
Pruessner, J.C., Champagne, F., Meaney, M.J., Dagher, A., 2004. Dopamine release in response to a psychological stress in humans and its relationship to early life maternal care: a positron emission tomography study using [11C]raclopride. J. Neurosci. 24 (11), 2825–2831.
Putnam, L.E., Rennert, M.P., 1984. Effects of obesity on autonomic responding in a mental arithmetic task. Psychophysiology 21, 594.
Richman, L.S., Bennett, G.G., Pek, J., Siegler, I., Williams Jr., R.B., 2007. Discrimination, dispositions, and cardiovascular responses to stress. Health Psychol. 26 (6), 675–683.
Roman, O., Seres, J., Pometlova, M., Jurcovicova, M., 2004. Neuroendocrine or behavioral effects of acute or chronic emotional stress in Wystar Kyoto (WKY) and spontaneously hypertensive (SHR) rats. Endocr. Regul. 38, 151–155.
Saab, P.G., Matthews, K.A., Stoney, C.M., McDonald, R.H., 1989. Premenopausal and post-menopausal women differ in their cardiovascular and neuroendocrine responses to behavioral stressors. Psychophysiology 26, 270–280.
Sack, M., Hopper, J.W., Lamprecht, F., 2004. Low respiratory sinus arrhythmia and prolonged psychophysiological arousal in posttraumatic stress disorder: heart rate dynamics and individual differences in arousal regulation. Biol. Psychiatry 55 (3), 284–290.
Salomon, K., Clift, A., Karlsdóttir, M., Rottenberg, J., 2009. Major depressive disorder is associated with attenuated cardiovascular reactivity and impaired recovery among those free of cardiovascular disease. Health Psychol. 28 (2), 157–165.
Santa Ana, E.J., Saladin, M.E., Back, S.E., Waldrop, A.E., Spratt, E.G., McRae, A.L., LaRowe, S.D., Timmerman, M.A., Upadhyaya, H., Brady, K.T., 2006. PTSD and the HPA axis: differences in response to the cold pressor task among individuals with child vs. adult trauma. Psychoneuroendocrinology 31, 501–509.
Schuler, J.L.H., O'Brien, W.H., 1997. Cardiovascular recovery from stress and hypertension risk factors: a meta-analytic review. Psychophysiology 34, 649–659.
Selye, H., 1955. Stress and disease. Science 122 (3171), 625–631.
Shapiro, A.P., Moutsos, S.E., Krifcher, E., 1963. Patterns of pressor response to noxious stimuli in normal hypertensive and diabetic subjects. J. Clin. Investig. 42, 1890–1898.

Siegrist, J., Klein, D., 1990. Occupational stress and cardiovascular reactivity in blue-collar workers. Work Stress. 4 (4), 295–304.
Stoney, C.M., Davis, M.C., Matthews, K.A., 1987. Sex differences in physiological responses to stress and in coronary heart disease: a causal link? Psychophysiology 24, 127–131.
Taylor, S.E., Lerner, J.S., Sage, R.M., Lehman, B.J., Seeman, T.E., 2004. Early environment, emotions, responses to stress, and health. J. Pers. 72 (6), 1365–1393.
Treiber, F.A., Kapuku, G.K., Davis, H., Pollock, J.S., Pollock, D.M., 2002. Plasma endothelin-1 release during acute stress: role of ethnicity and sex. Psychosom. Med. 64, 707–713.
Treiber, F.A., Kamarck, T., Schneiderman, N., Sheffield, D., Kapuku, G., Taylor, T., 2003. Cardiovascular reactivity and development of preclinical and clinical disease states. Psychosom. Med. 65 (1), 46–62.
Uchino, B.N., Kiecolt-Glaser, J.K., Cacioppo, J.T., 1992. Age-related changes in cardiovascular response as a function of a chronic stressor and social support. J. Pers. Soc. Psychol. 63 (5), 839–846.
Voellmin, A., Hug, E., Wilhelm, F.H., Schaefer, V., Gaab, J., La Marca, R., Pruessner, J.C., Bader, K., 2015. Blunted endocrine and cardiovascular reactivity in young healthy women reporting a history of childhood adversity. Psychneuroendocrinology 51, 58–67.
Wirtz, P.H., Ehlert, U., Kottwitz, M.U., La Marca, R., Semmer, N.K., 2013. Occupational role stress is associated with higher cortisol reactivity to acute stress. J. Occup. Health Psychol. 18, 121–131.

CHAPTER 6

Circulatory system alterations under stress

Paul D. Chantler[a], Jefferson C. Frisbee[b]
[a]Division of Exercise Physiology, Robert C. Byrd Health Sciences Center, School of Medicine, West Virginia University, Morgantown, WV, United States
[b]Department of Medical Biophysics, Schulich School of Medicine & Dentistry, Western University, London, ON, Canada

Contents

Psychological stress has long been known to be associated with CVD, with the first reported evidence that emotional stress being associated with the pathogenesis of acute myocardial infarction reported in 1910 (Obraztsov and Strazhesko, 1910). Subsequently, the international INTERHEART study confirmed that psychosocial factors were associated with acute myocardial infarction with an odds ratio (a measure of the association between an exposure and an outcome; >1 exposure associated with higher odds of outcome) of 2.67 (i.e., greater risk) (Yusuf et al., 2004). A recent meta-analysis revealed that severe mental diseases, including major depressive disorder, were associated with an increased risk of developing CVD compared with controls with a hazard ratio of 1.54 (De Hert et al., 2018). Furthermore, the association between stress and CVD is thought to be independent of traditional cardiovascular risk factors (e.g., age, sex, smoking, hypertension, elevated blood lipids, diabetes mellitus, and obesity) (Rosengren et al., 2004;

Cardiovascular Implications of Stress and Depression
https://doi.org/10.1016/B978-0-12-815015-3.00006-4

Kivimaki et al., 2006; Steptoe and Kivimaki, 2013). This chapter examines the adverse effects of psychological stress on the cardiovascular system, and how sex and disease states mediate these cardiovascular changes to stress.

Before evaluating the literature in this area, it is important to consider the definitions of stress and depression, and their dynamic interplay with CVD, as they pertain to the literature pertaining to the effects of stress on the circulatory system. As outlined in Chapter 1, the term "stress" refers to the specific stimuli or alterations in environmental contexts to which an organism is exposed. Chronic exposure to stress can lead to a maladaptive state of the bodily equilibrium as it adapts to the surrounding environment called allostasis. In contrast, "depression" refers to the diagnosed conditions of either Major Depressive or Persistent Depressive Disorders, and "depressive symptoms" refer to the underlying cognitive, behavioral, emotional, and physical symptoms upon which these diagnoses are made. The development of depression involves a complex interplay between psychological and neurobiological factors (Palazidou, 2012; Baum and Posluszny, 1999), which makes it difficult to separate the distinctive roles of stress vs depression on alterations within the circulatory system that lead to the development of CVD. Added difficulties involved in modeling the physiological phenotypes of stress-related disorders such as depression in rodents are discussed in Chapter 3. Because animals cannot be diagnosed with depression, research employing animal models need to rely on assessing behaviors that resemble a depressive phenotype (e.g., social withdrawal, restlessness, reduced grooming, inactivity). Most of the stress/depression literature detailing the pathophysiological changes to the heart and vasculature (both brain and systemic circulation) is derived from animal models.

The most appropriate rodent model for human clinical depression is the unpredictable chronic stress protocol described in Chapter 3, based on its ability to reproduce simultaneously the development of many human clinical depressive symptoms, including anhedonia and learned helplessness (Willner, 2005), and changes within the cardiovascular system that promote CVD. In this regard, most of the literature exploring how stress and depression are associated with the structural and functional changes in the cardiovascular system does not attempt to separate the effects of exposure to stress from the resulting depressive symptoms. Conceptually, these animal models assume that exposure to stress is the precipitant for both subsequent depressive behaviors and cardiovascular complications that are observed, but there has been very little attention paid to determining whether the altered cardiovascular system results directly from the body's response to stress or whether it is mediated indirectly through the depressive behavioral symptoms that result from exposure to stress. One exception to this is the work of Bergström et al. who used chronic mild stress to elicit an anhedonic phenotype in rats (Bergstrom et al., 2008). They noted that not all rats developed this anhedonic phenotype, and split the cohort into resilient and susceptible animals, allowing the researchers to identify molecular changes in these subphenotypes. The authors identified a significant difference in gene expression of brain derived neurotrophic factor (upregulated in resilient rats) and vascular endothelial growth factor (downregulated in the stress sensitive rats), in the CA3 region of the ventral hippocampus. Unfortunately, the resiliency studies have mainly focused on neurochemical differences and shed very little light on the modifications to the cardiovascular system that is the topic of this chapter. Nevertheless, it will be important for future research to examine differences in structural and/or functional changes in the circulatory system between stress-resilient animals and those with vulnerable phenotypes. Because the preponderance of studies examining the effects of exposure to stress upon modifications to the circulatory system have yet to distinguish the effects of stress from the influence of depressive behavioral symptoms, we will include studies that have explored either the stress-CVD link or the depression-CVD linkage. Studies evaluating the effects of stress and/or depression on vascular structures will be considered first, followed by those that have focused on the structure and function of the heart itself. Fig. 6.1

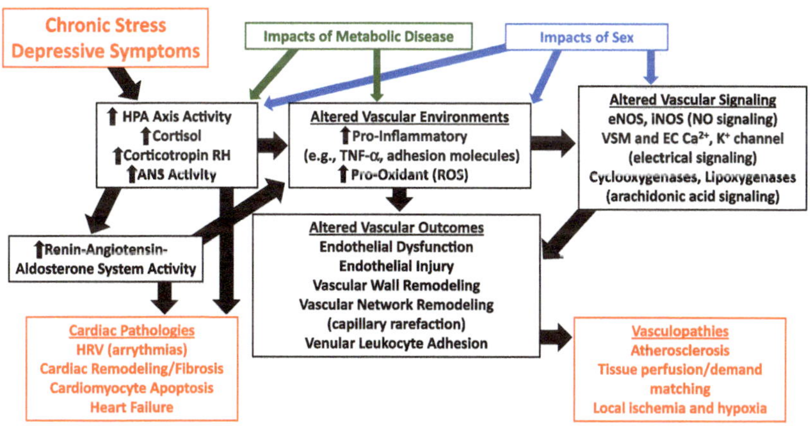

Fig. 6.1 Schematic representation of the impact and pathways linking chronic stress and depressive symptoms to poor cardiac and/or vascular outcomes. *ANS*, autonomic nervous system; *EC*, endothelial cell; *eNOS*, endothelial nitric oxide synthase; *HPA*, hypothalamic-pituitary-adrenal; *HRV*, heart rate variability; *iNOS*, inducible nitric oxide synthase; *NO*, nitric oxide; *RH*, releasing hormone; *ROS*, reactive oxygen species; *TNF-a*, tumor necrosis factor-alpha; *VSM*, vascular smooth muscle.

provides a brief schematic representation of the impact and pathways linking chronic stress and depressive symptoms to poor cardiac and/or vascular outcomes.

Vascular dysfunction

Endothelial function

The central arteries deliver blood from the heart to peripheral tissues and act as a "Windkessel" (translated as an elastic "air chamber") in which the shape of the arterial blood pressure waveform changes in response to the increased blood flow following heart contractions and the accommodation of this increased flow by the aorta and large elastic arteries to ensure a steady supply of blood to the peripheral organs. In contrast, resistance arterioles play an important role in regulating local blood flow and thus oxygenation and nutrient delivery. Both the large (aorta, carotid etc.) and small arteries are lined with endothelial cells (EC) that play a pivotal role in regulating several arterial properties, including vessel diameter, permeability, angiogenesis, anti-thrombosis, and the response to inflammation. One potential pathway through which chronic exposure to psychosocial stressors are hypothesized to influence CVD risk is through injury to the endothelium. Indeed, acute exposure to a mental stressor is significantly associated with prolonged endothelial dysfunction (Spieker et al., 2002), whereas frequent and repetitive (chronic) stresses lead to impairment in vascular reactivity (Black and Garbutt, 2002). It is well known that endothelial dysfunction plays a critical role in the initiation and progression of atherosclerosis, and thus, may be a major pathway linking chronic stress and CVD.

In human studies, Ghiadoni et al. (2000) was among the first to study the effects of acute mental stress on endothelial function as estimated by flow-mediated dilation (FMD). In this experiment, FMD in healthy individuals after mental stress was comparable with FMD at baseline in patients with diabetes. From a chronic perspective, the Multi-Ethnic Study of Atherosclerosis (MESA) identified that higher levels of chronic stress were associated with a lower absolute FMD response, which remained evident after adjusting for demographic/socioeconomic characteristics, and behavioral/biological CVD risk factors (Kershaw et al., 2017). In animal models using vasoactive substances (e.g., acetylcholine, phenylephrine etc.) to explore vascular health, reproducible results have confirmed the impairment of endothelium-dependent vasorelaxation of multiple vascular beds (aorta, middle cerebral artery, gracilis arteriole) in mice and rats exposed to at least 8 weeks of unpredictable chronic mild stress (UCMS) to induce a depression-like phenotype (Isingrini et al., 2011; Brooks et al., 2018a,c; Branyan et al., 2018). Despite the reproducible response of impaired endothelium-dependent vasorelaxation to the administration of vasoactive drugs, the actions of chronic stress on endothelium-independent vasorelaxation are less consistent. For example, 8 weeks of UCMS did not impair endothelium-independent vasorelaxation in the aorta of middle cerebral artery of lean mice or rats (Stanley et al., 2014; Brooks et al., 2018a) but an impaired endothelium-independent vasorelaxation was evident in the gracilis arteriole of lean rats (Branyan et al., 2018). Some studies, but not all (Isingrini et al., 2011; Stanley et al., 2014), have reported an altered contractile response after UCMS. Branyan et al. (2018) reported a reduced phenylephrine-induced vasoconstriction in the gracilis arteriole, whereas Brooks et al. (2018a) identified an increased serotonin-induced vasoconstriction response in the middle cerebral artery. Both studies were performed using lean rats that were exposed to 8 weeks of UCMS. These data confirm the pivotal role of endothelial functioning in explaining how stress-elicits vascular dysfunction (Fig. 6.1).

Arterial remodeling

Arterial stiffness

Arterial stiffness is a natural process of aging caused by structural and cellular change within vessel walls that accelerates with the co-occurrence of CVD risk factors (e.g., diabetes, hypertension etc.). Arterial stiffening has importance clinical consequences as it permits increased systolic blood pressure (SBP) and higher blood flow pulsations to be transmitted deeper down the vascular tree into the microvasculature of the brain and other peripheral organs. Arterial stiffness, as measured by pulse wave velocity or indirectly via augmentation index, has been found to be increased in individuals with depression and/or anxiety disorders (Bomhof-Roordink et al., 2015; Seldenrijk et al., 2011; Onete et al., 2018). Further, chronic depression is linked to a greater increase arterial stiffness over a 3-year period, compared to non-depressed individuals (Satoh et al., 2015). In contrast, antidepressant treatment among highly depressed women was associated with a reduction in arterial stiffness (Oulis et al., 2010). The interaction between arterial stiffness and antidepressant drug therapy is less clear among older adults. In a sample of adults aged 70 years or older with newly diagnosed depression, anti-depressant treatment with duloxetine (Cymbalta) but not escitalopram (Lexapro) significantly increased arterial stiffness after 12 months of treatment, which was attributed to, in part, the effects of duloxetine increasing heart rate and its actions on monoamines and catecholamine metabolism. However, the drug-specific effect remained after controlling for cardiovascular risk factors (Scuteri et al., 2013). These data suggest that exposure to stress, anxiety, and ultimately depression may therefore enhance the development and progression of arterial stiffness.

Evidence also suggests that arterial stiffness may accelerate the development of depressive disorders. The link between arterial stiffness and depression is not fully known, but the effects of arterial stiffness altering brain structure (i.e., small vessel lesions and endothelial permeability) may play an important role. According to the vascular depression hypothesis, arterial stiffness, which is associated with the accumulation of cerebral small vessel lesions in the mood regulatory centers of the brain (Mitchell et al., 2011), may occur as depression develops. Indeed, arterial stiffness has been associated with white matter hyperintensities, cerebral lacunar infarction, and cortical brain atrophy (Henry-Feugeas, 2009; Bateman et al., 2008). A previous study has identified the presence of supratentorial cerebral white matter hyperintensities and subcortical infarcts that occur with depression and could explain the association between arterial stiffness and depression. (van Sloten et al., 2015). It has been postulated that increased arterial stiffness can lead to greater oscillations in wall shear stress, leading to endothelial dysfunction and the production of low-grade inflammation (Chappell et al., 1998). In turn, the endothelial dysfunction can disturb neurogenesis (Zhao et al., 2008), lead to small vessel disease (Hoth et al., 2007), and limit cerebrovascular reactivity (Brooks et al., 2018a). Furthermore, a pro-inflammatory environment stimulates the production of pro-inflammatory markers by microglia, which could induce symptoms of depression (Miller et al., 2009). Irrespective of whether depression is a cause or effect of arterial stiffness, the relationships between arterial stiffness, CVD, and depression seem to be interactive and tightly coupled.

Atherosclerosis

Endothelial dysfunction is an initial step in the development of atherosclerosis. Atherosclerosis is a chronic inflammatory state of the arterial wall associated with the accumulation of macrophages, white blood cells, lipids, and the release of inflammatory proteins and cytokines (Woollard, 2013). Injury to the endothelium and the subsequent inflammatory response induces production of adhesion molecules and recruitment of immune cells, which in turn perpetuates the vicious cycle of macrophage infiltration, foam cell formation, atherosclerotic plaque formation, and adverse thrombotic events

(Gu et al., 2012). Numerous studies have reported a strong relationship between high levels of psychosocial stress and the development of coronary artery disease (Castro-Diehl et al., 2014; Zipursky et al., 2017). In a study using primates, coronary artery atherosclerosis was significantly increased in monkeys who developed depressive symptoms vs control monkeys (Shively et al., 2015). Similarly, in apolipoprotein E knockout mice, exposure to chronic stress resulted in increased atheroma formation (Kumari et al., 2003), more stenosis, and an increased degree of perivascular fibrosis (Roth et al., 2015). Histological analysis revealed that comparison to unstressed mice, mice exposed to chronic stress had profound changes in plaque composition, as evidenced by increased macrophage and T-cell infiltration and decreased smooth muscle cell mass, reflecting an unstable plaque phenotype (Zhang et al., 2010; Roth et al., 2015) and a higher frequency of myocardial infarctions (Roth et al., 2015). In addition to atherosclerosis in the coronary vessels, atherosclerosis also develops in the large vessels (carotid, aorta etc.) and plays an important role in accelerating arterial stiffness. For example, Razzoli et al. (2018) developed a model of lifelong chronic psychosocial stress that involved a number of weeks in which mice were exposed to daily defeats by a dominant mouse and sensory contact housing, and an aging phase lasting until spontaneous death or moribund status. During the aging phase, mice were housed in sensory contact and experienced a continued degree of threat stress related to the previous encounters. As such, the mice exposed to the lifelong chronic psychosocial stress developed spontaneous early-stage atherosclerotic lesions of the aortic sinuses characterized by significant immune cell infiltration and sporadic rupture and calcification, none of which was found in dominant mice. Stressed animals also show morphological vascular changes and metabolic derangements including hypertrophy of the intima and tunica media of the thoracic aorta, increased serum levels of total cholesterol, lipids, triglycerides, and atherogenic indexes (Neves et al., 2009). These data highlight an important role for social stress and low social status on shortening lifespan and increasing the risk of CVD.

Vessel density

The total length of capillaries in the human brain is approximately 600 km. We and others have demonstrated a significant reduction in cerebral microvessel density in rodents exposed to chronic stress (Brooks et al., 2018a; Czeh et al., 2010; Kiuchi et al., 2012). In contrast, using the social defeat model of chronic stress, Pearson-Leary et al. (2017) indicated that microvessel density in the brain was increased. It was postulated that the observed increase in microvessel density would provide metabolic support following increased neural activity. These differing results may be a consequence of the duration of stress imposed, for example 7 days (Pearson-Leary) vs 5–8 weeks (Brooks et al., 2015, 2018a,b,c,d; Czeh et al., 2010; Kiuchi et al., 2012). Of note, there is a link between suppressed blood flow, capillary density, and neurogenesis. For example, Heine et al. showed that a stress-induced decline in hippocampal neurogenesis mainly occurred near capillaries and suggested that the reduction in blood flow in response to chronic stress led to suppressed blood flow and capillary density in the hippocampus that was in turn followed by the occurrence of depressive behaviors (Heine et al., 2005). These findings suggest that the development and improvement of stress behavior may be closely related to changes in neurogenesis and angiogenesis in the hippocampus.

Sex specific differences in arterial function

Sex differences in stress-related hormonal secretion are well known, whereby females, regardless of age, have an altered hormonal secretion compared to males in response to stress (Heuser et al., 1994). In premenopausal women with an ovariectomy, a significant reduction of adrenocorticotropic hormone and cortisol are found in the absence of any response to stress (De Leo et al., 1998). Further, there are sex differences in the cognitive consequences of repeated stress, with males showing impairment of hippocampal-dependent memory, whereas females do not (Luine et al., 1994, 2007; Bowman et al., 2001). In men and women, neural activation patterns to the same tasks are quite different between the sexes even when performance is similar (Derntl et al., 2010). This leads to the hypothesis that men and women use different strategies to approach and deal with stress in their daily lives, in part because of the subtle differences in brain architecture and functioning.

Recent data from animal studies has demonstrated important sex-specific differences in the response to chronic stress. Using a mild UCMS protocol, female Wistar rats were found to have reduced serotonergic activity in the hippocampus and reduced dopaminergic activity in the prefrontal cortex, in contrast to male rats (Dalla et al., 2005). Sex-differences have also been reported in the dysregulation of the hypothalamic-pituitary-adrenal (HPA) axis, whereby higher serum corticosterone concentrations were reported in female vs male rats (Xing et al., 2013; Harpaz et al., 2013). These data would suggest that females may be more vulnerable to the UCMS model than males. Indeed, we have previously shown that female rats display a more phenotypically severe depressive state compared to male rats (Brooks et al., 2018c). At the neural level, exposure to chronic stress results in dendritic atrophy of neurons in the prefrontal cortex in male rats but results in dendritic hypertrophy in female rats (Garrett and Wellman, 2009). However, dendritic hypertrophy was absent in ovariectomized female rats, suggesting that ovarian hormones modulate the morphological changes in brain structure with stress.

Regarding aortic stiffness, sex differences in humans have been identified, in that aortic stiffness was associated with depression in men aged 60 years or younger, but, in women the association was not statistically significant (Onete et al., 2018). A possible explanation for the interaction might reflect the importance of hormonal fluctuations, as seen during the menstrual cycle and perimenopause, in predisposing women aged 60 years or younger to depression moreso than aortic stiffness. Reproductive hormones are thought to be vasoprotective, especially before menopause (Segers et al., 2007), thus further diminishing the association between arterial stiffness and depression in women. An alternative explanation suggested was that substantial heterogeneity among those with depressive disorders exists, which might result in important differences of this diagnosis between men and women (Onete et al., 2018).

Pre-clinical models have expanded our understanding on the role of sex-hormones, depression, and vascular health. Despite a more phenotypically severe depressive behavioral profile in female compared to male rats, female rats demonstrate a better cerebrovascular and aortic endothelial-dependent dilation (albeit lower than female non-stressed rats) compared to male rats exposed to UCMS (Brooks et al., 2018c). Such sex-differences in the vascular response to UCMS were also evident in a mouse model (Stanley et al., 2014), further supporting the sex-specific effects of stress upon CVD outcomes. The vascular protection from UCMS in females appeared to reflect a superior maintenance of endothelial function, with more normal levels of endothelium-dependent vasodilators (nitric oxide, hydrogen peroxide, and prostacyclin). However, the protective effect in female rats was dependent on the maintenance of a normal sex hormone profile, as ovariectomized females prior to UCMS failed to exhibit this protective effect and resulted in endothelial dysfunction being virtually identical to those in males (Brooks et al., 2018c). While the specific nature of these mechanisms, their relative importance, or how they may interact to produce an integrative outcome remain an area of active investigation, the beneficial impact of estrogen on the ability of the vascular endothelium to promote the bioavailability of nitric oxide, prostacyclin, and other dilator metabolites (Leblanc et al., 2009; Liochev and Fridovich, 2007), as well as the ability of estrogen to blunt a pro-oxidant or pro-inflammatory environment (Kim and Levin, 2006; Strehlow et al., 2003) are likely pathways. On a cautionary note, these data reflect the impact of ovariectomy at a young age (between 5 and 6 weeks of age) rather than natural alterations to sex hormone profiles that are associated with aging. Irrespective, these data add to our understanding of these relationships, providing support to the concept that a major contributor to the links between chronic stress/depressive symptoms and vascular dysfunction is mediated through the genesis of a pro-oxidant and pro-inflammatory condition that compromises the vascular environment.

The role of pre-existing CVD on arterial function

The presence of pre-existing cardiovascular and metabolic diseases (e.g., obesity, metabolic syndrome, diabetes etc.) has been shown to dysregulate the HPA axis so that neuroendocrine hyper-responsiveness to different neuropeptides and acute stress challenges is observed (Pasquali and Vicennati, 2000; Pasquali et al., 2006; Chan et al., 2002b). With diabetes, an increased plasma adrenocorticotrophic hormone, RNA expression of corticosterone, and hypothalamic corticotrophin-releasing hormone are observed (Chan et al., 2002a). Elevated levels of circulating corticosterone were noted in rats with metabolic syndrome compared to control rats and circulating corticosterone concentrations were further elevated with exposure to chronic stress (Brooks et al., 2018a,b). Such data point towards vascular dysfunction occurring in response to stress. Indeed, there is strong evidence linking existing vascular dysfunction, due to the presence of CVD, to the development of major depressive disorders especially with the presence of two or more classic risk factors (Serlin et al., 2011; Valkanova and Ebmeier, 2013). The metabolic syndrome affects approximately 20–25% of adults worldwide and has a strong association with depression (Pratt and Brody, 2014; Kahl et al., 2015). The normal development of metabolic syndrome is characterized by chronic low-grade inflammation and a pro-oxidative environment that negatively affects multiple vascular beds, including those located in the brain (Brooks et al., 2015; Phillips et al., 2005).

We have shown that the cerebrovascular and aortic dysfunction (impaired endothelial function and microvascular rarefaction) with UCMS was enhanced in the presence of metabolic syndrome (Brooks et al., 2018a,c,d). Interestingly, these cerebrovascular adaptions to UCMS in lean healthy rats reflect what we see in metabolic syndrome rats (without UCMS); i.e., 8 weeks of UCMS transformed a healthy cerebrovasculature to one with similar dysfunction seen in the metabolic syndrome. However, when chronic exposure to stress was performed in rats that were developing the metabolic syndrome phenotype, the negative effects on the aorta and brain vessels were amplified (Brooks et al., 2018a,d). Furthermore, unlike lean rats exposed to UCMS, the metabolic syndrome rats also developed impaired smooth muscle dilation (endothelial independent dilation) coupled with increased cerebrovascular stiffness (Brooks et al., 2018a). The profile of vasoactive metabolites was further altered with the combined presentation of metabolic syndrome and UCMS, such that the evolving pro-oxidant and pro-inflammatory environment within the cerebrovasculature nearly abolishes nitric oxide and prostacyclin levels (Brooks et al., 2018b). These results suggest that metabolic syndrome results in greater pathological cerebrovascular and conduit arterial alterations compared to lean rats when exposed to chronic stress. Interesting, sex differences reported above with regard to the protection of the cerebrovasculature and aorta with UCMS were abolished with the presence of the metabolic syndrome (Brooks et al., 2018d). Whereby, healthy female rats can somewhat buffer the impact of UCMS on the vasculature, female rats presenting with metabolic syndrome and exposed to 8 weeks of UCMS resulted in a severely impaired vasculature (aorta and cerebral vessels) (Brooks et al., 2018d).

Potential mechanisms leading to vascular dysfunction and remodeling

The route through which stress exerts its pathological effects on the vasculature continues to be the source of some debate. However, several physiological mechanisms explored in more detail in other chapters of this book have been hypothesized as having critical roles, including the role of oxidative stress, inflammation, and the autonomic nervous system (Fig. 6.1),

Oxidative stress

Although the mechanisms through which chronic stress leads to endothelial dysfunction and vascular remodeling remains a process of exploration, many of the studies above noted significant increases in the vascular production of reactive oxygen species (ROS) and a reduction in nitric oxide (NO) bioavailability (Branyan et al., 2018; Brooks et al., 2018a; Stanley et al., 2014). Low levels of ROS are powerful signaling molecules essential for maintaining normal vascular function, whereas uncontrolled overproduction of ROS exacerbates oxidative stress, resulting in vascular cell damage, the induction of proliferation and migration of vascular smooth muscle cells, recruitment of inflammatory cells, lipid peroxidation, activation of metalloproteinases and deposition of extracellular matrix, collectively causing vascular remodeling (Konior et al., 2014; Raaz et al., 2014). Nicotinamide adenine dinucleotide phosphate (NADPH) oxidase, xanthine oxidase, and uncoupled NO Synthase (NOS) are the most important sources of vascular ROS. One of the most important signaling molecules for vascular function is NO, and under healthy physiological conditions in the presence of endothelial NOS (eNOS), NO is produced from the conversion of L-arginine to L-citrulline, which ultimately stimulates NO-sensitive guanylyl cyclase and the generation of cyclic GMP resulting in a reuptake of calcium in the vascular smooth muscle cells causing vasodilation. However, in the presence of an increased production of superoxides, NO is quickly converted to peroxynitrite (a potent free radial), reducing the bioavailability of NO. Further, under pathological conditions, dysfunctional uncoupled eNOS no longer produces NO, but rather the dysfunctional eNOS produces more superoxides which further aggravates the oxidative stress burden (i.e., reduced NO and increased superoxides production) (Forstermann and Sessa, 2012). Indeed, pro-oxidative conditions, such as UCMS, can alter the balance between constricting and dilating metabolites, by shifting arachidonic acid metabolism through cyclooxygenase to the production of vasoconstricting metabolites (e.g., thromboxane) that can compete against the vasodilatory stimulus from NO, prostacyclin, or other vasodilators (Isingrini et al., 2011).

Impaired NO bioavailability, through inhibition of eNOS, can also lead to increased arterial stiffness (Zieman et al., 2005) and decreased angiogenesis (Fukumura et al., 2001). A previous study has shown that the balance between oxidant stress and endothelial function (e.g., NO bioavailability, altered arachidonic acid metabolism) were key factors involved in the progression and severity of microvascular rarefaction (Chantler et al., 2015). Both arteriogenesis and angiogenesis are tightly modulated by environmental cues and likely differ under physiologic and disease conditions. Further, these processes are critically dependent on expression of vascular endothelial growth factors (VEGF) (Audet et al., 2011; Hoier and Hellsten, 2014; Lloyd et al., 2005). While VEGF is an essential trigger to initiate angiogenesis, evidence suggests that thrombospondin-1 (TSP-1) is an important factor for capillary regression and/or pathologically-mediated rarefaction (Olfert et al., 2016). Indeed, a fine bi-directional interaction exists between TSP-1,

VEGF, and NO. For example, TSP-1 can negatively regulate NO signaling (Zhou et al., 2006; Isenberg et al., 2006), and in turn, decreased NO production can induce TSP-1 expression (Isenberg et al., 2008). TSP-1 can also interfere with Vegf binding to its receptor (Chu et al., 2013; Greenaway et al., 2007). When TSP-1 is elevated, the VEGF signal pathways can be endogenously inhibited, regardless of whether VEGF levels are elevated. Further, NO simultaneously induces vasodilatation and enhances VEGF expression (Benoit et al., 1999). Evidence suggests that the processes of oxidative stress and inflammation on microvascular rarefaction are interconnected. ROS stimulate the induction of VEGF expression in endothelial cells, smooth muscle cells, and macrophages (Kim and Byzova, 2014; Kim et al., 2013). In addition, up regulation of TSP-1 is mediated by superoxides (Xie et al., 2010). Our laboratory has made a similar observation in obese Zucker rats (OZR; a model for metabolic syndrome), where TSP-1 expression can be elevated up to twofold in OZR whole brain extracts, with either VEGF protein expression being unchanged or slightly elevated in OZR compared to similar-aged healthy lean controls (Chantler and Olfert unpublished data).

Inflammation

There is much evidence supporting a role for pro-inflammatory cytokines in stress-induced CVD, and chapter 8 provides a detailed discussion of the bi-directional associations between stress, depression, and inflammation. In this section, the role of inflammation on vascular function is considered. Bernberg et al. (2012) reported increased levels of pro-inflammatory mediators interleukin 6 (IL-6) and chemokine (C-X-C motif) ligand 1 in apolipoprotein E knockout mice following psychological stress (Bernberg et al., 2012). There are increased circulating levels of tumor necrosis factor alpha (TNF-α) and IL-1β in a rat model of anhedonia induced by UCMS (Grippo et al., 2005). Increased levels of pro-inflammatory cytokines have been consistently reported in several human studies as well (Dowlati et al., 2010; Hiles et al., 2012; Howren et al., 2009). It is well known that inflammatory cytokines affect endothelial function. For instance, TNF-α and IL-1 decrease eNOS activity along with reducing eNOS mRNA and protein content (Zhang et al., 1997). Other suggested mechanisms include alteration in calcium channel expression and activity (Tiwari et al., 2006), upregulation of Rho-kinase expression and function (Hiroki et al., 2004), increased ROS production (Zhang et al., 2008), and enhanced cyclooxygenase expression (Mitchell et al., 1995). Inflammation and oxidative stress are closely related phenomena, as ROS is produced during immune reactions to form an important part of the body's natural defense or repair mechanisms. Under pro-inflammatory conditions, simultaneous production of superoxide and NO can be strongly activated to increase production 1000-fold, which will increase the formation of peroxynitrite by a 1,000,000-fold (Pacher et al., 2007). Meanwhile, reactive radicals attract more immune cells and further exacerbate inflammatory reactions. Nuclear factor kappa of activated B cells (NF-κB), a protein involved in the transcription of DNA, is thought to link psychological stress- and oxidative stress-induced inflammatory organ dysfunction, including impaired cardiovascular regulation (Bierhaus et al., 2004). The NF-κB pathway is a critical component of the inflammatory processes that is activated by oxidative stress and has a role as a ubiquitous transcription factor having multiple roles such as mediating inflammatory responses to a variety of signals, regulating immune function, causing endothelial cell activation, and controlling cell growth (Liu et al., 2017). NF-κB is normally present in an inactive form in the cytoplasm by virtue of binding to a family of inhibitory IκB proteins. Stimulation by a variety of stimuli activates subtypes of these inhibitory proteins, IKK-α and IKK-β, resulting in the phosphorylation of IκB and its proteasomal degradation. IκB degradation liberates NF-κB, allowing it to translocate to the nucleus and induce gene expression of pro-inflammatory cytokines such as IL-1β, IL-6, TNF-α, cyclooxygenase-2, lipoxygenase, iNOS, and cell adhesion molecules (VCAM-1, ICAM-1, PCAM, and *E*-selectin). Increased activity of NF-κB occurs in peripheral blood mononuclear cells of healthy volunteers undergoing a brief laboratory test, which was proportional to the levels of catecholamines and cortisol secretion (Bierhaus et al., 2003). The Trier Social Stress Test, used as a form of mental stress that includes a speech and an arithmetic test in the presence of an audience, also increased NF-κB in participants (Kirschbaum et al., 1993).

Stress hormones and autonomic nervous system function

Chapter 5 provides a detailed description linking stress hormones and the autonomic function with the associated CVD phenotype induced by exposure to stress and the development of depression. In this section, changes in stress hormones with stress and depression on how they mediate cardiovascular physiology are considered (Fig. 6.1).

Stressful circumstances activate the HPA axis and the autonomic nervous system (ANS) to release stress hormones (Baune et al., 2012). Overstimulation of the ANS and HPA axis is associated with an over-activation of the renin-angiotensin system, which leads to increased levels of homocysteine and elevated cardiovascular activity accompanied by various degrees of endothelial damage. Further, glucocorticoids regulate vascular reactivity by acting on both endothelial and vascular smooth muscle cells. Increased cortisol levels decrease NO bioavailability directly by inhibiting eNOS directly (Liu et al., 2009), and indirectly through increasing the production of oxidative stress via cortisol (Iuchi et al., 2003). Cortisol-induced dysregulations of NO production in the endothelium may, therefore induce some of the deleterious effects associated with stress. Cortisol can also affect vascular function via the enhancement of endothelin-1 secretion from vascular cells (Nickel et al., 2009). This effect can be a direct result of cortisol or may be the consequence of glucocorticoid-induced oxidative stress in vascular tissue (Satoh et al., 2015). In a study of healthy human subjects, exposure to only 3 min of mental stress elicited endothelial dysfunction which was prevented in the presence of endothelin-A receptor antagonist (BQ-123), signifying the role of endothelin in vascular dysfunction in response to mental stress (Spieker et al., 2002). Furthermore, corticotropin-releasing hormone (CRH), another stress hormone, can increase secretion of endothelin-1 from cultured human endothelial cells in a time- and dose-dependent manner (Nickel et al., 2009). Corticotropin also causes $CD14^{+}$ cells to release TNF-α while also damaging endothelial barrier function by inducing apoptosis of endothelial cells (Sojka et al., 2006). Furthermore, corticotropin promotes monocyte adhesion to the endothelium (Wilbert-Lampen et al., 2006).

In addition to the effects of stress on the HPA system, hypersensitivity of the sympathetic nervous system (SNS) occurs during chronically stressful conditions. Animals exposed to chronic repetitive ("habituated") stress are able to synthesize and store higher amounts of catecholamines, although catecholamine release at rest and the surge after an episode of habituated stress are lower in stressed animals. However, stimulation by a novel stress elicited an exaggerated sympathetic response in these animals (Groeschel and Braam, 2011), suggesting increased hyper-responsiveness of the ANS and that the catecholamine surge may be regulated differently when stimulated by habitual versus novel stress. Hyperactivity of the SNS, especially on a chronic basis, can harm the cardiovascular system because the continuous pressor effects of released catecholamines may counteract vasodilatory effects of NO and can also accelerate the atherosclerotic process. Of note, cortisol can also increase SNS arousal, resulting in an increase in SBP and HR, which are known risk factors for arterial stiffness (Sutton-Tyrrell et al., 2005). Another mechanism is activation of the renin-angiotensin-aldosterone system (RAAS) resulting from sympathetic overactivity. Central activation of mineralocorticoid receptors leads to sympathetic overactivity and increased secretion of pro-inflammatory cytokines (Gomez-Sanchez, 2004). Rats stressed by a new environment (open field) or predator exposure show increased RAAS activity (Clamage et al., 1976). Angiotensin-converting enzyme (ACE) inhibitors reduce the despair behavior in rodents during swim tests (Martin et al., 1990), and mice lacking ACE demonstrate fewer signs of depression (Okuyama et al., 1999). Angiotensin Ii, a potent vasoconstrictor, can harm endothelial cells independently of its pressor effect (Aberg and Ferrer, 1990), by increasing cellular lipid peroxidation (Keidar et al., 1995) and reducing cholesterol efflux (Kaplan et al., 2002) in macrophages and scavenger cells; both actions lead to accelerated foam cell formation and a hastened atherosclerotic process. Angiotensin II stimulates uptake of oxidized LDL by macrophages, possibly mediated by IL-6 (Keidar et al., 2001). Margination, adhesion, and translocation of leukocytes, all prerequisites for atherosclerosis, are enhanced by angiotensin II type 1 receptors (AT_1Rs) (Mateo et al., 2006). Angiotensin II also promotes the expression and production of adhesive and proinflammatory molecules (vascular cell adhesion protein 1, intercellular adhesion molecule 1, monocyte chemoattractant protein-1, macrophage inflammatory protein-1α, and IL-8) on the endothelial and vascular smooth muscle cells (Liu et al., 2006). Angiotensin II affects vascular smooth muscle signal transduction and several growth factors (transforming growth factor-β, platelet-derived growth factor, and basic fibroblast growth factor) that stimulate smooth muscle proliferation (Kaplan et al., 1996). Central and peripheral blockade of AT_1Rs with the anti-hypertensive agent, candesartan, provides cardiovascular benefits in addition to modifying responses to psychological stress (Pavel et al., 2008). In addition, candesartan abolished hormonal and sympathetic responses mediated by the HPA axis in rodents undergoing isolation-induced stress while also reducing anxiety and protecting against gastric ulceration caused by cold-restraint stress (Schiavone et al., 1988).

Despite the untoward consequences of activation of AT_1Rs by angiotensin II, there is another counterregulatory cascade of reactions that is triggered by activating angiotensin II type 2 (AT_2)/Mas receptors, which has beneficial effects during stressful situations. Angiotensin 1–7, the product of ACEII interacting with angiotensinogen, is the principal agonist for AT_2 receptors. By acting preferentially on AT_2 receptors, angiotensin 1–7 has beneficial effects during the stress response. Walther et al. (1998) suggested that AT_2receptor-deficient mice showed increased anxiety in behavioral tests. Furthermore, a functional interaction between AT_2 and AT_1Rs occurs in the amygdala, an important nucleus in the limbic system regulating many psychological responses (Von Bohlen Und Halbach et al., 2000). Intravenous injections of an AT_2 receptor agonist or an ACEII activator reduced stress-induced tachycardia (Martins Lima et al., 2013). Intraventricular injection of these substances also diminished isoproterenol-induced tachycardia and renal sympathetic activation. Therefore, one can conclude that activation of AT_2 receptors modifies sympathetic nervous activity during stressful conditions.

Adverse effect of stress on the heart

As described in this chapter, numerous vascular consequences of chronic stress have been identified. Although these direct relations have been reported, modifications in vascular function are likely to have secondary effects on cardiac function. Such cardiac complications include stress-induced cardiac ischemia and arrhythmias. It has always seemed plausible that acute mental stress could "trigger" abnormal heart rhythms and potential for sudden cardiac death in patients with existing heart disease. Initial evidence from natural disasters and war supports the link between exposure to mental stress and onset of heart attacks (Suglia et al. 2015). Physiologically, acute mental stress is accompanied by activation of the SNS to the heart, which can cause ventricular arrhythmias. Increased adrenaline secretion by the adrenal medulla activates platelets, predisposing to thrombosis. Further, there is a rapid increase in BP, which can cause unstable coronary artery plaques to block coronary arteries resulting in a myocardial infarction. Further, the increased BP is commonly accompanied by an increased rate-pressure product and myocardial oxygen consumption. Although the increased oxygen demand requires increased blood flow through the coronary arteries, studies of patients during mental stress showed increased norepinephrine spillover from the heart and a reduction of blood flow through regions of high stenosis by an average of 27% (Steptoe and Brydon, 2009). The combination of hypoxia and sympathetic stimulation can cause arrhythmias. Cardiac troponin T (cTnT) (Thygesen et al., 2007), and more recently high-sensitivity assays (hs) cTnT (Collinson et al., 2012), are plasma proteins routinely measured in diagnosing acute myocardial infarction with higher levels associated with a greater incidence of heart disease, cardiovascular mortality, and all-cause mortality. Interestingly, a heightened cortisol response to mental stress was associated with detectable plasma levels of hs-cTnT in healthy participants, independent of coronary atherosclerosis (Lazzarino et al., 2013). Studies have consistently documented HPA hyperactivity in depressed patients (Arborelius et al., 1999; Ehlert et al., 2001). In addition, cardiac rhythmicity disorders, in particular, ventricular arrhythmias, are the main culprit for sudden cardiac deaths in patients with CVD (Harris and Lysitsas, 2016).

Disrupted autonomic function has been observed in the development of depression and its association with CVD. Indeed, depressed individuals display higher resting heart rate (HR), leading to increased cardiac output and altered basal and stressor-associated BP, higher basal levels of plasma norepinephrine, and greater sympathetic reactivity to stressors, compared to non-depressed controls (Kayano et al., 2015; Nahshoni et al., 2004; Sheffield et al., 1998; Veith et al., 1994). Furthermore, a specific reduction in cardiac parasympathetic regulation has been reported, with depressed individuals demonstrating reduced heart rate variability (HRV) and atypical parasympathetic reactivity to stressors. Of note, decreased HRV is a known risk factor for sudden death and ventricular arrhythmias in patients with CVD (Huikuri and Makikallio, 2001).

Animal models have provided strong evidence supporting the link between ANS dysfunction and depressive behaviors. Numerous studies have reported an increase in resting HR following exposure to chronic stressors (Grippo et al., 2008; Mercanoglu et al., 2008; Duarte et al., 2015). However, chronic stress has not been found to affect the intrinsic HR (examined with dual blockade of sympathetic and parasympathetic activities) (Duarte et al., 2015; McNeal et al., 2014; Almeida et al., 2015), suggesting the increased HR is a reflection of the stress stimuli affecting autonomic control. Autonomic activity, as assessed by HRV, noted an increase in the low frequency vs high frequency power spectral analysis oscillatory components of HR (Bundzikova-Osacka et al., 2015), suggesting a change in cardiac sympathovagal balance towards sympathetic predominance. Other studies, have identified an increase in cardiac sympathetic tone either with (Grippo and Johnson, 2009; Grippo et al., 2007; McNeal et al., 2014) or without (Grippo et al., 2002; Duarte et al., 2015) a change in parasympathetic tone. These differences can be attributed to the type of chronic stress stimulus deployed in these studies (i.e., chronic social isolation vs chronic variable stress). However, in general the effects of chronic stress result in sympathetic predominance, which is in line with the resting tachycardia induced by the chronic stress models. Of note, the increase in sympathetic contribution of cardiac autonomic balance corroborates the increased susceptibility to cardiac arrhythmias (Grippo et al., 2012), cardiac contractile dysfunction (Grippo et al., 2006; Xie et al., 2012), and increased severity of myocardial ischemia (Mercanoglu et al., 2008) induced by chronic stress.

Evidence also suggests that chronic stress impacts cardiac function and remodeling. In a mouse model, end systolic diameter was larger and fractional shortening (often used to represent cardiac contractility) was reduced with chronic stress (Roth et al., 2015). In addition, monkeys exhibiting depressive behaviors also presented with higher resting HRs, smaller chamber sizes, and a smaller left ventricle without evidence of systolic or diastolic dysfunction compared to monkeys with few depressive behaviors (Groban et al., 2014). Repeated social defeat episodes have been shown to cause accumulation of fibrous tissue in the left ventricular myocardium, maladaptive cardiac hypertrophy, and changes in electrical conduction system of the heart (e.g., reduced myocardial refractoriness and impaired conduction) (Carnevali et al., 2013, 2015; Costoli et al. 2004).

Previous clinical studies have indicated that depressive symptoms exacerbate chronic heart failure in human patients (Liu et al., 2015; Rutledge et al., 2006) and in mice (Lu et al., 2017). Indeed, 7 days of chronic stress combined with surgically induced heart failure led to an increased mortality rate, cardiac hypertrophy, BP, oxidative stress, and reduced ejection fraction (Lu et al., 2017). Furthermore, in a rat model, chronic stress aggravated pressure overload-induced cardiac hypertrophy, cardiac fibrosis, and cardiomyocytes apoptosis, which contributed to loss in cardiac function (Liu et al., 2015).

Summary and conclusions

Although it is generally accepted that psychological stress can lead to adverse cardiovascular consequences, the mechanisms that may underpin this are not well understood. However, the literature presented in this chapter provide evidence that stress does indeed make it way into the vascular structure of the circulatory system as summarized in Fig. 6.1. These cardiovascular dysfunctions are likely caused by a pro-oxidative and inflammatory environment in the tissues and vessels resulting from exposure to stress. Data from humans and animal studies of stress have highlighted the critical role for endothelial dysfunction in stress-induced cardiovascular disorders.

Chronic psychological stress also leads to significant cardiac dysfunction, which might reflect the direct effects of stress on the heart and the secondary effects of stress on the vasculature, which plays a key role in regulating and modifying cardiac function. Evidence from pre-clinical models suggest that the pathophysiological responses to chronic stress are somewhat sex-specific and dependent on the pre-existence of metabolic disease and CVD. Findings typically reveal that exposure to stress impacts the vasculature of males more than females, at least ovulating females, and persons diagnosed with CVD or metabolic syndrome more than healthy persons. However, further research is still needed to explore the sex and disease specific role on the cardiovascular system with chronic exposure to stress.

As noted previously in this chapter, the research comprising this literature sheds very little information on the distinctive contributions of the influences of stress and depression on vascular function or whether depression mediates the relation between stress and CVD outcomes. Although both exposure to stress and presence of diagnoses of depression have been associated with identifiable changes in vascular responses, very little effort has focused on how they interact in influencing vascular processes. In particular, because animal studies rely only on depressive-like behaviors or organisms, this is a limitation of the literature that will need to rely on more sophisticated studies of vascular processes in humans exposed to stress accompanied with and without depressive disorders.

One additional limitation of this body of work is the tendency to examine very narrow components of the overall physiological stress response. For example, studies examining the role of oxidative stress as a mechanism in translating exposure to stress to vascular outcomes are important, but reveal very little regarding the roles of co-occurring inflammatory or ANS responses to stress. Although this current literature provides important clues into how stress influences individual elements of the vascular stress response, it will be important for future research to examine multiple physiological processes to account for the complexity of the bodily stress response.

References

Aberg, G., Ferrer, P., 1990. Effects of captopril on atherosclerosis in cynomolgus monkeys. J. Cardiovasc. Pharmacol. 15 (Suppl 5), S65–S72.

Almeida, J., Duarte, J.O., Oliveira, L.A., Crestani, C.C., 2015. Effects of nitric oxide synthesis inhibitor or fluoxetine treatment on depression-like state and cardiovascular changes induced by chronic variable stress in rats. Stress 18, 462–474.

Arborelius, L., Owens, M.J., Plotsky, P.M., Nemeroff, C.B., 1999. The role of corticotropin-releasing factor in depression and anxiety disorders. J. Endocrinol. 160, 1–12.

Audet, G.N., Meek, T.H., Garland Jr., T., Olfert, I.M., 2011. Expression of angiogenic regulators and skeletal muscle capillarity in selectively bred high aerobic capacity mice. Exp. Physiol. 96, 1138–1150.

Bateman, G.A., Levi, C.R., Schofield, P., Wang, Y., Lovett, E.C., 2008. The venous manifestations of pulse wave encephalopathy: windkessel dysfunction in normal aging and senile dementia. Neuroradiology 50, 491–497.

Baum, A., Posluszny, D.M., 1999. Health psychology: mapping biobehavioral contributions to health and illness. Annu. Rev. Psychol. 50, 137–163.

Baune, B.T., Stuart, M., Gilmour, A., Wersching, H., Heindel, W., Arolt, V., Berger, K., 2012. The relationship between subtypes of depression and cardiovascular disease: a systematic review of biological models. Transl. Psychiatry 2, e92.

Benoit, H., Jordan, M., Wagner, H., Wagner, P.D., 1999. Effect of NO, vasodilator prostaglandins, and adenosine on skeletal muscle angiogenic growth factor gene expression. J. Appl. Physiol. (1985) 86, 1513–1518.

Bergstrom, A., Jayatissa, M.N., Mork, A., Wiborg, O., 2008. Stress sensitivity and resilience in the chronic mild stress rat model of depression; an in situ hybridization study. Brain Res. 1196, 41–52.

Bernberg, E., Ulleryd, M.A., Johansson, M.E., Bergstrom, G.M., 2012. Social disruption stress increases IL-6 levels and accelerates atherosclerosis in ApoE-/- mice. Atherosclerosis 221, 359–365.

Bierhaus, A., Wolf, J., Andrassy, M., Rohleder, N., Humpert, P.M., Petrov, D., Ferstl, R., Von Eynatten, M., Wendt, T., Rudofsky, G., Joswig, M., Morcos, M., Schwaninger, M., McEwen, B., Kirschbaum, C., Nawroth, P.P., 2003. A mechanism converting psychosocial stress into mononuclear cell activation. Proc. Natl. Acad. Sci. U. S. A. 100, 1920–1925.

Bierhaus, A., Humpert, P.M., Nawroth, P.P., 2004. NF-kappaB as a molecular link between psychosocial stress and organ dysfunction. Pediatr. Nephrol. 19, 1189–1191.

Black, P.H., Garbutt, L.D., 2002. Stress, inflammation and cardiovascular disease. J. Psychosom. Res. 52, 1–23.

Bomhof-Roordink, H., Seldenrijk, A., Van Hout, H.P., Van Marwijk, H.W., Diamant, M., Penninx, B.W., 2015. Associations between life stress and subclinical cardiovascular disease are partly mediated by depressive and anxiety symptoms. J. Psychosom. Res. 78, 332–339.

Bowman, R.E., Zrull, M.C., Luine, V.N., 2001. Chronic restraint stress enhances radial arm maze performance in female rats. Brain Res. 904, 279–289.

Branyan, K.W., Devallance, E.R., Lemaster, K.A., Skinner, R.C., Bryner, R.W., Olfert, I.M., Kelley, E.E., Frisbee, J.C., Chantler, P.D., 2018. Role of chronic stress and exercise on microvascular function in metabolic syndrome. Med. Sci. Sports Exerc. 50, 957–966.

Brooks, S.D., Devallance, E., D'audiffret, A.C., Frisbee, S.J., Tabone, L.E., Shrader, C.D., Frisbee, J.C., Chantler, P.D., 2015. Metabolic syndrome impairs reactivity and wall mechanics of cerebral resistance arteries in obese Zucker Rats. Am. J. Physiol. Heart Circ. Physiol. 309, H1846–H1859.

Brooks, S., Branyan, K., Devallance, E., Skinner, R., Lemaster, K., Sheets, J.W., Pitzer, C.R., Asano, S., Bryner, R., Olfert, I.M., Frisbee, J.C., Chantler, P.D., 2018a. Psychological stress induced cerebrovascular dysfunction: the role of metabolic syndrome and exercise. Exp. Physiol. 103 (5), 761–776.

Brooks, S.D., Hileman, S., Chantler, P.D., Milde, S., Lemaster, K.C., Frisbee, S., Shoemaker, J.K., Jackson, D.N., Frisbee, J.C., 2018b. Protection from chronic stress- and depressive symptom-induced vascular endothelial dysfunction in female rats is abolished by preexisting metabolic disease. Am. J. Physiol. Heart Circ. Physiol. 314 (5), H1085–H1097.

Brooks, S.D., Hileman, S., Chantler, P.D., Milde, S., Lemaster, K.C., Frisbee, S., Shoemaker, J.K., Jackson, D.N., Frisbee, J.C., 2018c. Protection from vascular dysfunction in female rats with chronic stress and depressive symptoms. Am. J. Physiol. Heart Circ. Physiol. 314 (5), H1070–H1084.

Brooks, S.D., Hileman, S.M., Chantler, P.D., Milde, S.A., Lemaster, K.A., Frisbee, S.J., Shoemaker, J.K., Jackson, D.N., Frisbee, J.C., 2018d. Protection from chronic stress- and depressive symptom-induced vascular endothelial dysfunction in female rats is abolished by preexisting metabolic disease. Am. J. Physiol. Heart Circ. Physiol. 314, H1085–H1097.

Bundzikova-Osacka, J., Ghosal, S., Packard, B.A., Ulrich-Lai, Y.M., Herman, J.P., 2015. Role of nucleus of the solitary tract noradrenergic neurons in post-stress cardiovascular and hormonal control in male rats. Stress 18, 221–232.

Carnevali, L., Trombini, M., Rossi, S., Graiani, G., Manghi, M., Koolhaas, J.M., Quaini, F., Macchi, E., Nalivaiko, E., Sgoifo, A., 2013. Structural and electrical myocardial remodeling in a rodent model of depression. Psychosom. Med. 75, 42–51.

Carnevali, L., Vacondio, F., Rossi, S., Callegari, S., Macchi, E., Spadoni, G., Bedini, A., Rivara, S., Mor, M., Sgoifo, A., 2015. Antidepressant-like activity and cardioprotective effects of fatty acid amide hydrolase inhibitor Urb694 in socially stressed Wistar Kyoto rats. Eur. Neuropsychopharmacol. 25, 2157–2169.

Castro-Diehl, C., Diez Roux, A.V., Seeman, T., Shea, S., Shrager, S., Tadros, S., 2014. Associations of socioeconomic and psychosocial factors with urinary measures of cortisol and catecholamines in the multi-ethnic study of atherosclerosis (MESA). Psychoneuroendocrinology 41, 132–141.

Chan, O., Chan, S., Inouye, K., Shum, K., Matthews, S.G., Vranic, M., 2002a. Diabetes impairs hypothalamo-pituitary-adrenal (HPA) responses to hypoglycemia, and insulin treatment normalizes HPA but not epinephrine responses. Diabetes 51, 1681–1689.

Chan, O., Inouye, K., Vranic, M., Matthews, S.G., 2002b. Hyperactivation of the hypothalamo-pituitary-adrenocortical axis in streptozotocin-diabetes is associated with reduced stress responsiveness and decreased pituitary and adrenal sensitivity. Endocrinology 143, 1761–1768.

Chantler, P.D., Shrader, C.D., Tabone, L.E., D'audiffret, A.C., Huseynova, K., Brooks, S.D., Branyan, K.W., Grogg, K.A., Frisbee, J.C., 2015. Cerebral cortical microvascular rarefaction in metabolic syndrome is dependent on insulin resistance and loss of Nitric Oxide bioavailability. Microcirculation 22, 435–445.

Chappell, D.C., Varner, S.E., Nerem, R.M., Medford, R.M., Alexander, R.W., 1998. Oscillatory shear stress stimulates adhesion molecule expression in cultured human endothelium. Circ. Res. 82, 532–539.

Chu, L.Y., Ramakrishnan, D.P., Silverstein, R.L., 2013. Thrombospondin-1 modulates VEGF signaling via CD36 by recruiting Shp-1 to Vegfr2 complex in microvascular endothelial cells. Blood 122, 1822–1832.

Clamage, D.M., Sanford, C.S., Vander, A.J., Mouw, D.R., 1976. Effects of psychosocial stimuli on plasma renin activity in rats. Am. J. Phys. 231, 1290–1294.

Collinson, P.O., Heung, Y.M., Gaze, D., Boa, F., Senior, R., Christenson, R., Apple, F.S., 2012. Influence of population selection on the 99th percentile reference value for cardiac troponin assays. Clin. Chem. 58, 219–225.

Costoli, T., Bartolomucci, A., Graiani, G., Stilli, D., Laviola, G., Sgoifo, A., 2004. Effects of chronic psychosocial stress on cardiac autonomic responsiveness and myocardial structure in mice. Am. J. Physiol. Heart Circ. Physiol. 286, H2133–H2140.

Czeh, B., Abumaria, N., Rygula, R., Fuchs, E., 2010. Quantitative changes in hippocampal microvasculature of chronically stressed rats: no effect of fluoxetine treatment. Hippocampus 20, 174–185.

Dalla, C., Antoniou, K., Drossopoulou, G., Xagoraris, M., Kokras, N., Sfikakis, A., Papadopoulou-Daifoti, Z., 2005. Chronic mild stress impact: are females more vulnerable? Neuroscience 135, 703–714.

De Hert, M., Detraux, J., Vancampfort, D., 2018. The intriguing relationship between coronary heart disease and mental disorders. Dialogues Clin. Neurosci. 20, 31–40.

De Leo, V., La Marca, A., Talluri, B., D'antona, D., Morgante, G., 1998. Hypothalamo-pituitary-adrenal axis and adrenal function before and after ovariectomy in premenopausal women. Eur. J. Endocrinol. 138, 430–435.
Derntl, B., Finkelmeyer, A., Eickhoff, S., Kellermann, T., Falkenberg, D.I., Schneider, F., Habel, U., 2010. Multidimensional assessment of empathic abilities: neural correlates and gender differences. Psychoneuroendocrinology 35, 67–82.
Dowlati, Y., Herrmann, N., Swardfager, W., Liu, H., Sham, L., Reim, E.K., Lanctôt, K.L., 2010. A meta-analysis of cytokines in major depression. Biol. Psychiatry 67, 446–457.
Duarte, J.O., Cruz, F.C., Leao, R.M., Planeta, C.S., Crestani, C.C., 2015. Stress vulnerability during adolescence: comparison of chronic stressors in adolescent and adult rats. Psychosom. Med. 77, 186–199.
Ehlert, U., Gaab, J., Heinrichs, M., 2001. Psychoneuroendocrinological contributions to the etiology of depression, posttraumatic stress disorder, and stress-related bodily disorders: the role of the hypothalamus-pituitary-adrenal axis. Biol. Psychol. 57, 141–152.
Forstermann, U., Sessa, W.C., 2012. Nitric oxide synthases: regulation and function. Eur. Heart J. 33, 829–837. 837a-837d.
Fukumura, D., Gohongi, T., Kadambi, A., Izumi, Y., Ang, J., Yun, C.O., Buerk, D.G., Huang, P.L., Jain, R.K., 2001. Predominant role of endothelial nitric oxide synthase in vascular endothelial growth factor-induced angiogenesis and vascular permeability. Proc. Natl. Acad. Sci. U. S. A. 98, 2604–2609.
Garrett, J.E., Wellman, C.L., 2009. Chronic stress effects on dendritic morphology in medial prefrontal cortex: sex differences and estrogen dependence. Neuroscience 162, 195–207.
Ghiadoni, L., Donald, A.E., Cropley, M., Mullen, M.J., Oakley, G., Taylor, M., O'connor, G., Betteridge, J., Klein, N., Steptoe, A., Deanfield, J.E., 2000. Mental stress induces transient endothelial dysfunction in humans. Circulation 102, 2473–2478.
Gomez-Sanchez, E.P., 2004. Brain mineralocorticoid receptors: orchestrators of hypertension and end-organ disease. Curr. Opin. Nephrol. Hypertens. 13, 191–196.
Greenaway, J., Lawler, J., Moorehead, R., Bornstein, P., Lamarre, J., Petrik, J., 2007. Thrombospondin-1 inhibits VEGF levels in the ovary directly by binding and internalization via the low density lipoprotein receptor-related protein-1 (LRP-1). J. Cell. Physiol. 210, 807–818.
Grippo, A.J., Johnson, A.K., 2009. Stress, depression and cardiovascular dysregulation: a review of neurobiological mechanisms and the integration of research from preclinical disease models. Stress 12, 1–21.
Grippo, A.J., Moffitt, J.A., Johnson, A.K., 2002. Cardiovascular alterations and autonomic imbalance in an experimental model of depression. Am. J. Phys. Regul. Integr. Comp. Phys. 282, R1333–R1341.
Grippo, A.J., Francis, J., Beltz, T.G., Felder, R.B., Johnson, A.K., 2005. Neuroendocrine and cytokine profile of chronic mild stress-induced anhedonia. Physiol. Behav. 84, 697–706.
Grippo, A.J., Beltz, T.G., Weiss, R.M., Johnson, A.K., 2006. The effects of chronic fluoxetine treatment on chronic mild stress-induced cardiovascular changes and anhedonia. Biol. Psychiatry 59, 309–316.
Grippo, A.J., Lamb, D.G., Carter, C.S., Porges, S.W., 2007. Social isolation disrupts autonomic regulation of the heart and influences negative affective behaviors. Biol. Psychiatry 62, 1162–1170.
Grippo, A.J., Moffitt, J.A., Johnson, A.K., 2008. Evaluation of baroreceptor reflex function in the chronic mild stress rodent model of depression. Psychosom. Med. 70, 435–443.
Grippo, A.J., Moffitt, J.A., Sgoifo, A., Jepson, A.J., Bates, S.L., Chandler, D.L., McNeal, N., Preihs, K., 2012. The integration of depressive behaviors and cardiac dysfunction during an operational measure of depression: investigating the role of negative social experiences in an animal model. Psychosom. Med. 74, 612–619.
Groban, L., Kitzman, D.W., Register, T.C., Shively, C.A., 2014. Effect of depression and sertraline treatment on cardiac function in female nonhuman primates. Psychosom. Med. 76, 137–146.

Groeschel, M., Braam, B., 2011. Connecting chronic and recurrent stress to vascular dysfunction: no relaxed role for the renin-angiotensin system. Am. J. Phys. Renal Phys. 300, F1–10.

Gu, H.F., Tang, C.K., Yang, Y.Z., 2012. Psychological stress, immune response, and atherosclerosis. Atherosclerosis 223, 69–77.

Harpaz, I., Abutbul, S., Nemirovsky, A., Gal, R., Cohen, H., Monsonego, A., 2013. Chronic exposure to stress predisposes to higher autoimmune susceptibility in C57BL/6 mice: glucocorticoids as a double-edged sword. Eur. J. Immunol. 43, 758–769.

Harris, P., Lysitsas, D., 2016. Ventricular arrhythmias and sudden cardiac death. BJA Education 16, 221–229.

Heine, V.M., Zareno, J., Maslam, S., Joels, M., Lucassen, P.J., 2005. Chronic stress in the adult dentate gyrus reduces cell proliferation near the vasculature and VEGF and Flk-1 protein expression. Eur. J. Neurosci. 21, 1304–1314.

Henry-Feugeas, M.C., 2009. Intracranial MR dynamics in clinically diagnosed Alzheimer's disease: the emerging concept of "pulse wave encephalopathy". Curr. Alzheimer Res. 6, 488–502.

Heuser, I.J., Gotthardt, U., Schweiger, U., Schmider, J., Lammers, C.H., Dettling, M., Holsboer, F., 1994. Age-associated changes of pituitary-adrenocortical hormone regulation in humans: importance of gender. Neurobiol. Aging 15, 227–231.

Hiles, S.A., Baker, A.L., De Malmanche, T., Attia, J., 2012. A meta-analysis of differences in IL-6 and IL-10 between people with and without depression: exploring the causes of heterogeneity. Brain Behav. Immun. 26, 1180–1188.

Hiroki, J., Shimokawa, H., Higashi, M., Morikawa, K., Kandabashi, T., Kawamura, N., Kubota, T., Ichiki, T., Amano, M., Kaibuchi, K., Takeshita, A., 2004. Inflammatory stimuli upregulate Rho-kinase in human coronary vascular smooth muscle cells. J. Mol. Cell. Cardiol. 37, 537–546.

Hoier, B., Hellsten, Y., 2014. Exercise-induced capillary growth in human skeletal muscle and the dynamics of VEGF. Microcirculation 21, 301–314.

Hoth, K.F., Tate, D.F., Poppas, A., Forman, D.E., Gunstad, J., Moser, D.J., Paul, R.H., Jefferson, A.L., Haley, A.P., Cohen, R.A., 2007. Endothelial function and white matter hyperintensities in older adults with cardiovascular disease. Stroke 38, 308–312.

Howren, M.B., Lamkin, D.M., Suls, J., 2009. Associations of depression with C-reactive protein, IL-1, and IL-6: a meta-analysis. Psychosom. Med. 71, 171–186.

Huikuri, H.V., Makikallio, T.H., 2001. Heart rate variability in ischemic heart disease. Auton. Neurosci. 90, 95–101.

Isenberg, J.S., Wink, D.A., Roberts, D.D., 2006. Thrombospondin-1 antagonizes nitric oxide-stimulated vascular smooth muscle cell responses. Cardiovasc. Res. 71, 785–793.

Isenberg, J.S., Frazier, W.A., Roberts, D.D., 2008. Thrombospondin-1: a physiological regulator of nitric oxide signaling. Cell. Mol. Life Sci. 65, 728–742.

Isingrini, E., Surget, A., Belzung, C., Freslon, J.L., Frisbee, J., O'donnell, J., Camus, V., D'audiffret, A., 2011. Altered aortic vascular reactivity in the unpredictable chronic mild stress model of depression in mice: UCMS causes relaxation impairment to ACh. Physiol. Behav. 103, 540–546.

Iuchi, T., Akaike, M., Mitsui, T., Ohshima, Y., Shintani, Y., Azuma, H., Matsumoto, T., 2003. Glucocorticoid excess induces superoxide production in vascular endothelial cells and elicits vascular endothelial dysfunction. Circ. Res. 92, 81–87.

Kahl, K.G., Schweiger, U., Correll, C., Muller, C., Busch, M.L., Bauer, M., Schwarz, P., 2015. Depression, anxiety disorders, and metabolic syndrome in a population at risk for type 2 diabetes mellitus. Brain Behav 5, e00306.

Kaplan, J.R., Adams, M.R., Clarkson, T.B., Manuck, S.B., Shively, C.A., Williams, J.K., 1996. Psychosocial factors, sex differences, and atherosclerosis: lessons from animal models. Psychosom. Med. 58, 598–611.

Kaplan, M., Aviram, M., Knopf, C., Keidar, S., 2002. Angiotensin II reduces macrophage cholesterol efflux: a role for the AT-1 receptor but not for the ABC1 transporter. Biochem. Biophys. Res. Commun. 290, 1529–1534.
Kayano, H., Koba, S., Matsui, T., Fukuoka, H., Kaneko, K., Shoji, M., Toshida, T., Watanabe, N., Geshi, E., Kobayashi, Y., 2015. Impact of depression on masked hypertension and variability in home blood pressure in treated hypertensive patients. Hypertens. Res. 38, 751–757.
Keidar, S., Kaplan, M., Hoffman, A., Aviram, M., 1995. Angiotensin II stimulates macrophage-mediated oxidation of low density lipoproteins. Atherosclerosis 115, 201–215.
Keidar, S., Heinrich, R., Kaplan, M., Hayek, T., Aviram, M., 2001. Angiotensin II administration to atherosclerotic mice increases macrophage uptake of oxidized ldl: a possible role for interleukin-6. Arterioscler. Thromb. Vasc. Biol. 21, 1464–1469.
Kershaw, K.N., Lane-Cordova, A.D., Carnethon, M.R., Tindle, H.A., Liu, K., 2017. Chronic stress and endothelial dysfunction: the multi-ethnic study of atherosclerosis (MESA). Am. J. Hypertens. 30, 75–80.
Kim, Y.W., Byzova, T.V., 2014. Oxidative stress in angiogenesis and vascular disease. Blood 123, 625–631.
Kim, J.K., Levin, E.R., 2006. Estrogen signaling in the cardiovascular system. Nucl. Recept. Signal. 4, e013.
Kim, Y.W., West, X.Z., Byzova, T.V., 2013. Inflammation and oxidative stress in angiogenesis and vascular disease. J. Mol. Med. (Berl) 91, 323–328.
Kirschbaum, C., Pirke, K.M., Hellhammer, D.H., 1993. The 'trier social stress test'—a tool for investigating psychobiological stress responses in a laboratory setting. Neuropsychobiology 28, 76–81.
Kiuchi, T., Lee, H., Mikami, T., 2012. Regular exercise cures depression-like behavior via VEGF-Flk-1 signaling in chronically stressed mice. Neuroscience 207, 208–217.
Kivimaki, M., Virtanen, M., Elovainio, M., Kouvonen, A., Vaananen, A., Vahtera, J., 2006. Work stress in the etiology of coronary heart disease—a meta-analysis. Scand. J. Work Environ. Health 32, 431–442.
Konior, A., Schramm, A., Czesnikiewicz-Guzik, M., Guzik, T.J., 2014. NADPH oxidases in vascular pathology. Antioxid. Redox Signal. 20, 2794–2814.
Kumari, M., Grahame-Clarke, C., Shanks, N., Marmot, M., Lightman, S., Vallance, P., 2003. Chronic stress accelerates atherosclerosis in the apolipoprotein E deficient mouse. Stress 6, 297–299.
Lazzarino, A.I., Hamer, M., Gaze, D., Collinson, P., Steptoe, A., 2013. The association between cortisol response to mental stress and high-sensitivity cardiac troponin T plasma concentration in healthy adults. J. Am. Coll. Cardiol. 62, 1694–1701.
Leblanc, A.J., Reyes, R., Kang, L.S., Dailey, R.A., Stallone, J.N., Moningka, N.C., Muller-Delp, J.M., 2009. Estrogen replacement restores flow-induced vasodilation in coronary arterioles of aged and ovariectomized rats. Am. J. Physiol. Regul. Integr. Comp. Physiol. 297, R1713–R1723.
Liochev, S.I., Fridovich, I., 2007. The effects of superoxide dismutase on H2O2 formation. Free Radic. Biol. Med. 42, 1465–1469.
Liu, H.Q., Wei, X.B., Sun, R., Cai, Y.W., Lou, H.Y., Wang, J.W., Chen, A.F., Zhang, X.M., 2006. Angiotensin II stimulates intercellular adhesion molecule-1 via an AT1 receptor/nuclear factor-kappaB pathway in brain microvascular endothelial cells. Life Sci. 78, 1293–1298.
Liu, Y., Mladinov, D., Pietrusz, J.L., Usa, K., Liang, M., 2009. Glucocorticoid response elements and 11 beta-hydroxysteroid dehydrogenases in the regulation of endothelial nitric oxide synthase expression. Cardiovasc. Res. 81, 140–147.
Liu, W., Wang, X., Mei, Z., Gong, J., Gao, X., Zhao, Y., Ma, J., Xie, F., Qian, L., 2015. Chronic stress promotes the progression of pressure overload-induced cardiac dysfunction through inducing more apoptosis and fibrosis. Physiol. Res. 64, 325–334.

Liu, T., Zhang, L., Joo, D., Sun, S.C., 2017. NF-kappaB signaling in inflammation. Signal Transduction Targeted Ther. 2.
Lloyd, P.G., Prior, B.M., Li, H., Yang, H.T., Terjung, R.L., 2005. VEGF receptor antagonism blocks arteriogenesis, but only partially inhibits angiogenesis, in skeletal muscle of exercise-trained rats. Am. J. Physiol. Heart Circ. Physiol. 288, H759–H768.
Lu, X.T., Liu, X.Q., Wang, B., Sun, Y.Y., Yang, R.X., Xing, Y.F., Sun, P., Wang, Y.B., Zhao, Y.X., 2017. The role of psychological stress on heart autophagy in mice with heart failure. Psychosom. Med. 79, 1036–1044.
Luine, V., Villegas, M., Martinez, C., McEwen, B.S., 1994. Repeated stress causes reversible impairments of spatial memory performance. Brain Res. 639, 167–170.
Luine, V.N., Beck, K.D., Bowman, R.E., Frankfurt, M., Maclusky, N.J., 2007. Chronic stress and neural function: accounting for sex and age. J. Neuroendocrinol. 19, 743–751.
Martin, P., Massol, J., Scalbert, E., Puech, A.J., 1990. Involvement of angiotensin-converting enzyme inhibition in reversal of helpless behavior evoked by perindopril in rats. Eur. J. Pharmacol. 187, 165–170.
Martins Lima, A., Xavier, C.H., Ferreira, A.J., Raizada, M.K., Wallukat, G., Velloso, E.P., Dos Santos, R.A., Fontes, M.A., 2013. Activation of angiotensin-converting enzyme 2/angiotensin-(1-7)/Mas axis attenuates the cardiac reactivity to acute emotional stress. Am. J. Physiol. Heart Circ. Physiol. 305, H1057–H1067.
Mateo, T., Abu Nabah, Y.N., Abu Taha, M., Mata, M., Cerda-Nicolas, M., Proudfoot, A.E., Stahl, R.A., Issekutz, A.C., Cortijo, J., Morcillo, E.J., Jose, P.J., Sanz, M.J., 2006. Angiotensin II-induced mononuclear leukocyte interactions with arteriolar and venular endothelium are mediated by the release of different CC chemokines. J. Immunol. 176, 5577–5586.
McNeal, N., Scotti, M.A., Wardwell, J., Chandler, D.L., Bates, S.L., Larocca, M., Trahanas, D.M., Grippo, A.J., 2014. Disruption of social bonds induces behavioral and physiological dysregulation in male and female prairie voles. Auton. Neurosci. 180, 9–16.
Mercanoglu, G., Safran, N., Uzun, H., Eroglu, L., 2008. Chronic emotional stress exposure increases infarct size in rats: the role of oxidative and nitrosative damage in response to sympathetic hyperactivity. Methods Find. Exp. Clin. Pharmacol. 30, 745–752.
Miller, A.H., Maletic, V., Raison, C.L., 2009. Inflammation and its discontents: the role of cytokines in the pathophysiology of major depression. Biol. Psychiatry 65, 732–741.
Mitchell, J.A., Larkin, S., Williams, T.J., 1995. Cyclooxygenase-2: regulation and relevance in inflammation. Biochem. Pharmacol. 50, 1535–1542.
Mitchell, G.F., Van Buchem, M.A., Sigurdsson, S., Gotal, J.D., Jonsdottir, M.K., Kjartansson, O., Garcia, M., Aspelund, T., Harris, T.B., Gudnason, V., Launer, L.J., 2011. Arterial stiffness, pressure and flow pulsatility and brain structure and function: the age, gene/environment susceptibility—reykjavik study. Brain 134, 3398–3407.
Nahshoni, E., Aravot, D., Aizenberg, D., Sigler, M., Zalsman, G., Strasberg, B., Imbar, S., Adler, E., Weizman, A., 2004. Heart rate variability in patients with major depression. Psychosomatics 45, 129–134.
Neves, V.J., Moura, M.J., Tamascia, M.L., Ferreira, R., Silva, N.S., Costa, R., Montemor, P.L., Narvaes, E.A., Bernardes, C.F., Novaes, P.D., Marcondes, F.K., 2009. Proatherosclerotic effects of chronic stress in male rats: altered phenylephrine sensitivity and nitric oxide synthase activity of aorta and circulating lipids. Stress 12, 320–327.
Nickel, T., Deutschmann, A., Hanssen, H., Summo, C., Wilbert-Lampen, U., 2009. Modification of endothelial biology by acute and chronic stress hormones. Microvasc. Res. 78, 364–369.
Obraztsov, V.P., Strazhesko, N., 1910. The symptomatology and diagnosis of coronary thrombosis. In: Vorobeva, V.A., Konchalovski, M. (Eds.), Works of the First Congress of Russian Therapists. Comradeship Typography of A.E Mamontov, Moscow.
Okuyama, S., Sakagawa, T., Sugiyama, F., Fukamizu, A., Murakami, K., 1999. Reduction of depressive-like behavior in mice lacking angiotensinogen. Neurosci. Lett. 261, 167–170.

Olfert, I.M., Baum, O., Hellsten, Y., Egginton, S., 2016. Advances and challenges in skeletal muscle angiogenesis. Am. J. Physiol. Heart Circ. Physiol. 310, H326–H336.
Onete, V., Henry, R.M., Sep, S.J.S., Koster, A., Van Der Kallen, C.J., Dagnelie, P.C., Schaper, N., Kohler, S., Reesink, K., Stehouwer, C.D.A., Schram, M.T., 2018. Arterial stiffness is associated with depression in middle-aged men - the Maastricht Study. J. Psychiatry Neurosci. 43, 111–119.
Oulis, P., Kouzoupis, A., Kyrkou, K., Masdrakis, V.G., Georgiopoulos, G., Karapoulios, E., Georgiou, S., Karakatsanis, N.A., Lykka, M., Papadimitriou, G.N., Papamichael, C., Stamatelopoulos, K., 2010. Reversal of increased arterial stiffness in severely depressed women after 6-week antidepressant treatment. J. Affect. Disord. 122, 164–166.
Pacher, P., Beckman, J.S., Liaudet, L., 2007. Nitric oxide and peroxynitrite in health and disease. Physiol. Rev. 87, 315–424.
Palazidou, E., 2012. The neurobiology of depression. Br. Med. Bull. 101, 127–145.
Pasquali, R., Vicennati, V., 2000. The abdominal obesity phenotype and insulin resistance are associated with abnormalities of the hypothalamic-pituitary-adrenal axis in humans. Horm. Metab. Res. 32, 521–525.
Pasquali, R., Vicennati, V., Cacciari, M., Pagotto, U., 2006. The hypothalamic-pituitary-adrenal axis activity in obesity and the metabolic syndrome. Ann. N.Y. Acad. Sci. 1083, 111–128.
Pavel, J., Benicky, J., Murakami, Y., Sanchez-Lemus, E., Saavedra, J.M., 2008. Peripherally administered angiotensin II AT1 receptor antagonists are anti-stress compounds in vivo. Ann. N.Y. Acad. Sci. 1148, 360–366.
Pearson-Leary, J., Eacret, D., Chen, R., Takano, H., Nicholas, B., Bhatnagar, S., 2017. Inflammation and vascular remodeling in the ventral hippocampus contributes to vulnerability to stress. Transl. Psychiatry 7, e1160.
Phillips, S.A., Sylvester, F.A., Frisbee, J.C., 2005. Oxidant stress and constrictor reactivity impair cerebral artery dilation in obese Zucker rats. Am. J. Phys. Regul. Integr. Comp. Phys. 288, R522–R530.
Pratt, L.A., Brody, D.J., 2014. Depression and obesity in the U.S. Adult household population, 2005–2010. In: National Health and Nutrition Examination Survey Data. Nchs Data Brief No. 167. https://www.cdc.gov/nchs/products/databriefs/db167.htm.
Raaz, U., Toh, R., Maegdefessel, L., Adam, M., Nakagami, F., EMRICH, F.C., Spin, J.M., Tsao, P.S., 2014. Hemodynamic regulation of reactive oxygen species: implications for vascular diseases. Antioxid. Redox Signal. 20, 914–928.
Razzoli, M., Nyuyki-Dufe, K., Gurney, A., Erickson, C., McCallum, J., Spielman, N., Marzullo, M., Patricelli, J., Kurata, M., Pope, E.A., Touma, C., Palme, R., Largaespada, D.A., Allison, D.B., Bartolomucci, A., 2018. Social stress shortens lifespan in mice. Aging Cell 17 (4), e12778.
Rosengren, A., Hawken, S., Ounpuu, S., Sliwa, K., Zubaid, M., Almahmeed, W.A., Blackett, K.N., Sitthi-Amorn, C., Sato, H., Yusuf, S., Investigators, I., 2004. Association of psychosocial risk factors with risk of acute myocardial infarction in 11119 cases and 13648 controls from 52 countries (the INTERHEART study): case-control study. Lancet 364, 953–962.
Roth, L., Rombouts, M., Schrijvers, D.M., Lemmens, K., De Keulenaer, G.W., Martinet, W., De Meyer, G.R., 2015. Chronic intermittent mental stress promotes atherosclerotic plaque vulnerability, myocardial infarction and sudden death in mice. Atherosclerosis 242, 288–294.
Rutledge, T., Reis, V.A., Linke, S.E., Greenberg, B.H., Mills, P.J., 2006. Depression in heart failure a meta-analytic review of prevalence, intervention effects, and associations with clinical outcomes. J. Am. Coll. Cardiol. 48, 1527–1537.
Satoh, H., Fujii, S., Tsutsui, H., 2015. Persistent depression is a significant risk factor for the development of arteriosclerosis in middle-aged Japanese male subjects. Hypertens. Res. 38, 84–88.

Schiavone, M.T., Santos, R.A., Brosnihan, K.B., Khosla, M.C., Ferrario, C.M., 1988. Release of vasopressin from the rat hypothalamo-neurohypophysial system by angiotensin-(1-7) heptapeptide. Proc. Natl. Acad. Sci. U. S. A. 85, 4095–4098.
Scuteri, A., Modestino, A., Fedullo, F., Assisi, A.P., Gianni, W., 2013. Depression treatment selectively modifies arterial stiffness in older participants. J. Gerontol. A Biol. Sci. Med. Sci. 68, 719–725.
Segers, P., Rietzschel, E.R., De Buyzere, M.L., Vermeersch, S.J., De Bacquer, D., Van Bortel, L.M., De Backer, G., Gillebert, T.C., Verdonck, P.R., Asklepios, I., 2007. Noninvasive (input) impedance, pulse wave velocity, and wave reflection in healthy middle-aged men and women. Hypertension 49, 1248–1255.
Seldenrijk, A., Van Hout, H.P., Van Marwijk, H.W., De Groot, E., Gort, J., Rustemeijer, C., Diamant, M., Penninx, B.W., 2011. Depression, anxiety, and arterial stiffness. Biol. Psychiatry 69, 795–803.
Serlin, Y., Levy, J., Shalev, H., 2011. Vascular pathology and blood-brain barrier disruption in cognitive and psychiatric complications of type 2 diabetes mellitus. Cardiovasc. Psychiatry Neurol. 2011, 609202.
Sheffield, D., Krittayaphong, R., Cascio, W.E., Light, K.C., Golden, R.N., Finkel, J.B., Glekas, G., Koch, G.G., Sheps, D.S., 1998. Heart rate variability at rest and during mental stress in patients with coronary artery disease: differences in patients with high and low depression scores. Int. J. Behav. Med. 5, 31–47.
Shively, C.A., Register, T.C., Appt, S.E., Clarkson, T.B., 2015. Effects of long-term sertraline treatment and depression on coronary artery atherosclerosis in premenopausal female primates. Psychosom. Med. 77, 267–278.
Sojka, P., Stalnacke, B.M., Bjornstig, U., Karlsson, K., 2006. One-year follow-up of patients with mild traumatic brain injury: occurrence of post-traumatic stress-related symptoms at follow-up and serum levels of cortisol, S-100B and neuron-specific enolase in acute phase. Brain Inj. 20, 613–620.
Spieker, L.E., Hurlimann, D., Ruschitzka, F., Corti, R., Enseleit, F., Shaw, S., Hayoz, D., Deanfield, J.E., Luscher, T.F., Noll, G., 2002. Mental stress induces prolonged endothelial dysfunction via endothelin-A receptors. Circulation 105, 2817–2820.
Stanley, S.C., Brooks, S.D., Butcher, J.T., D'audiffret, A.C., Frisbee, S.J., Frisbee, J.C., 2014. Protective effect of sex on chronic stress- and depressive behavior-induced vascular dysfunction in Balb/cJ mice. J. Appl. Physiol. (1985) 117, 959–970.
Steptoe, A., Brydon, L., 2009. Emotional triggering of cardiac events. Neurosci. Biobehav. Rev. 33, 63–70.
Steptoe, A., Kivimaki, M., 2013. Stress and cardiovascular disease: an update on current knowledge. Annu. Rev. Public Health 34, 337–354.
Strehlow, K., Rotter, S., Wassmann, S., Adam, O., Grohe, C., Laufs, K., Bohm, M., Nickenig, G., 2003. Modulation of antioxidant enzyme expression and function by estrogen. Circ. Res. 93, 170–177.
Suglia, S.F., Sapra, K.J., Koenen, K.C., 2015. Violence and cardiovascular health: a systematic review. Am. J. Prev. Med. 48, 205–212.
Sutton-Tyrrell, K., Najjar, S.S., Boudreau, R.M., Venkitachalam, L., Kupelian, V., Simonsick, E.M., Havlik, R., Lakatta, E.G., Spurgeon, H., Kritchevsky, S., Pahor, M., Bauer, D., Newman, A., Health, A.B.C.S., 2005. Elevated aortic pulse wave velocity, a marker of arterial stiffness, predicts cardiovascular events in well-functioning older adults. Circulation 111, 3384–3390.
Thygesen, K., Alpert, J.S., White, H.D., Joint ESC/ACCF/AHA/WHF Task Force for the Redefinition of Myocardial Infarction, 2007. Universal definition of myocardial infarction. J. Am. Coll. Cardiol. 50, 2173–2195.
Tiwari, S., Zhang, Y., Heller, J., Abernethy, D.R., Soldatov, N.M., 2006. Atherosclerosis-related molecular alteration of the human CaV1.2 calcium channel alpha1C subunit. Proc. Natl. Acad. Sci. U. S. A. 103, 17024–17029.

Valkanova, V., Ebmeier, K.P., 2013. Vascular risk factors and depression in later life: a systematic review and meta-analysis. Biol. Psychiatry 73, 406–413.
Van Sloten, T.T., Protogerou, A.D., Henry, R.M., Schram, M.T., Launer, L.J., Stehouwer, C.D., 2015. Association between arterial stiffness, cerebral small vessel disease and cognitive impairment: a systematic review and meta-analysis. Neurosci. Biobehav. Rev. 53, 121–130.
Veith, R.C., Lewis, N., Linares, O.A., Barnes, R.F., Raskind, M.A., Villacres, E.C., Murburg, M.M., Ashleigh, E.A., Castillo, S., Peskind, E.R., et al., 1994. Sympathetic nervous system activity in major depression. Basal and desipramine-induced alterations in plasma norepinephrine kinetics. Arch. Gen. Psychiatry 51, 411–422.
Von Bohlen Und Halbach, O., Walther, T., Bader, M., Albrecht, D., 2000. Interaction between Mas and the angiotensin AT1 receptor in the amygdala. J. Neurophysiol. 83, 2012–2021.
Walther, T., Balschun, D., Voigt, J.P., Fink, H., Zuschratter, W., Birchmeier, C., Ganten, D., Bader, M., 1998. Sustained long term potentiation and anxiety in mice lacking the Mas protooncogene. J. Biol. Chem. 273, 11867–11873.
Wilbert-Lampen, U., Straube, F., Trapp, A., Deutschmann, A., Plasse, A., Steinbeck, G., 2006. Effects of corticotropin-releasing hormone (CRH) on monocyte function, mediated by Crh-receptor subtype R1 and R2: a potential link between mood disorders and endothelial dysfunction? J. Cardiovasc. Pharmacol. 47, 110–116.
Willner, P., 2005. Chronic mild stress (CMS) revisited: consistency and behavioural-neurobiological concordance in the effects of CMS. Neuropsychobiology 52, 90–110.
Woollard, K.J., 2013. Immunological aspects of atherosclerosis. Clin. Sci. (Lond.) 125, 221–235.
Xie, H.H., Zhou, S., Chen, D.D., Channon, K.M., Su, D.F., Chen, A.F., 2010. GTP cyclohydrolase I/BH4 pathway protects EPCs via suppressing oxidative stress and thrombospondin-1 in salt-sensitive hypertension. Hypertension 56, 1137–1144.
Xie, F., Sun, L., Su, X., Wang, Y., Liu, J., Zhang, R., Wang, N., Zhao, J., Ban, T., Niu, H., Ai, J., 2012. Neuropeptide Y reverses chronic stress-induced baroreflex hypersensitivity in rats. Cell. Physiol. Biochem. 29, 463–474.
Xing, Y., He, J., Hou, J., Lin, F., Tian, J., Kurihara, H., 2013. Gender differences in CMS and the effects of antidepressant venlafaxine in rats. Neurochem. Int. 63, 570 575.
Yusuf, S., Hawken, S., Ounpuu, S., Dans, T., Avezum, A., Lanas, F., McQueen, M., Budaj, A., Pais, P., Varigos, J., Lisheng, L., Investigators, I.S., 2004. Effect of potentially modifiable risk factors associated with myocardial infarction in 52 countries (the INTERHEART study): case-control study. Lancet 364, 937–952.
Zhang, J., Patel, J.M., Li, Y.D., Block, E.R., 1997. Proinflammatory cytokines downregulate gene expression and activity of constitutive nitric oxide synthase in porcine pulmonary artery endothelial cells. Res. Commun. Mol. Pathol. Pharmacol. 96, 71–87.
Zhang, C., Park, Y., Picchi, A., Potter, B.J., 2008. Maturation-induces endothelial dysfunction via vascular inflammation in diabetic mice. Basic Res. Cardiol. 103, 407–416.
Zhang, T., Chen, Y., Liu, H., Zhou, Z., Zhai, Y., Yang, J., 2010. Chronic unpredictable stress accelerates atherosclerosis through promoting inflammation in apolipoprotein E knockout mice. Thromb. Res. 126, 386–392.
Zhao, C., Deng, W., Gage, F.H., 2008. Mechanisms and functional implications of adult neurogenesis. Cell 132, 645–660.
Zhou, L., Isenberg, J.S., Cao, Z., Roberts, D.D., 2006. Type I collagen is a molecular target for inhibition of angiogenesis by endogenous thrombospondin-1. Oncogene 25, 536–545.
Zieman, S.J., Melenovsky, V., Kass, D.A., 2005. Mechanisms, pathophysiology, and therapy of arterial stiffness. Arterioscler. Thromb. Vasc. Biol. 25, 932–943.
Zipursky, R.T., Press, M.C., Srikanthan, P., Gornbein, J., McClelland, R., Watson, K., Horwich, T.B., 2017. Relation of stress hormones (urinary catecholamines/cortisol) to coronary artery calcium in men versus women (from the multi-ethnic study of atherosclerosis [MESA]). Am. J. Cardiol. 119, 1963–1971.

CHAPTER 7

Human neuroimaging-based connections between stress, cardiovascular disease and depression

Julie A. Brefczynski-Lewis
Department of Neuroscience, West Virginia University, Morgantown, WV, United States

Contents

Introduction

The intimate relationship between the brain and cardiovascular system has long been established (Cannon, 1957; Richter, 1957; Engel, 1971). Although progress has been made in understanding aspects of this relationship, the field is still relatively young, with unresolved complexities and tantalizing recent discoveries. This chapter will provide an overview of neuroimaging

Cardiovascular Implications of Stress and Depression
https://doi.org/10.1016/B978-0-12-815015-3.00007-6

studies that have examined the relations between stress, depression, and cardiovascular health. Because the brain is generally the first organ in the body to detect environmental stress, the neural responses to stress arguably reflect some of the first stress-related physiological changes in the body. A better understanding of these cortical responses to stress is likely to lead to improved development, testing, and monitoring of new treatments and/or prevention efforts for reducing cardiovascular disease (CVD). It would also assist with our understanding of the brain-based obstacles that may pose challenges to implementing effective treatments in this large and growing population. The necessary steps to address the global CVD epidemic need to go beyond the core instructions of taking medication, reducing stress, improving diet, and exercising.

In this chapter, we begin with a brief overview of human neuroimaging methods and then discuss the early brain responses related to exposure to different types of stress, the development of depression, and the role of CVD. Finally, neuroimaging studies of common treatments and therapies to reduce risk for CVD will be examined with respect to how they impact the brain.

Neuroimaging methods

This section provides a brief overview of the typical methods used for measuring the cortical stress response. First, differences in regional brain structure and volume can be readily measured with Magnetic Resonance Imaging (MRI) using voxel-wide brain morphometry (VBM) or hand traced measurements of individual structures. VBM is an analysis that looks for volume differences of grey matter between groups (Whitwell, 2009). Second, a different MRI brain sequence called Diffusion Tensor Imaging (DTI) can look at the integrity of brain structures and their connectivity by measuring how easily water flows along known or abnormal pathways in the white matter of the brain.

Next, baseline brain function is a measure of brain activity that occurs while a person is at rest and not assigned a specific task. Again, using MRI, Resting State MRI (rsMRI) is the least invasive of such measures, as it uses a functional MRI (fMRI) image collection sequence, but without a task. In rsMRI, one can observe the regions of the brain that are most frequently active as a correlated network during an individual's resting state, with multiple networks including the most well studied Default Mode Network (DMN) (Raichle et al. 2001). In addition, a baseline of metabolic activity in the brain can be obtained using a radioactively labeled tracers, such as the sugar tracer fluorodeoxyglucose (FDG) and the water tracer (5-O-H_2O) with the imaging tool, Positron Emission Tomography (PET). PET reveals an average, "long-exposure" picture of what regions are more versus less metabolically active, but not necessarily as part of the same network as in rsMRI. PET can also trace other markers of activity such as inflammation, genetic transcription and neurotransmitter activity (Wang et al., 2014). However, these methods are relatively novel, and yielded contradictory results with depression, likely due to small sample sizes (Smith and Jakobsen, 2013).

Brain functional imaging refers to measuring brain activity in response to specific stimuli or tasks and requires comparisons of activity over time (Bandettini, 2012). fMRI combines the spatial specificity of MRI with a moderate degree of temporal resolution, usually on the order of a few seconds in response to a stimulus type or task, such as exposure to facial expressions or negatively/positively-valanced images or sounds. The stimuli or task can elicit Blood Oxygen-Level Dependent (BOLD) changes in the MRI response that reflect altered metabolic demand due to the ensuing depletions of metabolites (fMRI dip) and subsequent dilation of vessels to bring oxygenated blood to the necessary brain regions (Glover, 2011). fMRI can also be further analyzed using task-related functional connectivity (fcMRI) in which the strengths of correlations of metabolic activity in various brain regions can be examined to determine how well they are synchronized. Functional PET scans using a short half-life water tracer, 15-O-H_2O, can also show task-related activity, as can FDG PET scans if the tracer is delivered prior to the activity or delivered via infusions during multiple tasks (Villien et al., 2014; Rischka et al., 2018).

All the above methodologies can be used to examine brain differences between groups (e.g., CVD vs. healthy controls) or examine the same set of participants longitudinally over time, (i.e., pre-and post-intervention). Other imaging modalities such as electrophysiological (EEG), functional near infrared spectroscopy (fNIRS), magnoencephalography (MEG), single-photon emission computed tomography (SPECT), and neuromodulatory methods, will only be peripherally addressed due to the magnitude of the literature examining cortical responses to stress. This chapter will focus on neurophysiological mechanisms for explaining relations among CVD, depression, and stress from meta-analyses, and several particularly revealing studies. The results are summarized in a table in the discussion summary.

Cortical responses to stress and the brain regions affected

Acute stress exposure and the brain

Necessary for survival of most species, including our own, is the development of adaptive responses to threat and uncertainty. Response to stress has both physiological and neurological components that are intimately linked. Some of these responses to stress affect behavior in potentially lifesaving ways. Repetitive exposure to acute stressors can lead to long term adaptive brain changes. If there are ongoing threats to life, these responses can be beneficial, at least in terms of short-term survival. For example, one may acquire a heightened fear response, an anxious disposition, and associated enhanced physiological reactivity to, for instance, loud noises or sudden movements which leads to a quicker behavioral response to confront or avoid dangerous situations or signal potential danger. Indeed, individuals with elevated levels of anxiety, which can accompany increased stress (Bremner et al., 1996), were found to have faster reaction times to threat-related stimuli than individuals with low trait anxiety, which was correlated with increased amygdala activity (Fig. 7.1A) (Reidel et al., 2016). Survival-based adaptations to

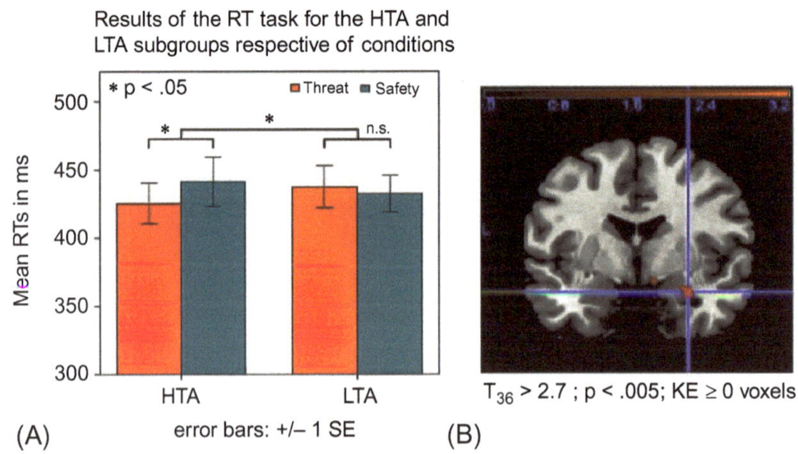

Fig. 7.1 Adaptive response to stress. (A) Chart of reaction times (RT) to "Threat" *(Red)* versus "Safety" *(Blue)* cues in those with high trait anxiety (HTA) versus those with low trait anxiety (LTA), labeled below on the X axis. Those with high anxiety have a significantly faster reaction times to threat stimuli compared to those with LTA. (B) This faster reaction time was coupled with higher amygdala activity in HTA subjects. A coronal brain slice showing amygdala, reveals increased activation in the right amygdala during presentation of threat stimuli in HTA versus LTA subjects (*red* activation at focus of the *blue* cross-hair). *(Reprinted from Reidel, P., Jacob, M.J., Muller, D.K., Vetter, N.C., Smolka, M.N., Marxen, M., 2016. Amygdala fMRI signal as a predictor of reaction time. Front. Hum. Neurosci. 10, 516.)*

stress likely include altered response properties of the amygdala (Fig. 7.1B). The amygdala is located bilaterally on the temporal lobes with diverse roles including early detection of salient threat-related objects, emotions, and danger, as well as mediation of affective expression. The amygdala obtains input from the sensory modalities and can be modulated (especially inhibited) via input from the prefrontal cortex (Davidson, 2002). The amygdala also has influence over the autonomic nervous systems and behavioral systems via connections to the hypothalamus, brain stem regions, and reward regions like the nucleus accumbens, which work together to establish autonomic tone.

The amygdala may serve as an early gateway for stress detection and the initiation of potentially longer-term stress related brain changes. Even in healthy individuals, the amygdala can be activated by affective stimuli, such as facial expressions portraying emotions and by pictures of weapons and vicious animals (Bzdok et al., 2013). An aspect of amygdala function seems hardwired for fear; for example, brain activity in this region preferentially increases in response to viewed eye stimuli with increased eye-white area that characterizes human fear expression (Morris et al., 1999; Hardee et al., 2008). A fear-evoked state of hypervigilance and hypersensitivity to affective stimuli, however, does not last indefinitely in the healthy brain, and it would be maladaptive for those who live in a relatively safe society for the brain to respond acutely with fear responses to a broad range of everyday stimuli.

Regions of the brain that regulate emotion (e.g., medial prefrontal cortex) may help temper amygdala reactivity when such a response is not necessary. A high amygdala response of a healthy, secure person is elicited by a narrower range of stimuli than amygdala responses of a fearful person. This may be due in part to early childhood learning that occurs when caregivers do not respond to every noise, dropped object, or sudden appearance of novel stimuli. For example, an fMRI study in young children showed that a picture of the child's mother vs. a picture of a stranger (a mother of one of the other participants) buffered amygdala reactivity to a scary stimulus (Gee et al., 2014; Gee and Casey, 2015). Thus, it appears that young children with developing prefrontal cortex rely on their caregivers to shape this development, such that the prefrontal cortex-amygdala activation pattern looks like that of a healthy adult (inverse relationship), only when a trusted caregiver is present. The protective presence of a caregiver and how one's caregivers respond to potentially dangerous or stressful stimuli may shape how the brain learns to respond to stress and threat. This process can be altered either when parents are stressed or in situations of childhood adversity such as neglect, in which evidence showed that exposure to these sorts of childhood environments altered ones' autonomic and neuroendocrine responses to stress. Even in adulthood, exposure to various stressors and mental disorders results in higher than normal activation of the prefrontal cortex-amygdala connection (Gianaros and Sheu, 2009). This has been proposed to be either a compensatory mechanism in which inhibitory neurons in the prefrontal cortex are working harder to control amygdala reactivity, or resulting from feed-forward connections from the amygdala to the prefrontal cortex resulting from its over reactivity (Gianaros and Sheu, 2009).

Regarding the reactivity of the cardiovascular system and stress, an earlier review (Gianaros and Sheu, 2009) cited converging evidence from multiple neuroimaging studies on the critical relationship of the error monitoring and emotion/behavioral regulation system, including cingulate cortex and insula, in affecting reactivity of the cardiovascular system to stress. Briefly, the Anterior Cingulate Cortex (ACC) has been shown in fMRI studies to be involved in tasks related to the stress response. Such as detection of errors and cognitive control over emotional reactions (Carter et al., 1998; Bush et al., 2000; Kerns et al., 2004), with the perigenual region of the ACC involved with threat evaluation, negative mood induction, and anticipatory anxiety (Straube et al., 2009; Wager et al., 2009a,b). The insula has also been active in error detection and, in contrast with ACC, seems to be associated with awareness of the error (Klein et al., 2007) and internal body states (Critchley et al., 2004). The ACC and insula have been shown to be connected in a similar manor via efferent and afferent connections to affective and physiological control regions (nucleus tractus solitarius and other brainstem regions), which have a strong influence on the Hypothalamic-Pituitary-Adrenal (HPA) axis response to stress (Gianaros and Sheu, 2009).

Farther posterior in the cingulate gyrus, the Posterior Cingulate Cortex (PCC) is a brain region with self-referential functions that are indirectly involved in regulating cardiovascular reactivity to stress (Gianaros and Sheu, 2009). Although the PCC has less direct connections to HPA axis regions, it is associated with stress-related thought processes like rumination, which elicits altered connectivity of PCC to brain regions that more directly control bodily responses to stress (Cooney et al., 2010; Nejad et al., 2013). Hypermetabolism (FDG PET) of the PCC is found in depression, and hyperconnectivity with perigenual ACC is associated with rumination in depression, Attention Deficit and Hyperactivity Disorder (ADHD), traumatic brain injury (TBI) and Alzheimer's Disease (Leech and Sharp, 2014).

An fMRI study found that the degree of stress reactivity of these regions (ACC, prefrontal cortex, insula, PCC, as well as cerebellum) was correlated with the intensity of an individual's blood pressure response in a challenging cognitive task, portraying a potential "neural phenotype" of a highly reactive person who may be at risk for CVD (Gianaros et al., 2007). In other words, having a more excitable ACC-PCC-Insula response may predispose an individual to CVD risk, or be an early indicator of such risk.

In sum, if danger is real and ongoing (e.g., war and violent conflict), it may beneficial to have a robust reaction to every novel sound, movement, or disagreeable facial expression, as there is a greater chance that these stimuli may be life-threatening. The evolutionary drive to survive and produce and raise offspring reinforces the quick and accurate detection of threats in dangerous environments. Consequently, changes in brain activity are entirely predictable to increase the probability of survival in high stress environments. However, for most humans, ongoing hyperreactivity to acute stress is maladaptive, both socially and for long-term survival. Many contemporary sources of environmental stress are not life-threatening, although human physiology has been programmed to respond to them as though they were. In sum, hyperreactivity to acute environmental stressors involves numerous brain regions including the amygdala, insula, ACC, as well as the PCC. These regions can influence the physical and mental responses to stress, including exaggerated autonomic nervous system and HPA activation, as well as perpetual rumination or hyper self-awareness. This activation is associated with a less regulation by the medial prefrontal cortex, a key structure that normally would inhibit or dampen the magnitude of these stress responses. This pattern of cortical stress responding could create pathological changes to the brain that provides the foundation for conditions like CVD and depression. These patterns do appear to be malleable, as further addressed in a later section.

Chronic stress exposure and the brain

All humans inevitably experience periodic circumstances in life in which they are exposed to a higher frequency and intensity of life stressors. When exposed to repetitive stress and high negative emotion, different regions of the brain begin to change, resulting in behavioral effects and a direct physiological response in peripheral regions of the body, which can result in "burnout" or "stress fatigue" (Demerouti et al., 2001; Tei et al., 2014). Chronic stress can lead to chronic inflammation, which is known to affect brain function and functional connectivity (Kraynak et al., 2018)

(Chapter 8). Imaging studies have shown that the brain begins to change in these highly stressful environments, resulting in decreased baseline prefrontal and increased amygdala activity. These changes, however, appear to be fully or partially reversible upon alleviation of stress or with therapy or cognitive training (Gavelin et al., 2017; Savic et al., 2017).

With chronic stress, one also begins to see changes in cognitive, memory, and attention systems. Liston et al. (2009) utilized a predictable temporary chronic high stress paradigm and examined the changes in brain and behavior during and after a one-month stressful student exam period. During an attention shifting task, the fMRI scans showed reductions in dorsal lateral prefrontal cortex (DLPFC), a region responsible for attentional control, as well as higher order cognitive functions. They additional saw widespread diminished connectivity between this prefrontal region and regions involved in stress and affective responses, including medial prefrontal, insula, and basal ganglia regions (Liston et al., 2009). After the stress period was over and perceived stress was no longer elevated, activity and connectivity within these regions returned to normal, showing no differences between the formerly-stressed students and control participants. These results indicate that stable changes in affective, cognitive, and attention-related brain function can occur with chronic stress exposure, but in healthy individuals these are typically reversible after the stress period lifts.

However, chronic stress can often extend into longer periods of time. Ongoing chronic stress can come in the form of lack of perceived control, lack of purpose, and social isolation/loneliness, all of which are risk factors for both CVD and depression (Harlow et al., 1986; Penninx et al., 1997; Friedmann et al., 2006; Koizumi et al., 2008; Lindsay, 2009). First, lack of perceived control can come from situations that seem inescapable, such as certain types of pain, ongoing stressors in work and life, as well as socioeconomic challenges, discrimination, and as a result of other environmental determinants (Kurian and Cardarelli, 2007; Yan et al., 2017). For example, poverty is associated with greater perceived loss of control and stress, and a MRI study of over 800 children and adolescents found structural brain differences correlating with poverty, including smaller frontal lobes, temporal lobes, and hippocampi (Hair et al., 2015). However, it is difficult to determine causality due to potential dietary differences and other factors between those living in poverty-stricken areas and those in more affluent communities. A smaller controlled study used fMRI to study perceived control of a pain stressor, such that subjects were delivered the same degree of pain but either accurately told they had no control or falsely told they had control via a faux button. Perceived lack of control resulted in activation in the ACC and right insula, which correlated with a higher perceived unpleasantness of the pain (Salomons et al., 2004). Such avoidable stressors over time may lead to longer term brain changes in such regions that influence cardiovascular responses to stress, including affective regions like ACC and insula, as well as memory and cognition brain regions such as frontal and temporal cortex and hippocampus.

Another stressor related to the perception of control is an ongoing sense of lack of purpose in life. The Japanese have a term Ikigai that suggests an optimal overlap between finding a way to live a life that is beneficial to others, utilizing one's talents, and is financially viable (Koizumi et al., 2008). Having this Ikigai or sense of purpose can be protective against depression and CVD risk (Koizumi et al., 2008; Schaefer et al., 2013; Cohen et al., 2016). In terms of brain mechanisms, PCC has been found to correlate in size (using VBM) with both perceived sense of purpose and increased positive versus negative emotions, which have been proposed as potential indicators of overall happiness (Sato et al., 2015). In contrast, a small, overactive, and hyper-frontal connectivity of PCC (small VBM, but overactive metabolism) is characteristic of depression and high rumination, and is generally thought to be associated with hyper self-focus (Cooney et al., 2010). In addition, higher activity in PCC is correlated with high blood pressure reactivity (Gianaros et al., 2007). This suggests that perceived sense of purpose and corresponding PCC size and function, may be important to monitor in relation to both CVD and depression-related rehabilitation and prevention.

The last of these complex factors associated with chronic stress, loneliness or lack of social support, is a major risk factor for CVD and depression, and in a comprehensive meta-analysis, exceeding the risk of even hypertension and obesity (Holt-Lunstad et al., 2010). Loneliness has been found to affect the limbic reward system. For example, individuals who are lonely versus non-lonely have decreased brain reward region responses (ventral striatum) to pleasant pictures of social enjoyment of strangers (Cacioppo et al., 2009). However, they have a hyperreactivity response in these regions to close familiar persons (Inagaki et al., 2016). This indicates a potential reward-related brain network dysfunction that may play a role in CVD recovery treatment plans and considerations related to social support.

In summary, in terms of chronic stress, finding ways to decrease intensity or shorten exposure, empower people and reduce loneliness is likely critical in reducing the brain signatures from chronic stress that can become CVD and depression risk factors. The point at which ongoing chronic stress leads to longer term, and potential irreversible changes in neural function and activity remains unknown. It appears, however, that as exposure to stress becomes more chronic, the amount of brain networks involved increases. In addition to alterations of activity in the amygdala, insula, ACC, PCC, and medial prefrontal cortex detected during exposure to acute stress, chronic stress deleteriously implicates function and structure of regions related to cognitive performance and attention such as DLPFC, as well as altered reward pathways in the brain. Further research naturally will be necessary to understand how brain changes evolve with long-term moderate/high stress situations such as ongoing occupational, economic, or caregiving stress (Demerouti et al., 2001;Violanti et al., 2006; Capistrant et al., 2012).

Exposure to extreme stress and the brain

Exposure to even short periods of very high intensity stress can create brain, behavioral, and physiological changes that are lasting and often challenging to treat. Research has shown that Post-Traumatic Stress Disorder (PTSD) arising from exposure to severe stress such as abuse, accidents, and intense occupational stress (e.g., military, first responders), has a robust link with brain dysfunction and CVD (Violanti et al., 2009; Edmondson and von Kanel, 2017). A meta-analysis of 26 PET and fMRI neuroimaging studies of patients diagnosed with PTSD, revealed common hyperactivity in amygdala and mid-cingulate cortex, and hypoactivity in emotional regulation, cognitive and memory regions, including prefrontal cortex, inferior frontal gyrus, and hippocampus (Hayes et al., 2012). As a specific example, a study of women with PTSD resulting from sexual assault, showed a reduced ability to downregulate responses to negative stimuli that was correlated with reduced prefrontal cortex activity (New et al., 2009). These results suggest that experiencing extreme stress can make long-term, stable (but still potentially reversible) changes in the brain, such that the brain's baseline function becomes more like the brain's response to acute or temporary stress. Increased amygdala and cingulate activity, coupled with reductions in size and function of the regions that normally temper activity in these regions such as prefrontal cortex, likely underlie reduced affective control and other emotion regulation symptoms common in PTSD, which in turn have influence over physiological functions.

Exposure to early life stress and the brain

Early life adversity is a better predictor of developing depression in later life than genetics (Risch et al., 2009) and is also a risk factor for CVD

(Gassen et al., 2017). Children exposed to adverse early life stressors can have altered brain development, which is linked to future risk for CVD and depression (Chapman et al., 2004; Anda et al., 2006). These data are supported by rodent and human studies, such that how young are treated has been shown to affect more than 900 genes, some of which lead to long term changes in the brain (Weaver et al., 2006). For example, rats born to mothers who experienced high levels of stress, such as inadequate maternal care, have long term changes in the genetic expression related to glucocorticoid receptor density in the hippocampus (Weaver et al., 2004). In addition, in a human post-mortem study of suicide victims, those with a history of childhood abuse also showed differential expression of a neuron-specific glucocorticoid receptor (NR3C1) promoter in the hippocampus (McGowan et al., 2009), which would directly change the function of this memory-related brain region, with downstream physiological influence.

Genetic expression that affects the density of brain receptors might underlie a host of documented stress-induced behaviors, such as reduced fear extinction, increased awareness of potential threats, spending less time caring for off-spring, and having a decreased threshold for aggressive behaviors (Cozolino, 2017). Children raised in situations characterized by neglect, high stress, and trauma have long term brain changes that are associated with increased anxiety, depression, and addiction (Frodl, 2016). An example of these brain changes comes from a neuroimaging study of children raised in an orphanage versus a foster home. This study showed that the functional connectivity of the prefrontal-amygdala brain regions was altered among young children raised in the orphanage and appeared to have prematurely manifested an inverse "adult" connectivity pattern, in which prefrontal cortex shows less activation when amygdala is active (Gee et al., 2014; Gee and Casey, 2015). In contrast, as discussed earlier, typically young children only show a mature (adult-like) pattern of connectivity when viewing their caregiver (Gee and Casey, 2015). Although the direct long-term effects of a premature adult-like pattern in emotional regulation circuitry is unknown, in a larger study using VBM, the medial prefrontal cortex was reduced in size among those with early childhood adversity compared to those without (van Harmelen et al., 2010). Further, children who experienced such adversity were at risk for an abnormal emotional regulation and stress-related health disorders.

Overall, early childhood adversity must be addressed as a major risk factor for depression and CVD (Felitti et al., 1998; Batten et al., 2004; Infurna et al., 2016). Part of the mechanisms comprising these risks may involve abnormal brain changes that are shaped via disrupted or compensatory developmental trajectories. Reduced or abnormal caregiving may affect genetic regulation of memory and cognitive brain circuitry, as well as affective and emotion regulation regions. Abnormal or lack of external guidance during childhood may alter how important stress-related brain circuitry is established, such as the connectivity between the frontal cortex and amygdala.

Cortical responses to stress and CVD

A consistent pattern of cortical changes and abnormalities has been observed in response to a broad range of stressful life experiences, including exposure to acute and chronic life stressors, traumatic events, and aversive childhoods. If the cortical stress response is involved in mediating the link between environmental stress and CVD, it is also important to examine cortical responses to stress among patients who have manifested CVD and pre-clinical CVD symptomologies, in comparison to healthy controls.

Anazodo et al. (2013) found structural grey matter volume decreases in CVD patients vs. age-matched controls, with atrophy in brain regions related to affective control and cognitive functions, including the superior frontal, medial frontal, and superior temporal gyri (Fig. 7.2, left). A review of 9 structural MRI studies of heart failure patients found converging evidence of structural differences in cingulate, insula, and amygdala, as well as decreased size and an increase of detected white matter microstructural alterations (as measured by DTI) in the hippocampus, thalamus, cerebellum, and subcortical brain regions (Alosco and Hayes, 2015). The combination of abnormalities in both grey matter and white matter integrity suggests that not only have these regions experienced atrophy, but their ability to properly communicate via the brain white matter system is impaired.

In terms of function, it has been shown that CVD patients have abnormal brain responses in affective and emotion regulation networks. For example, a PET study using O15-H_2O (a perfusion marker that reveals increased task-related blood flow), showed that CVD patients with a demonstrated stress-related vascular response (increased vasoconstriction) had increased brain response in insula and parietal cortex, including PCC and sub-regions, and decreased activation in medial prefrontal cortex (Shah et al., 2019). Emotion regulation changes have also been identified in individuals with pre-clinical CVD (Gianaros et al., 2009). For example, in response to fearful and angry expression faces, CVD risk patients vs. healthy

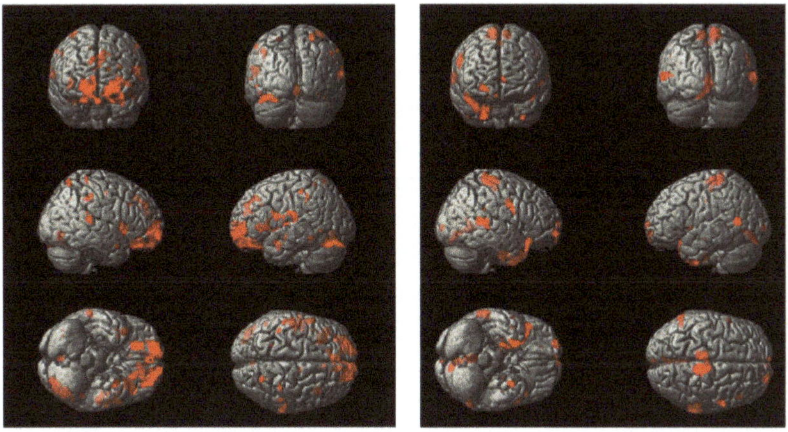

Fig. 7.2 Brain changes in CVD patients versus Healthy Controls are partially ameliorated with a aerobic exercise regime. The left panel shows in *red*, regions of the brain that are smaller (VBM) in CVD patients healthy age-matched controls. The brain is shown rotated in all directions for viewing including forward facing, backward facing, right side, left side, from the bottom, and from the top. The right panel shows in *red*, the regions within the CVD patients that gained grey matter volume after a 6 month aerobic exercise rehabilitation. *(Adapted from Anazodo, U.C., Shoemaker, J.K., Suskin, N., Lawrence, K.S., 2013. An investigation of changes in regional gray matter volume in cardiovascular disease patients, pre and post cardiovascular rehabilitation. NeuroImage: Clinical 3, 388–395.)*

controls exhibited increased dorsal right amygdala response, as well as the abnormal positive amygdala/prefrontal/ACC connectivity pattern, similar to that seen in other stress dysregulation studies (Gianaros et al., 2009).

In summary, CVD-related brain changes overlap with major brain systems affected by stress, with functional deficits related to affective processing and cognition. There are functional deficits in brain regions mediating healthy vs. unhealthy stress responses, including amygdala reactivity and connectivity, as well as insula and ACC changes. Similar changes were observed in these brain regions for healthy participants with hyper-reactive blood pressure increases in response to stress, as discussed in the acute stress sub-section (Gianaros et al., 2007) as well as those with pre-clinical CVD indicators, indicative of a potential mechanism for how these brain changes formed and became stable over time. These functional deficits occur robustly in brain regions having critical input to physiological control regions, that are simultaneously affected by circulating physiological hormones. This suggests a potential negative feedback loop, in which those who currently have, or are at risk for CVD, have an increased stress response that could put them at even further risk for CVD events over time.

Cortical responses to stress and depression

Depression has high co-morbidity and potentially common underlying mechanisms with dysregulated stress responses and the development of CVD (Musselman et al., 1998). Those with depression can have deficits in behavioral function related to "emotion regulation and processing, cognitive functioning, attention, salience, and sensorimotor processing" (Brakowski et al., 2017). Brain imaging data have corroborated such behavioral differences that are, in turn, linked to poor cardiovascular function and to CVD risk, both *directly* in terms of endocrine function and hormone disease, and *indirectly* in terms of engagement in risky behaviors.

Meta-analytic reviews show common anatomical differences between depressed patients and non-depressed controls, including reduced grey matter volume in the threat/emotion and regulation regions (i.e., amygdala and medial prefrontal cortex, also parahippocampal cortex) and in cognitive region DLPFC (Orange/Yellow regions of Fig. 7.3) (Bora et al., 2012; Lai, 2013). In addition, focused regional differences in small volume structures have been found among depressed patients, including a smaller hippocampus volume (Fig. 7.4A) (Bremner et al., 1996; Sheline et al., 2003) and a smaller amygdala volume (Hamilton et al., 2008). Both structures show volume loss that is less severe or not present in individuals treated for depression (Nezafati et al., 2015).

In terms of baseline brain function, a meta-analysis of 10 metabolic PET studies of depressed individuals using sugar tracer, FDG, revealed a common pattern of lower-than-typical metabolism in affective and reward

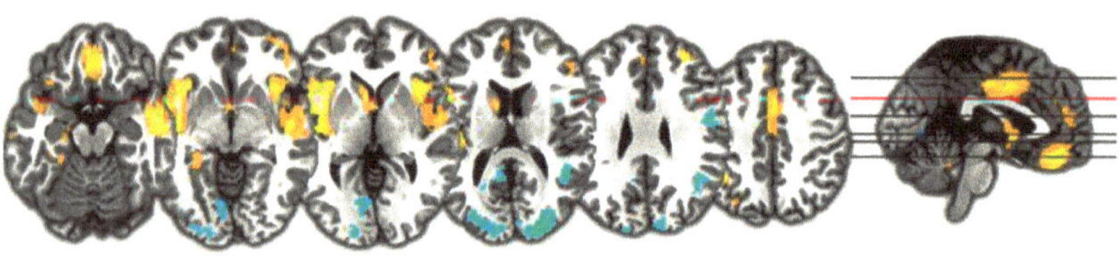

Fig. 7.3 Brain volume difference in depressed patients versus non-depressed controls. Data from a grey matter comparison (VBM) between these groups are overlaid onto a series of axial brain slices starting from an inferior slice to a superior slice, with a sagittal slice at right showing *black gridlines* that indicate the slice placement of the axial slices. Warm colors (*yellow/orange*) show decreased volume in depressed subjects in ventral medial frontal/orbital frontal, insula, cingulate, and basal ganglia regions. Cool colors (*blues*) show increased volume differences for depressed subjects in visual cortex, precuneus and temporal regions. *(Adapted from Wise, T., Radua, J., Via, E., Cardoner, N., Abe, O., Adams, T.M., Amico, F., Cheng, Y., Cole, J.H., Perico, C.D.A.M., Dickstein, D.P., 2017. Common and distinct patterns of grey-matter volume alteration in major depression and bipolar disorder: evidence from voxel-based meta-analysis. Mol. Psychiatry 22(10), 1455.)*

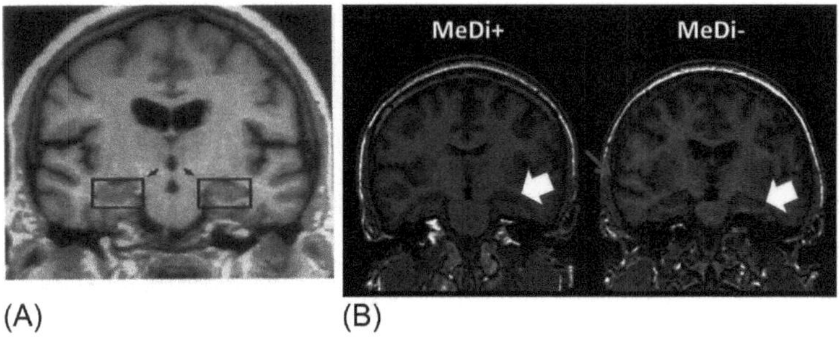

Fig. 7.4 Coronal MRI brain slices showing hippocampal atrophy in depression and with poor versus good diet. (A) Hippocampal atrophy in Major depression. *Black boxes* show hippocampal regions have *black* gaps near the top that are filled with cerebral spinal fluid instead of brain grey matter. (B) White arrows point to absence of age-related atrophy in hippocampus for older adults on a Mediterranean diet (MeDi+) versus atrophy (*black gaps*) in those with non-adherence to the diet (MeDi−). See increased *black* in MeDi− brain (cerebral spinal fluid versus brain) in hippocampus (*white arrows*), and other regions. *(From (A) Sheline, Y.I., Wang, P.W., Gado, M.H., Csernansky, J.G., Vannier, M.W., 1996. Hippocampal atrophy in recurrent major depression. Proc. Natl. Acad. Sci. USA 93(9), 3908–3913. Copyright (1996) National Academy of Sciences U.S.A. From (B) Mosconi, L., Murray, J., Tsui, W.H., Li, Y., Davies, M., Williams, S., Pirraglia, E., Spector, N., Osorio, R.S., Glodzik, L., McHugh, P., de Leon, M.J., 2014. Mediterranean diet and magnetic resonance imaging-assessed brain atrophy in cognitively normal individuals at risk for Alzheimer's disease. J. Prev. Alzheimers Dis. 1, 23–32.)*

regions of the brain including insula, cingulate, basal ganglia, and striatum, along with higher-than-typical metabolism in the thalamus and cerebellum (Su et al., 2018). Resting-state fMRI studies have found altered connectivity between these regions and the default mode network, in which frontal regions become decoupled with reward regions and limbic regions that regulate physiological response, in favor of regions that mediate self-focus/rumination and cognition (Mulders et al., 2015). For example, depressed individuals were observed to have reduced frontal connectivity (ACC) to the striatum (reward/motivation) and insula (affective/physiological feedback loop), which was coupled with increased connectivity of these ACC regions to rumination and cognition regions, including PCC and DLPFC (Philippi et al., 2015; Brakowski et al., 2017). The chronic decrease in metabolism and reduced frontal connectivity to limbic and reward regions (especially basal ganglia and striatal regions) may contribute to the anhedonia that is often reported in depression, with insula being a link to internal body perception changes and regulation (Critchley et al., 2004). However, the decreased size and increased frontal connectivity to PCC may indicate a dysfunction related to hyper self-focus and rumination (Leech and Sharp, 2014).

In a study specifically examining one of the hallmarks of depression, rumination, depressed patients (vs. controls) showed that rumination vs. other types of conceptual thought, resulted in increased activity in amygdala, medial prefrontal regions, ACC, DLPFD, parahippocampus, and PCC (Cooney et al., 2010). In general, those with depression can have exaggerated responses to affective (especially negative) stimuli, as well as a sustained response in the amygdala to negatively-valanced stimuli (Hamilton et al., 2012). In addition, a study showing decreased functional connectivity between amygdala and left prefrontal cortex in response to fearful faces, indicates the potential for executive system and emotional processing dysfunctions (Kong et al., 2013).

In summary, depression-related brain changes potentially explain specific symptoms of depression including anhedonia (reduced reward regions such as basal ganglia and ventral striatum), increased negativity and self-dwelling (increased and sustained amygdala response and decreased PCC size, and altered connectivity/function), decreased memory (smaller hippocampus), and lower cognitive control (decreased activity, size and connectivity of prefrontal regions, plus reduced connectivity of prefrontal cortex with the amygdala). The overlap of these brain abnormalities with stress-induced and CVD-related changes, suggests a potential common mechanism, with depression vs. CVD showing more reliable disfunction of reward/motivation related brain regions. Indeed, it has been shown that dysregulation of the HPA axis, in regions like amygdala, insula, and ACC are affected by and also can contribute to risk for both CVD and depression (Joynt et al., 2003). However, the direct link may be complex, as treatment for depression is linked to overall survival yet so far has not been strongly linked to reducing cardiac events (Zuidersma et al. 2013).

Therapies and treatments: Effect on brain

There is evidence that the cortical changes that occur in response to stress are not permanent (Weaver et al., 2005, 2006; Whittle and Singewald, 2014). It is therefore reasonable to hypothesize that treatments for stress that are known to have positive health effects on CVD and/or mental health outcomes, might also result in adaptive alterations of brain structure and function. Evidence for this hypothesis is considered in the following section.

Physical exercise

A prospective study of CVD patients engaged in a 6 month long aerobic exercise-based rehabilitation program, showed restoration of normalized volume in a subset of brain regions that had shown atrophy relative to controls prior to rehabilitation (Fig. 7.2, right). This included regions in the medial prefrontal, lateral prefrontal, superior frontal, and superior temporal gyri (Anazodo et al., 2013). Furthermore, another study found that in heart failure patients, the frequency of exercise correlated with behavioral and cognitive measures associated with the above regions, along with corresponding brain volume recovery of sub-cortical regions including thalamus and hypothalamus (Alosco and Hayes, 2015). Therefore, exercise has potentially beneficial effects on brain regions that directly regulate the cortical response to stress and downstream responses in the periphery.

Such direct benefits of physical exercise in the brain may be mediated via improved vascular function and structure (Branyan et al., 2018), as well as the release of anti-inflammatory and growth factors such as brain-derived neurotrophic factor into the bloodstream (Anderson-Hanley et al., 2018). For example, a rodent study showed that physical exercise increases norepinephrine output from the highly vascularized brain structure, the locus coeruleus (Brown et al., 1979). The locus coeruleus has strong influences on the stress response, as demonstrated by human imaging studies in which beta-blockers dampen or eliminate cortisol-dependent differences in amygdala reactivity (van Stegeren et al., 2007; Hurlemann et al., 2010). Exercise may also have indirect benefits, as it requires effort and brain involvement of task-related cortical networks incompatible with deleterious negative thought dwelling and rumination, such as the hyperactivity in the PCC (Wong et al., 2007). Indeed, physical exercise has been shown to be beneficial for reducing symptoms not just for CVD, but for depression, ADHD, anxiety, and other psychiatric disorders characterized by negative thought patterns (Penedo and Dahn, 2005; Archer and Kostrzewa, 2012).

Dietary interventions

Another standard lifestyle change for CVD patients is to consume a "heart healthy" diet that is lower in calories, saturated fats, sodium, and sugars and contains more fruits, vegetables, and other healthy foods. However, this can be difficult as simply viewing unhealthy versus healthy foods activates reward, motivation, and memory networks in the limbic system and its reduced connectivity to the prefrontal cortex (Val-Laillet et al., 2015; van Meer et al., 2016). If one already has a sluggish limbic system, as can happen as a result of chronic stress and depression, as reviewed in earlier sections of this chapter, such high reward stimulation may be difficult to resist. Thus, adherence to healthy diets is particularly difficult for this population.

Furthermore, a poor diet may feel therapeutic to these patients, as research in rats has shown that eating hyperpalatable foods (i.e., high sugar, fat, calorie-dense) results in a temporary dampening of HPA activity and the restoration of dopaminergic activity in a reward region, nucleus accumbens, that is typically suppressed during stress (Yau and Potenza, 2013). Unfortunately, this potential for self-medication with food is ultimately deleterious to the body in that it induces inflammation, including neuroinflammation of the memory-related region, hypothalamus (Gao et al., 2017), as well as the down regulation of dopaminergic receptors so that more and more food is necessary for satiation (Alsio et al., 2010).

Despite these brain-based challenges, it is worth finding ways to overcome these obstacles, as consuming a good diet can be beneficial for patients at risk for CVD or who have had a cardiac event (Sacks et al., 2001; Estruch, 2014). Structural MRI studies have shown that eating a Mediterranean-type diet is associated with maintenance of healthy cortical thickness in multiple brain regions including the orbital/medial prefrontal cortex, insula, PCC, and hippocampus (Mosconi et al., 2014). It was also found that the older adults with poor diets (highest deviation from Mediterranean diet) showed atrophy in the hippocampus similar to that seen in depression and Alzheimer's Disease (Fig. 7.4B, compare presence of black regions indicating atrophy in hippocampi marked by arrow in MeDi− with Fig. 7.4A depressed patient hippocampi in black boxes). This atrophy was not observed among those who adhered to the diet (Fig. 7.4B, See full hippocampal volume in MeDi+ subjects indicating no atrophy) (Mosconi et al., 2014). Since this study was not prospective, it is difficult to know whether the brain changes occurred in response to the diet or caused poor adherence to the diet. However, prospective studies with probiotics have shown the importance of dietary factors in promoting brain health, such as, in this case, structural increases in the hippocampus and reward regions like the nucleus accumbens, in addition to functional reductions in amygdala reactivity and improved integrity of prefrontal connections in response to emotional face stimuli (Tillisch et al., 2013, 2017; Bagga et al., 2018).

Dietary effects on the brain and body may be mediated directly by the nutritional improvement supplied to the brain and/or from changes in the microbiome of the gut (Chapter 11). Probiotics and nutrition are likely an important consideration in treatment of both mental health and CVD conditions, including potential restoration of brain structure and function that had been previously altered by stress and/or early life trauma (Cowan et al., 2016; Dinan and Cryan, 2017).

Stress management

An increasingly common intervention shown to improve stress responding and health of CVD patients is to provide training in stress coping strategies, including engagement in mind-body exercises like relaxation and mindfulness-based interventions (van Dixhoorn and White, 2005; Abbott et al., 2014; Goyal et al., 2014). Many complementary and alternative medicine treatments (e.g., progressive muscle relaxation, mindfulness meditation, yoga, tai chi, and prayer) induce what has been termed the relaxation response (Bhasin et al., 2013). Indeed, within just a 20-min period, stress management treatments have been shown to induce a complex array of up and down regulation of healthy versus unhealthy genes, respectively, resulting in a range of cellular activity changes. Examples of downstream affects of these genetic regulation changes include improvements in oxidative stress management, mitochondrial energy efficiency, inflammation control, and insulin regulation, many of which affect brain function and structure (Raison et al., 2006; Bhasin et al., 2013).

One such practice, Mindfulness Based Stress Reduction (MBSR) was developed by Jon Kabat-Zinn to aid in the reduction of stress for cardiac patients (Kabat-Zinn and Hanh, 2009). A review of 22 neuroimaging studies of this mindfulness practice and related variations, termed mindfulness-based inductions, revealed common increases in gray matter in the insula, anterior cingulate, medial/orbital frontal cortices, hippocampus, and the somatosensory cortex, as well as white matter integrity improvements (Fig. 7.5) (Fox et al., 2014). In addition, as with other therapies, mindfulness practice can result in the reduction of amygdala overreactivity (Holzel et al., 2011; Doll et al., 2016), which may result from improved connectivity of amygdala with frontal brain regions that regulate affective response (Holzel et al., 2011; Taren et al., 2015; Doll et al., 2016). In addition, mindfulness studies have that such training can result in reduced PCC overreactivity and increased PCC size, indicating that one potential mechanism for the cardiac benefits of mindfulness, may be reductions in negative thoughts, rumination, or self-dwelling; i.e., a restoration of a PCC signature more similar to the "happiness marker" previously discussed in the acute stress section (Brefczynski-Lewis et al., 2007; Holzel et al., 2011).

Mindfulness may also address CVD treatment indirectly by facilitating adherence to treatments and engaging in healthy behaviors (avoiding risk, adopting health behaviors). Researchers have observed beneficial effects of mindfulness on self-regulatory brain regions (prefrontal cortex, insula), for example in patients using a mindfulness program (Craving to Quit) for smoking cessation (Brewer et al., 2011, 2013), and such results could be

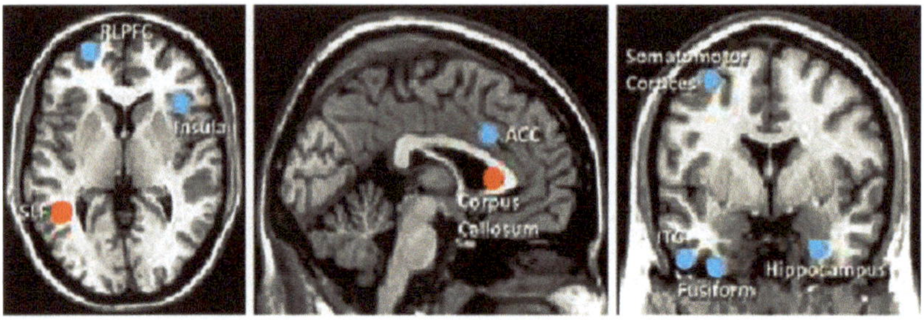

Fig. 7.5 Common grey matter and white matter differences in mindfulness-based interventions. Results of a metaanalysis of VBM studies of mindfulness are displayed on example axial, sagittal and coronal brain slices. *Blue* regions show grey matter volume increases in hippocampus, anterior cingulate, insula, lateral prefrontal cortex and sensory regions. *Red* regions show *white* matter integrity differences in the anterior corpus callosum and lateral fascicularus. *(Adapted from Fox, K.C., Nijeboer, S., Dixon, M.L., Floman, J.L., Ellamil, M., Rumak, S.P., Sedlmeier, P., Christoff, K., 2014. Is meditation associated with altered brain structure? A systematic review and meta-analysis of morphometric neuroimaging in meditation practitioners. Neurosci. Biobehav. Rev. 43, 48–73.)*

extrapolated to other habit-related challenges or addictions. Mindfulness, including the inherently mindful movement practices such as yoga, has been shown in multiple studies to increase functional efficacy and grey matter size in emotion regulation regions (prefrontal cortex) and the attention network (frontal-parietal cortex, thalamus, basal ganglia) (Brefczynski-Lewis et al., 2007; Jha et al., 2007). Changes related to metacognitive awareness of both attention and emotion (Teasdale, 1999) may be beneficial towards quitting deleterious health habits and negative thought habits, as well as subsequent adoption of healthy thought and behavioral habits.

Behavior therapy

The last therapy to be discussed, cognitive behavior therapy, is one often recommended for stress burnout, PTSD, CVD, and depression (Beck, 2006). A prospective FDG PET study of cognitive behavior therapy for depression showed restoration of more normalized metabolic brain activity post-therapy, such as increases in the hippocampus and dorsal mid-cingulate, ACC, ventral and dorsal lateral prefrontal cortex, parietal regions including PCC, and temporal regions (Goldapple et al., 2004). A meta-analysis of 17 neuroimaging prospective studies using various methods (e.g., MRI, PET, SPECT, and Magnetic Resonance Spectroscopy (MRS) showed increased anterior cingulate activity among depressed patients following cognitive behavior therapy, which is thought to be related to reinstating stable emotional regulation (Sankar et al., 2018).

In addition, for those with a history of either PTSD or adverse childhood events, therapy has the potential of improving cognitive control by strengthening the integrity of prefrontal cortex connections, reducing PCC-induced rumination, and potentially reversing some of the epigenetic methylation changes that have altered brain function due to such stressors, allowing these CVD-involved brain regions to function better (Weaver et al., 2005, 2006; Whittle and Singewald, 2014). As mentioned previously, both deleterious stress exposure and healing may be mediated by changes in genetic expression. Studies in humans have shown that for adults with PTSD, combining behavioral therapy with pharmaceuticals that can help facilitate genetic regulation changes, called histone deacetylase (HDAC) inhibitors, can lead to greater improvements in outcome than therapy alone (Whittle and Singewald, 2014). This treatment avenue, as well as determining mechanisms behind both adaptive and maladaptive brain changes, may yield important neuroimaging findings in the near future, as a novel PET neuroimaging tracer for HDAC activity has been developed that could help elucidate how epigenetic brain changes occur (Wang et al., 2014).

Based on these findings, it seems prudent that CVD patients be evaluated for behavioral factors, early life or adult trauma exposure, and depression. Strategies for overcoming the challenges of getting CVD patients to accept therapy need to be devised and tested, such as emphasizing that CVD is a brain as well as body disease, and that therapy may be a critical part of recovery. Engaging in exercise, consuming a balanced low-fat diet, learning how to better manage stress and engaging in therapy are all methods that can assist the brain in making physical and mental recovery easier or be used in prevention.

Summary

In examining the influence of stress on the brain through neuroimaging studies, commonalities in disturbances in brain structure and function become apparent across studies of acute and chronic stress exposure, PTSD and early life stress exposure, depression, and current and pre-clinical CVD. These changes are summarized in Table 7.1. Dysregulation of the functions of these brain networks could lead to compounding risk factors for CVD, especially exaggerated stress response via untempered amygdala, cingulate, and insula regional reactivity which leads to top down dysregulation of the HPA axis. This physiological dysregulation results in the release of cortisol and inflammatory factors, which in turn affect brain structure and function in a feedback cycle (Sah et al., 2003; Urry et al., 2006; Dedovic et al., 2009).

Table 7.1 Summary of reviewed brain system relationships with stress/CVD/depression dysfunction and benefit

Brain system	Regions involved	Dysfunction	Benefit
Emotion and threat monitoring/ regulation system	**Amygdala:** emotion, threat, directs HPA influence **Medial prefrontal:** emotion regulation, directs HPA influence **Connectivity between regions:** emotion regulation	**Amygdala**: hyper-reactive in all stress, CVD, negative stimuli/ smaller in depression **Medial prefrontal:** Smaller in depression, CVD Hyperactive in depression, low perceived control, decreased baseline in PTSD **Abnormal connectivity:** premature inverse in neglected children, positive in stress, CVD	**Amygdala:** less reactive with diet, mindfulness, beta-blockers **Medial prefrontal:** increased size after exercise rehabilitation, diet, mindfulness **Connectivity:** normalized with diet, mindfulness
Error-monitoring, self-monitoring, internal regulation system	**ACC:** cognitive and affective error monitoring and behavioral influence—direct input HPA axis **Insula:** body monitoring, emotion mediation, rumination, direct input HPA axis **PCC:** self-monitoring, episodic memory, rumination, potential signature of happiness	**ACC:** increased response to negative stimuli PTSD, stress, smaller in CVD **Insula:** smaller in depression, CVD, decreased metabolism in depression **PCC:** smaller in depression, CVD, decreased activity to positive versus negative stimuli if low sense of purpose, increased activity in depressed rumination	**ACC:** restored metabolic with therapy **Insula:** increased size with Mediterranean diet, mindfulness **PCC:** increased metabolism with therapy, increased size with Mediterranean diet, happiness factors

Cognitive, memory and attention-related systems	**Hippocampus/ parahippocampus:** memory **Attention network:** including DLPFC, parietal, temporal attention, cognitive functions, **temporal** -sensory integration **Thalamus:** attention, sensory filtering	**Hippocampus:** smaller in depression, CVD **Attention network:** reduced activity and connectivity to affective regions in chronic stress, reduced size in CVD **Thalamus:** decreased size in CVD, increased metabolism in depression	**Hippocampus:** size increases with exercise, diet, mindfulness, increased metabolism with therapy **Attention network:** increased size (frontal only) for exercise, mindfulness, therapy **Thalamus:** increased activation in mindfulness
Limbic and reward system	Basal ganglia, striatal regions and nucleus accumbens	Smaller, decreased frontal connectivity and decreased metabolism in depression, lower response to social pictures for high loneliness, Increased response unhealthy versus healthy foods	Mindfulness increases activity Can be temporarily restored with junk food

In addition, stress, CVD, and depression may lead to an increased likelihood of risk behaviors and thoughts that can further perpetuate these disorders, as one may be more likely to ruminate over life stressors, (ACC, insula, PCC), less able to halt negative self-referential thought patterns (PCC), and less likely to remember to engage in healthy habits or recognize triggers for falling into poor habits (hippocampus, attention regions). Lastly, an underactive reward system may result in difficulties related to motivation for healthy habits and increased likelihood in engaging in unhealthy habits to induce a temporary feeling of relief, as seen in the dietary section of this chapter (e.g., temporary restoration of dopamine activity in nucleus accumbens). With these brain-based complications in mind, it is clear that there are multiple challenges to consider when implementing changes for either prevention or treatment of CVD.

Neuroimaging use in informing CVD/depression treatment and risk abatement

If these brain-based challenges are overcome, there is considerable promise that most of these deficits can be reversed or at least ameliorated with treatment. Unfortunately, cardiac patients often are resistant to acknowledging or adhering to treatment recommendations for mental health issues, as such patients may view themselves as having a problem with their body, not their brain (Mensah and Collins, 2015). They are, however, likely to acknowledge the high stress in their lives, which may present an avenue for introducing therapeutic changes via stress management or mindfulness, which can at least start the healing process via positive brain changes in decreasing stress reactivity and strengthened ability to adhere to better habits. In addition, showing patients brain images that show positive brain changes with therapies could be useful in communicating the importance of addressing brain health, and inspire confidence in treatment avenues.

Brain-based research reviewed in earlier sections of this chapter, indicate that those at risk for CVD and depression often have less responsive reward systems and deficits in structure and function in brain regions related to memory and attention. Therefore, it is imperative that necessary treatment plans and lifestyle changes are rewarding and that a healthy lifestyle is made to be an easy choice, with factors that are optimized based on results of complementary or concurrent neuroimaging studies (Puska et al., 2016). In terms of individual treatment plans, beginning with exercise and easy dietary changes like probiotics, might present as a more acceptable early rehabilitation strategy for many patients. More comprehensive and challenging changes could be implemented over time as brain regions like prefrontal cortex and limbic regions heal, but future research is needed to verify such approaches. In addition, therapeutic approaches should take into consideration obstacles related to loneliness (for example in considering group versus home-based exercise therapies (Dalal et al., 2010), as well as underlying complex issues such as a lack of perceived purpose or control over one's life.

Conclusion

Overall, the pervasiveness, deadliness, and expense of CVD, and the effects of CVD risk factors and treatments on the brain, point towards the need for neuroscience-informed, comprehensive social or community-supported programs that support engagement in health behaviors, and amelioration of stress-risk conditions. More challenging issues are worth addressing, such as finding ways to increase the control or mastery people have over their lives, as well as their sense of purpose and social connection, as these are important factors influencing brain-based changes in CVD and depression. It is imperative that the public, policy makers, and medical professionals be made aware that CVD presents as a brain disease, as well as a body disease, so that it is understood that treatment plans should reflect strategies to overcome brain-based challenges to healthy lifestyle adoption. It is also imperative to show the positive side: that these brain changes originally occurred to help adapt to stress, and that even stable, deleterious brain-based challenges related to CVD and depression are changeable, as demonstrated via scientifically-vetted treatment plans. Responsible use of neuroimaging results may be an avenue towards this education, due the relatability of brain-based data images, even to non-scientists (Brefczynski-Lewis, 2011).

References

Abbott, R.A., Whear, R., Rodgers, L.R., Bethel, A., Thompson Coon, J., Kuyken, W., Stein, K., Dickens, C., 2014. Effectiveness of mindfulness-based stress reduction and mindfulness based cognitive therapy in vascular disease: a systematic review and meta-analysis of randomised controlled trials. J. Psychosom. Res. 76, 341–351.

Alosco, M.L., Hayes, S.M., 2015. Structural brain alterations in heart failure: a review of the literature and implications for risk of Alzheimer's disease. Heart Fail. Rev. 20, 561–571.

Alsio, J., Olszewski, P.K., Norback, A.H., Gunnarsson, Z.E., Levine, A.S., Pickering, C., Schioth, H.B., 2010. Dopamine D1 receptor gene expression decreases in the nucleus accumbens upon long-term exposure to palatable food and differs depending on diet-induced obesity phenotype in rats. Neuroscience 171, 779–787.

Anazodo, U.C., Shoemaker, J.K., Suskin, N., Lawrence, K.S., 2013. An investigation of changes in regional gray matter volume in cardiovascular disease patients, pre and post cardiovascular rehabilitation. NeuroImage Clin. 3, 388–395.

Anda, R.F., Felitti, V.J., Bremner, J.D., Walker, J.D., Whitfield, C., Perry, B.D., Dube, S.R., Giles, W.H., 2006. The enduring effects of abuse and related adverse experiences in childhood. A convergence of evidence from neurobiology and epidemiology. Eur. Arch. Psychiatry Clin. Neurosci. 256, 174–186.

Anderson-Hanley, C., Barcelos, N.M., Zimmerman, E.A., Gillen, R.W., Dunnam, M., Cohen, B.D., Yerokhin, V., Miller, K.E., Hayes, D.J., Arciero, P.J., Maloney, M., Kramer, A.F., 2018. The Aerobic and Cognitive Exercise Study (ACES) for community-dwelling older adults with or at-risk for mild cognitive impairment (MCI): neuropsychological, neurobiological and neuroimaging outcomes of a randomized clinical trial. Front. Aging Neurosci. 10, 76.

Archer, T., Kostrzewa, R.M., 2012. Physical exercise alleviates ADHD symptoms: regional deficits and development trajectory. Neurotox. Res. 21, 195–209.

Bagga, D., Reichert, J.L., Koschutnig, K., Aigner, C.S., Holzer, P., Koskinen, K., Moissl-Eichinger, C., Schopf, V., 2018. Probiotics drive gut microbiome triggering emotional brain signatures. Gut Microbes 9, 486–496.

Bandettini, P.A., 2012. Twenty years of functional MRI: the science and the stories. NeuroImage 62, 575–588.

Batten, S.V., Aslan, M., Maciejewski, P.K., Mazure, C.M., 2004. Childhood maltreatment as a risk factor for adult cardiovascular disease and depression. J. Clin. Psychiatry 65, 249–254.

Beck JSC, J., 2006. Cognitive Therapy. American Psychological Association.

Bhasin, M.K., Dusek, J.A., Chang, B.H., Joseph, M.G., Denninger, J.W., Fricchione, G.L., Benson, H., Libermann, T.A., 2013. Relaxation response induces temporal transcriptome changes in energy metabolism, insulin secretion and inflammatory pathways. PLoS One 8, e62817.

Bora, E., Fornito, A., Pantelis, C., Yucel, M., 2012. Gray matter abnormalities in major depressive disorder: a meta-analysis of voxel based morphometry studies. J. Affect. Disord. 138, 9–18.

Brakowski, J., Spinelli, S., Dorig, N., Bosch, O.G., Manoliu, A., Holtforth, M.G., Seifritz, E., 2017. Resting state brain network function in major depression—depression symptomatology, antidepressant treatment effects, future research. J. Psychiatr. Res. 92, 147–159.

Branyan, K.W., Devallance, E.R., Lemaster, K.A., Skinner, R.C., Bryner, R.W., Olfert, I.M., Kelley, E.E., Frisbee, J.C., Chantler, P.D., 2018. Role of chronic stress and exercise on microvascular function in metabolic syndrome. Med. Sci. Sports Exerc. 50, 957–966.

Brefczynski-Lewis, J.A., 2011. Neuroimaging: directions and potentials for communication research. Commun. Res. Rep. 28, 196–204.

Brefczynski-Lewis, J.A., Lutz, A., Schaefer, H.S., Levinson, D.B., Davidson, R.J., 2007. Neural correlates of attentional expertise in long-term meditation practitioners. Proc. Natl. Acad. Sci. U. S. A. 104, 11483–11488.

Bremner, J.D., Krystal, J.H., Southwick, S.M., Charney, D.S., 1996. Noradrenergic mechanisms in stress and anxiety: II. Clinical studies. Synapse 23, 39–51.

Brewer, J.A., Mallik, S., Babuscio, T.A., Nich, C., Johnson, H.E., Deleone, C.M., Minnix-Cotton, C.A., Byrne, S.A., Kober, H., Weinstein, A.J., Carroll, K.M., Rounsaville, B.J., 2011. Mindfulness training for smoking cessation: results from a randomized controlled trial. Drug Alcohol Depend. 119, 72–80.

Brewer, J.A., Elwafi, H.M., Davis, J.H., 2013. Craving to quit: psychological models and neurobiological mechanisms of mindfulness training as treatment for addictions. Psychol. Addict. Behav. 27, 366–379.

Brown, B.S., Payne, T., Kim, C., Moore, G., Krebs, P., Martin, W., 1979. Chronic response of rat brain norepinephrine and serotonin levels to endurance training. J. Appl. Physiol. Respir. Environ. Exerc. Physiol. 46, 19–23.

Bush, G., Luu, P., Posner, M.I., 2000. Cognitive and emotional influences in anterior cingulate cortex. Trends Cogn. Sci. 4, 215–222.

Bzdok, D., Laird, A.R., Zilles, K., Fox, P.T., Eickhoff, S.B., 2013. An investigation of the structural, connectional, and functional subspecialization in the human amygdala. Hum. Brain Mapp. 34, 3247–3266.

Cacioppo, J.T., Norris, C.J., Decety, J., Monteleone, G., Nusbaum, H., 2009. In the eye of the beholder: individual differences in perceived social isolation predict regional brain activation to social stimuli. J. Cogn. Neurosci. 21, 83–92.

Cannon, W.B., 1957. Voodoo death. Psychosom. Med. 19, 182–190.

Capistrant, B.D., Moon, J.R., Berkman, L.F., Glymour, M.M., 2012. Current and long-term spousal caregiving and onset of cardiovascular disease. J. Epidemiol. Community Health 66, 951–956.

Carter, C.S., Braver, T.S., Barch, D.M., Botvinick, M.M., Noll, D., Cohen, J.D., 1998. Anterior cingulate cortex, error detection, and the online monitoring of performance. Science 280, 747–749.

Chapman, D.P., Whitfield, C.L., Felitti, V.J., Dube, S.R., Edwards, V.J., Anda, R.F., 2004. Adverse childhood experiences and the risk of depressive disorders in adulthood. J. Affect. Disord. 82, 217–225.

Cohen, R., Bavishi, C., Rozanski, A., 2016. Purpose in life and its relationship to all-cause mortality and cardiovascular events: a meta-analysis. Psychosom. Med. 78, 122–133.

Cooney, R.E., Joormann, J., Eugene, F., Dennis, E.L., Gotlib, I.H., 2010. Neural correlates of rumination in depression. Cogn. Affect. Behav. Neurosci. 10, 470–478.

Cowan, C.S., Callaghan, B.L., Kan, J.M., Richardson, R., 2016. The lasting impact of early-life adversity on individuals and their descendants: potential mechanisms and hope for intervention. Genes Brain Behav. 15, 155–168.

Cozolino, L., 2017. The Neuroscience of Psychotherapy: Healing the Social Brain. WW Norton & Company.

Critchley, H.D., Wiens, S., Rotshtein, P., Ohman, A., Dolan, R.J., 2004. Neural systems supporting interoceptive awareness. Nat. Neurosci. 7, 189–195.

Dalal, H.M., Zawada, A., Jolly, K., Moxham, T., Taylor, R.S., 2010. Home based versus centre based cardiac rehabilitation: cochrane systematic review and meta-analysis. BMJ 340, b5631.

Davidson, R.J., 2002. Anxiety and affective style: role of prefrontal cortex and amygdala. Biol. Psychiatry 51 (1), 68–80.

Dedovic, K., Duchesne, A., Andrews, J., Engert, V., Pruessner, J.C., 2009. The brain and the stress axis: the neural correlates of cortisol regulation in response to stress. NeuroImage 47, 864–871.

Demerouti, E., Bakker, A.B., Nachreiner, F., Schaufeli, W.B., 2001. The job demands-resources model of burnout. J. Appl. Psychol. 86, 499–512.

Dinan, T.G., Cryan, J.F., 2017. The microbiome-gut-brain axis in health and disease. Gastroenterol. Clin. N. Am. 46, 77–89.

Doll, A., Holzel, B.K., Mulej Bratec, S., Boucard, C.C., Xie, X., Wohlschlager, A.M., Sorg, C., 2016. Mindful attention to breath regulates emotions via increased amygdala-prefrontal cortex connectivity. NeuroImage 134, 305–313.

Edmondson, D., von Kanel, R., 2017. Post-traumatic stress disorder and cardiovascular disease. Lancet Psychiatry 4, 320–329.

Engel, G.L., 1971. Sudden and rapid death during psychological stress. Folklore or folk wisdom? Ann. Intern. Med. 74, 771–782.

Estruch, R., 2014. Cardiovascular mortality: how can it be prevented? Nefrologia 34, 561–569.

Felitti, V.J., Anda, R.F., Nordenberg, D., Williamson, D.F., Spitz, A.M., Edwards, V., Koss, M.P., Marks, J.S., 1998. Relationship of childhood abuse and household dysfunction to many of the leading causes of death in adults. The Adverse Childhood Experiences (ACE) Study. Am. J. Prev. Med. 14, 245–258.

Fox, K.C., Nijeboer, S., Dixon, M.L., Floman, J.L., Ellamil, M., Rumak, S.P., Sedlmeier, P., Christoff, K., 2014. Is meditation associated with altered brain structure? A systematic review and meta-analysis of morphometric neuroimaging in meditation practitioners. Neurosci. Biobehav. Rev. 43, 48–73.

Friedmann, E., Thomas, S.A., Liu, F., Morton, P.G., Chapa, D., Gottlieb, S.S., Sudden Cardiac Death in Heart Failure Trial I, 2006. Relationship of depression, anxiety, and social isolation to chronic heart failure outpatient mortality. Am. Heart J. 152. 940 e941-948.

Frodl, T., O'Keane, V., 2016. Integrating the stress systems and neuroimaging in depression. In: Systems Neuroscience and Depression. Academic Press, pp. 269–308.

Gao, Y., Bielohuby, M., Fleming, T., Grabner, G.F., Foppen, E., Bernhard, W., Guzman-Ruiz, M., Layritz, C., Legutko, B., Zinser, E., Garcia-Caceres, C., Buijs, R.M., Woods, S.C., Kalsbeek, A., Seeley, R.J., Nawroth, P.P., Bidlingmaier, M., Tschop, M.H., Yi, C.X., 2017. Dietary sugars, not lipids, drive hypothalamic inflammation. Mol. Metab. 6, 897–908.

Gassen, N.C., Chrousos, G.P., Binder, E.B., Zannas, A.S., 2017. Life stress, glucocorticoid signaling, and the aging epigenome: implications for aging-related diseases. Neurosci. Biobehav. Rev. 74, 356–365.

Gavelin, H.M., Neely, A.S., Andersson, M., Eskilsson, T., Jarvholm, L.S., Boraxbekk, C.J., 2017. Neural activation in stress-related exhaustion: cross-sectional observations and interventional effects. Psychiatry Res. Neuroimaging 269, 17–25.

Gee, D.G., Casey, B.J., 2015. The impact of developmental timing for stress and recovery. Neurobiol. Stress 1, 184–194.

Gee, D.G., Gabard-Durnam, L., Telzer, E.H., Humphreys, K.L., Goff, B., Shapiro, M., Flannery, J., Lumian, D.S., Fareri, D.S., Caldera, C., Tottenham, N., 2014. Maternal buffering of human amygdala-prefrontal circuitry during childhood but not during adolescence. Psychol. Sci. 25, 2067–2078.

Gianaros, P.J., Sheu, L.K., 2009. A review of neuroimaging studies of stressor-evoked blood pressure reactivity: emerging evidence for a brain-body pathway to coronary heart disease risk. NeuroImage 47, 922–936.

Gianaros, P.J., Jennings, J.R., Sheu, L.K., Derbyshire, S.W., Matthews, K.A., 2007. Heightened functional neural activation to psychological stress covaries with exaggerated blood pressure reactivity. Hypertension 49, 134–140.

Gianaros, P.J., Hariri, A.R., Sheu, L.K., Muldoon, M.F., Sutton-Tyrrell, K., Manuck, S.B., 2009. Preclinical atherosclerosis covaries with individual differences in reactivity and functional connectivity of the amygdala. Biol. Psychiatry 65, 943–950.

Glover, G.H., 2011. Overview of functional magnetic resonance imaging. Neurosurg. Clin. N. Am. 22, 133–139. vii.

Goldapple, K., Segal, Z., Garson, C., Lau, M., Bieling, P., Kennedy, S., Mayberg, H., 2004. Modulation of cortical-limbic pathways in major depression: treatment-specific effects of cognitive behavior therapy. Arch. Gen. Psychiatry 61, 34–41.

Goyal, M., Singh, S., Sibinga, E.M., Gould, N.F., Rowland-Seymour, A., Sharma, R., Berger, Z., Sleicher, D., Maron, D.D., Shihab, H.M., Ranasinghe, P.D., Linn, S., Saha, S., Bass, E.B., Haythornthwaite, J.A., 2014. Meditation programs for psychological stress and well-being: a systematic review and meta-analysis. JAMA Intern. Med. 174, 357–368.

Hair, N.L., Hanson, J.L., Wolfe, B.L., Pollak, S.D., 2015. Association of child poverty, brain development, and academic achievement. JAMA Pediatr. 169, 822–829.

Hamilton, J.P., Siemer, M., Gotlib, I.H., 2008. Amygdala volume in major depressive disorder: a meta-analysis of magnetic resonance imaging studies. Mol. Psychiatry 13, 993–1000.

Hamilton, J.P., Etkin, A., Furman, D.J., Lemus, M.G., Johnson, R.F., Gotlib, I.H., 2012. Functional neuroimaging of major depressive disorder: a meta-analysis and new integration of base line activation and neural response data. Am. J. Psychiatry 169, 693–703.

Hardee, J.E., Thompson, J.C., Puce, A., 2008. The left amygdala knows fear: laterality in the amygdala response to fearful eyes. Soc. Cogn. Affect. Neurosci. 3, 47–54.

Harlow, L.L., Newcomb, M.D., Bentler, P.M., 1986. Depression, self-derogation, substance use, and suicide ideation: lack of purpose in life as a mediational factor. J. Clin. Psychol. 42, 5–21.
Hayes, J.P., Vanelzakker, M.B., Shin, L.M., 2012. Emotion and cognition interactions in PTSD: a review of neurocognitive and neuroimaging studies. Front. Integr. Neurosci. 6, 89.
Holt-Lunstad, J., Smith, T.B., Layton, J.B., 2010. Social relationships and mortality risk: a meta-analytic review. PLoS Med. 7, e1000316.
Holzel, B.K., Lazar, S.W., Gard, T., Schuman-Olivier, Z., Vago, D.R., Ott, U., 2011. How does mindfulness meditation work? Proposing mechanisms of action from a conceptual and neural perspective. Perspect. Psychol. Sci. 6, 537–559.
Hurlemann, R., Walter, H., Rehme, A.K., Kukolja, J., Santoro, S.C., Schmidt, C., Schnell, K., Musshoff, F., Keysers, C., Maier, W., Kendrick, K.M., Onur, O.A., 2010. Human amygdala reactivity is diminished by the beta-noradrenergic antagonist propranolol. Psychol. Med. 40, 1839–1848.
Inagaki, T.K., Muscatell, K.A., Moieni, M., Dutcher, J.M., Jevtic, I., Irwin, M.R., Eisenberger, N.I., 2016. Yearning for connection? Loneliness is associated with increased ventral striatum activity to close others. Soc. Cogn. Affect. Neurosci. 11, 1096–1101.
Infurna, M.R., Reichl, C., Parzer, P., Schimmenti, A., Bifulco, A., Kaess, M., 2016. Associations between depression and specific childhood experiences of abuse and neglect: a meta-analysis. J. Affect. Disord. 190, 47–55.
Jha, A.P., Krompinger, J., Baime, M.J., 2007. Mindfulness training modifies subsystems of attention. Cogn. Affect. Behav. Neurosci. 7, 109–119.
Joynt, K.E., Whellan, D.J., O'Connor, C.M., 2003. Depression and cardiovascular disease: mechanisms of interaction. Biol. Psychiatry 54, 248–261.
Kabat-Zinn, J., Hanh, T.N., 2009. Full Catastrophe Living: Using the Wisdom of Your Body and Mind to Face Stress, Pain, and Illness. Delta.
Kerns, J.G., Cohen, J.D., AW III, M.D., Cho, R.Y., Stenger, V.A., Carter, C.S., 2004. Anterior cingulate conflict monitoring and adjustments in control. Science 303, 1023–1026.
Klein, T.A., Endrass, T., Kathmann, N., Neumann, J., von Cramon, D.Y., Ullsperger, M., 2007. Neural correlates of error awareness. NeuroImage 34, 1774–1781.
Koizumi, M., Ito, H., Kaneko, Y., Motohashi, Y., 2008. Effect of having a sense of purpose in life on the risk of death from cardiovascular diseases. J. Epidemiol. 18, 191–196.
Kong, L., Chen, K., Tang, Y., Wu, F., Driesen, N., Womer, F., Fan, G., Ren, L., Jiang, W., Cao, Y., Blumberg, H.P., Xu, K., Wang, F., 2013. Functional connectivity between the amygdala and prefrontal cortex in medication-naive individuals with major depressive disorder. J. Psychiatry Neurosci. 38, 417–422.
Kraynak, T.E., Marsland, A.L., Gianaros, P.J., 2018. Neural mechanisms linking emotion with cardiovascular disease. Curr. Cardiol. Rep. 20, 128.
Kurian, A.K., Cardarelli, K.M., 2007. Racial and ethnic differences in cardiovascular disease risk factors: a systematic review. Ethn. Dis. 17, 143–152.
Lai, C.H., 2013. Gray matter volume in major depressive disorder: a meta-analysis of voxel-based morphometry studies. Psychiatry Res. 211, 37–46.
Leech, R., Sharp, D.J., 2014. The role of the posterior cingulate cortex in cognition and disease. Brain 137, 12–32.
Lindsay, S., 2009. The influence of childhood poverty on the self-management of heart disease in later life. In: Social Sources of Disparities in Health and Health Care and Linkages to Policy, Population Concerns and Providers of Care. Emerald Group Publishing Limited, pp. 191–196.
Liston, C., McEwen, B.S., Casey, B.J., 2009. Psychosocial stress reversibly disrupts prefrontal processing and attentional control. Proc. Natl. Acad. Sci. U. S. A. 106, 912–917.
McGowan, P.O., Sasaki, A., D'Alessio, A.C., Dymov, S., Labonte, B., Szyf, M., Turecki, G., Meaney, M.J., 2009. Epigenetic regulation of the glucocorticoid receptor in human brain associates with childhood abuse. Nat. Neurosci. 12, 342–348.

Mensah, G.A., Collins, P.Y., 2015. Understanding mental health for the prevention and control of cardiovascular diseases. Glob. Heart 10, 221–224.
Morris, J.S., Ohman, A., Dolan, R.J., 1999. A subcortical pathway to the right amygdala mediating "unseen" fear. Proc. Natl. Acad. Sci. U. S. A. 96, 1680–1685.
Mosconi, L., Murray, J., Tsui, W.H., Li, Y., Davies, M., Williams, S., Pirraglia, E., Spector, N., Osorio, R.S., Glodzik, L., McHugh, P., de Leon, M.J., 2014. Mediterranean diet and magnetic resonance imaging-assessed brain atrophy in cognitively normal individuals at risk for Alzheimer's disease. J. Prev. Alzheimers Dis. 1, 23–32.
Mulders, P.C., van Eijndhoven, P.F., Schene, A.H., Beckmann, C.F., Tendolkar, I., 2015. Resting-state functional connectivity in major depressive disorder: a review. Neurosci. Biobehav. Rev. 56, 330–344.
Musselman, D.L., Evans, D.L., Nemeroff, C.B., 1998. The relationship of depression to cardiovascular disease: epidemiology, biology, and treatment. Arch. Gen. Psychiatry 55, 580–592.
Nejad, A.B., Fossati, P., Lemogne, C., 2013. Self-referential processing, rumination, and cortical midline structures in major depression. Front. Hum. Neurosci. 7, 666.
New, A.S., Fan, J., Murrough, J.W., Liu, X., Liebman, R.E., Guise, K.G., Tang, C.Y., Charney, D.S., 2009. A functional magnetic resonance imaging study of deliberate emotion regulation in resilience and posttraumatic stress disorder. Biol. Psychiatry 66, 656–664.
Nezafati, M.H., Vojdanparast, M., Nezafati, P., 2015. Antidepressants and cardiovascular adverse events: a narrative review. ARYA Atheroscler. 11, 295–304.
Penedo, F.J., Dahn, J.R., 2005. Exercise and well-being: a review of mental and physical health benefits associated with physical activity. Curr. Opin. Psychiatry 18, 189–193.
Penninx, B.W., van Tilburg, T., Kriegsman, D.M., Deeg, D.J., Boeke, A.J., van Eijk, J.T., 1997. Effects of social support and personal coping resources on mortality in older age: the Longitudinal Aging Study Amsterdam. Am. J. Epidemiol. 146, 510–519.
Philippi, C.L., Motzkin, J.C., Pujara, M.S., Koenigs, M., 2015. Subclinical depression severity is associated with distinct patterns of functional connectivity for subregions of anterior cingulate cortex. J. Psychiatr. Res. 71, 103–111.
Puska, P., Vartiainen, E., Nissinen, A., Laatikainen, T., Jousilahti, P., 2016. Background, principles, implementation, and general experiences of the North Karelia Project. Glob. Heart 11, 173–178.
Raichle, M.E., MacLeod, A.M., Snyder, A.Z., Powers, W.J., Gusnard, D.A., Shulman, G.L., 2001. A default mode of brain function. Proc. Nat. Acad. Sci. U. S. A. 98, 676–682.
Raison, C.L., Capuron, L., Miller, A.H., 2006. Cytokines sing the blues: inflammation and the pathogenesis of depression. Trends Immunol. 27, 24–31.
Richter, C.P., 1957. On the phenomenon of sudden death in animals and man. Psychosom. Med. 19, 191–198.
Reidel, P., Jacob, M.J., Muller, D.K., Vetter, N.C., Smolka, M.N., Marxen, M., 2016. Amygdala fMRI signal as a predictor of reaction time. Front. Hum. Neurosci. 10, 516.
Risch, N., Herrell, R., Lehner, T., Liang, K.Y., Eaves, L., Hoh, J., Griem, A., Kovacs, M., Ott, J., Merikangas, K.R., 2009. Interaction between the serotonin transporter gene (5-HTTLPR), stressful life events, and risk of depression: a meta-analysis. JAMA 301, 2462–2471.
Rischka, L., Gryglewski, G., Pfaff, S., Vanicek, T., Hienert, M., Klobl, M., Hartenbach, M., Haug, A., Wadsak, W., Mitterhauser, M., Hacker, M., Kasper, S., Lanzenberger, R., Hahn, A., 2018. Reduced task durations in functional PET imaging with [(18)F]FDG approaching that of functional MRI. NeuroImage 181, 323–330.
Sacks, F.M., Svetkey, L.P., Vollmer, W.M., Appel, L.J., Bray, G.A., Harsha, D., Obarzanek, E., Conlin, P.R., Miller III, E.R., Simons-Morton, D.G., Karanja, N., Lin, P.H., Group DASCR, 2001. Effects on blood pressure of reduced dietary sodium and the dietary approaches to stop hypertension (DASH) diet. DASH-Sodium Collaborative Research Group. N. Engl. J. Med. 344, 3–10.

Sah, P., Faber, E.S., Lopez De Armentia, M., Power, J., 2003. The amygdaloid complex: anatomy and physiology. Physiol. Rev. 83, 803–834.
Salomons, T.V., Johnstone, T., Backonja, M.M., Davidson, R.J., 2004. Perceived controllability modulates the neural response to pain. J. Neurosci. 24, 7199–7203.
Sankar, A., Melin, A., Lorenzetti, V., Horton, P., Costafreda, S.G., Fu, C.H.Y., 2018. A systematic review and meta-analysis of the neural correlates of psychological therapies in major depression. Psychiatry Res. Neuroimaging 279, 31–39.
Sato, W., Kochiyama, T., Uono, S., Kubota, Y., Sawada, R., Yoshimura, S., Toichi, M., 2015. The structural neural substrate of subjective happiness. Sci. Rep. 5, 16891.
Savic, I., Perski, A., Osika, W., 2017. MRI shows that exhaustion syndrome due to chronic occupational stress is associated with partially reversible cerebral changes. Cereb. Cortex 28, 894–906.
Schaefer, S.M., Morozink Boylan, J., van Reekum, C.M., Lapate, R.C., Norris, C.J., Ryff, C.D., Davidson, R.J., 2013. Purpose in life predicts better emotional recovery from negative stimuli. PLoS One 8. e80329.
Shah, A., et al., 2019. Brain correlates of stress-induced peripheral vasoconstriction in patients with cardiovascular disease. Psychophysiology 56. e13291.
Sheline, Y.I., Gado, M.H., Kraemer, H.C., 2003. Untreated depression and hippocampal volume loss. Am. J. Psychiatry 160, 1516–1518.
Smith, D.F., Jakobsen, S., 2013. Molecular neurobiology of depression: PET findings on the elusive correlation with symptom severity. Front. Psychiatry 4, 8.
Straube, T., Schmidt, S., Weiss, T., Mentzel, H.J., Miltner, W.H., 2009. Dynamic activation of the anterior cingulate cortex during anticipatory anxiety. NeuroImage 44, 975–981.
Su, H., Zuo, C.T., Zhang, H.W., Jiao, F.Y., Zhang, B., Tang, W.J., Geng, D.Y., Guan, Y.H., Shi, S.X., 2018. Regional cerebral metabolism alterations affect resting-state functional connectivity in major depressive disorder. Quant. Imag. Med. Surg. 8, 910–924.
Taren, A.A., Gianaros, P.J., Greco, C.M., Lindsay, E.K., Fairgrieve, A., Brown, K.W., Rosen, R.K., Ferris, J.L., Julson, E., Marsland, A.L., Bursley, J.K., Ramsburg, J., Creswell, J.D., 2015. Mindfulness meditation training alters stress-related amygdala resting state functional connectivity: a randomized controlled trial. Soc. Cogn. Affect. Neurosci. 10, 1758–1768.
Teasdale, J.D., 1999. Metacognition, mindfulness and the modification of mood disorders. Clin. Psychol. Psychother. 6, 142–155.
Tei, S., Becker, C., Kawada, R., Fujino, J., Jankowski, K.F., Sugihara, G., Murai, T., Takahashi, H., 2014. Can we predict burnout severity from empathy-related brain activity? Transl. Psychiatry 4, e393.
Tillisch, K., Labus, J., Kilpatrick, L., Jiang, Z., Stains, J., Ebrat, B., Guyonnet, D., Legrain-Raspaud, S., Trotin, B., Naliboff, B., Mayer, E.A., 2013. Consumption of fermented milk product with probiotic modulates brain activity. Gastroenterology 144, 1394–U1136.
Tillisch, K., Mayer, E.A., Gupta, A., Gill, Z., Brazeilles, R., Le Neve, B., Vlieg, J.E.T.V., Guyonnet, D., Derrien, M., Labus, J.S., 2017. Brain structure and response to emotional stimuli as related to gut microbial profiles in healthy women. Psychosom. Med. 79, 905–913.
Urry, H.L., van Reekum, C.M., Johnstone, T., Kalin, N.H., Thurow, M.E., Schaefer, H.S., Jackson, C.A., Frye, C.J., Greischar, L.L., Alexander, A.L., Davidson, R.J., 2006. Amygdala and ventromedial prefrontal cortex are inversely coupled during regulation of negative affect and predict the diurnal pattern of cortisol secretion among older adults. J. Neurosci. 26, 4415–4425.
Val-Laillet, D., Aarts, E., Weber, B., Ferrari, M., Quaresima, V., Stoeckel, L.E., Alonso-Alonso, M., Audette, M., Malbert, C.H., Stice, E., 2015. Neuroimaging and neuromodulation approaches to study eating behavior and prevent and treat eating disorders and obesity. Neuroimage Clin. 8, 1–31.

van Dixhoorn, J., White, A., 2005. Relaxation therapy for rehabilitation and prevention in ischaemic heart disease: a systematic review and meta-analysis. Eur. J. Cardiovasc. Prev. Rehabil. 12, 193–202.
van Harmelen, A.L., van Tol, M.J., van der Wee, N.J., Veltman, D.J., Aleman, A., Spinhoven, P., van Buchem, M.A., Zitman, F.G., Penninx, B.W., Elzinga, B.M., 2010. Reduced medial prefrontal cortex volume in adults reporting childhood emotional maltreatment. Biol. Psychiatry 68, 832–838.
van Meer, F., van der Laan, L.N., Charbonnier, L., Viergever, M.A., Adan, R.A., Smeets, P.A., Consortium IF, 2016. Developmental differences in the brain response to unhealthy food cues: an fMRI study of children and adults. Am. J. Clin. Nutr. 104, 1515–1522.
van Stegeren, A.H., Wolf, O.T., Everaerd, W., Scheltens, P., Barkhof, F., Rombouts, S.A., 2007. Endogenous cortisol level interacts with noradrenergic activation in the human amygdala. Neurobiol. Learn. Mem. 87, 57–66.
Villien, M., Wey, H.Y., Mandeville, J.B., Catana, C., Polimeni, J.R., Sander, C.Y., Zurcher, N.R., Chonde, D.B., Fowler, J.S., Rosen, B.R., Hooker, J.M., 2014. Dynamic functional imaging of brain glucose utilization using fPET-FDG. NeuroImage 100, 192–199.
Violanti, J.M., Burchfiel, C.M., Miller, D.B., Andrew, M.E., Dorn, J., Wactawski-Wende, J., Beighley, C.M., Pierino, K., Joseph, P.N., Vena, J.E., Sharp, D.S., Trevisan, M., 2006. The buffalo cardio-metabolic occupational police stress (BCOPS) pilot study: methods and participant characteristics. Ann. Epidemiol. 16, 148–156.
Violanti, J.M., Burchfiel, C.M., Fekedulegn, D., Andrew, M.E., Dorn, J., Hartley, T.A., Charles, L.E., Miller, D.B., 2009. Cortisol patterns and brachial artery reactivity in a high stress environment. Psychiatry Res. 169, 75–81.
Wager, T.D., Waugh, C.E., Lindquist, M., Noll, D.C., Fredrickson, B.L., Taylor, S.F., 2009a. Brain mediators of cardiovascular responses to social threat: part I: reciprocal dorsal and ventral sub-regions of the medial prefrontal cortex and heart-rate reactivity. NeuroImage 47, 821–835.
Wager, T.D., van Ast, V.A., Hughes, B.L., Davidson, M.L., Lindquist, M.A., Ochsner, K.N., 2009b. Brain mediators of cardiovascular responses to social threat, part II: Prefrontal-subcortical pathways and relationship with anxiety. NeuroImage 47, 836–851.
Wang, C., Schroeder, F.A., Hooker, J.M., 2014. Visualizing epigenetics: current advances and advantages in HDAC PET imaging techniques. Neuroscience 264, 186–197.
Weaver, I.C., Cervoni, N., Champagne, F.A., D'Alessio, A.C., Sharma, S., Seckl, J.R., Dymov, S., Szyf, M., Meaney, M.J., 2004. Epigenetic programming by maternal behavior. Nat. Neurosci. 7, 847–854.
Weaver, I.C., Champagne, F.A., Brown, S.E., Dymov, S., Sharma, S., Meaney, M.J., Szyf, M., 2005. Reversal of maternal programming of stress responses in adult offspring through methyl supplementation: altering epigenetic marking later in life. J. Neurosci. 25, 11045–11054.
Weaver, I.C., Meaney, M.J., Szyf, M., 2006. Maternal care effects on the hippocampal transcriptome and anxiety-mediated behaviors in the offspring that are reversible in adulthood. Proc. Natl. Acad. Sci. U. S. A. 103, 3480–3485.
Whittle, N., Singewald, N., 2014. HDAC inhibitors as cognitive enhancers in fear, anxiety and trauma therapy: where do we stand? Biochem. Soc. Trans. 42, 569–581.
Whitwell, J.L., 2009. Voxel-based morphometry: an automated technique for assessing structural changes in the brain. J. Neurosci. 29, 9661–9664.
Wong, S.W., Masse, N., Kimmerly, D.S., Menon, R.S., Shoemaker, J.K., 2007. Ventral medial prefrontal cortex and cardiovagal control in conscious humans. NeuroImage 35, 698–708.
Yan, R., Li, W., Yin, L., Wang, Y., Bo, J., Investigators, P.U.-C., 2017. Cardiovascular diseases and risk-factor burden in urban and rural communities in high-, middle-, and low-income regions of China: a Large Community-Based Epidemiological Study. J. Am. Heart Assoc. 6, .

Yau, Y.H., Potenza, M.N., 2013. Stress and eating behaviors. Minerva Endocrinol. 38, 255–267.
Zuidersma, M., Conradi, H.J., van Melle, J.P., Ormel, J., de Jonge, P., 2013. Depression treatment after myocardial infarction and long-term risk of subsequent cardiovascular events and mortality: a randomized controlled trial. J. Psychosom. Res. 74, 25–30.

Further reading

Buettner, D., 2012. The Blue Zones: 9 Lessons for Living Longer From People Who've Lived the Longest. National Geographic Books.
Chrysohoou, C., Pitsavos, C., Lazaros, G., Skoumas, J., Tousoulis, D., Stefanadis, C., Ikaria Study, I., 2016. Determinants of all-cause mortality and incidence of cardiovascular disease (2009 to 2013) in older adults: the Ikaria Study of the Blue Zones. Angiology 67, 541–548.
Sheline, Y.I., Wang, P.W., Gado, M.H., Csernansky, J.G., Vannier, M.W., 1996. Hippocampal atrophy in recurrent major depression. Proc. Natl. Acad. Sci. USA 93 (9), 3908–3913.
Shonkoff, J.P., Boyce, W.T., McEwen, B.S., 2009. Neuroscience, molecular biology, and the childhood roots of health disparities: building a new framework for health promotion and disease prevention. JAMA 301, 2252–2259.
Wise, T., Radua, J., Via, E., Cardoner, N., Abe, O., Adams, T.M., Amico, F., Cheng, Y., Cole, J.H., Perico, C.D.A.M., Dickstein, D.P., 2017. Common and distinct patterns of grey-matter volume alteration in major depression and bipolar disorder: evidence from voxel-based meta-analysis. Mol. Psychiatry 22 (10), 1455.

CHAPTER 8

The role of inflammation and oxidative stress in depression and cardiovascular disease

Susan K. Wood
Pharmacology, Physiology and Neuroscience, University of South Carolina School of Medicine, Columbia, SC, United States

Contents

Introduction

The immune system regulates physical, psychiatric and cardiovascular health, and when immune dysfunction ensues, so does cardiovascular disease (CVD) and psychiatric disorders. The immune system is comprised of two arms; the innate and the adaptive immune response, both of which

Cardiovascular Implications of Stress and Depression
https://doi.org/10.1016/B978-0-12-815015-3.00008-8

can be impacted by stress (Elenkov and Chrousos, 1999; Haapakoski et al., 2016; Medina-Rodriguez et al., 2018). Characterized by an acute and rapid activation as a first-line response to a stressful challenge, the innate immune response in the brain is regulated by activation of antigen presenting cells of myeloid lineage such as macrophage/microglia, dendritic cells, and neutrophils. If this first line of response is not sufficient, the adaptive immune response is activated. In contrast to the innate system, adaptive immunity is a progressive response that relies on antigen-specific T cells and B cells (Den Haan et al., 2014). Both immune responses result in the release of pro- and anti-inflammatory cytokines and chemokines and there is significant communication between innate and adaptive immune system regulators (reviewed in: Haapakoski et al., 2016). This chapter will highlight evidence of a role for innate and adaptive immune dysregulation following chronic exposure to stress as a conduit through which stress exposure may initiate the pathogenesis of comorbidity between psychiatric disorders such as depression, and CVD.

Stress as a catalyst for inflammation

While stress was not traditionally recognized as an immune system activator, more recently it has become well characterized that stress exposure, especially stressors of a social nature, stimulates cytokine release. For example, in rodents social stress increases the number of macrophages and stimulates the expression of cytokine messenger RNA (mRNA) and the release of proinflammatory cytokines in the plasma (Wohleb et al., 2012, 2014b; Powell et al., 2013). In fact, preclinical studies that take individual differences in stress susceptibility into account have revealed that immune activation in the plasma in response to stress is more pronounced in susceptible rats and mice as compared with the resilient phenotype (Hodes et al., 2014; Stewart et al., 2015; Wood et al., 2015). Stress in humans also parallels these findings as modeled by the Trier social stress test; subjects reporting enhanced feelings of anxiety and depression are correlated with increases in stress-induced inflammation in plasma (Carroll et al., 2011; Christian et al., 2013). In addition to changes in circulating cytokines following stress, increases in neuroinflammation have also been demonstrated in stress susceptible rats. As will be discussed in the next section, central cytokine expression is poised to serve as a critical link between stress and psychiatric disorders and it is important to understand stress sensitive pathways involving both the innate and adaptive immune systems that are capable of increasing inflammation in the brain. Two of the most likely stress sensitive mechanisms of increased neuroinflammation include disruption of the blood brain barrier (BBB) and the activation and sensitization (i.e., priming) of microglia (Fig. 8.1).

Stress-induced breakdown of the BBB as a source of neuroinflammation

Stress is consistently reported to increase circulating proinflammatory cytokines, and recent work has demonstrated that cytokines circulating in the periphery could impact the brain. The BBB is a meshwork of specialized endothelial cells along blood vessels that surround the brain and regulates the movement of cytokines and chemokines (along with other substances) between the peripheral circulation and the brain. In a healthy system with an intact BBB, cytokines are considered too large and hydrophilic to passively diffuse across the BBB (Banks, 2005). However, a few families of cytokines (i.e., Interleukin-1, TNF, and Interleukin-6) exhibit saturable transport mechanisms to efficiently pass from the blood to the brain (Banks et al., 1989, 1991). Moreover, a feed-forward mechanism has also been reported by which pro-inflammatory cytokines can also disrupt the integrity of the BBB (Muramatsu et al., 2012). In line with these findings, stress exposure that is known to promote proinflammatory cytokine release is also recognized to disrupt the BBB. For example, male rats that are susceptible to social defeat stress selectively demonstrated increased permeability of the BBB in the ventral hippocampus while the resilient subset of rats did not (Pearson-Leary et al., 2017). Moreover, administration of vascular endothelial growth factor-164, a proinflammatory cytokine, increased permeability of the BBB and induced vulnerability in socially stressed rats (Pearson-Leary et al., 2017). Stress-induced disruption of the BBB has also been reported in a mouse model of social defeat; susceptible subsets of male mice demonstrated stress-induced suppression of claudin-5, an endothelial cell-specific tight junction protein, in the hippocampus and nucleus accumbens compared with the resilient subset of mice and controls. Furthermore, exaggerated permeability of the BBB was also confirmed in the susceptible subset of mice (Menard et al., 2017). Beyond vascular endothelial growth factor-164 and claudin-5, several other stress-sensitive molecular cascades are poised to disrupt the BBB in the face of stress. For example the danger, damage, and disease signal-high mobility group box-1 (HMGB-1) is a chromatin protein that is upregulated by social defeat stress (Finnell et al., 2017b) and neutralizing HMGB-1 using a monoclonal antibody protects against ischemia-induced BBB disruption in rats

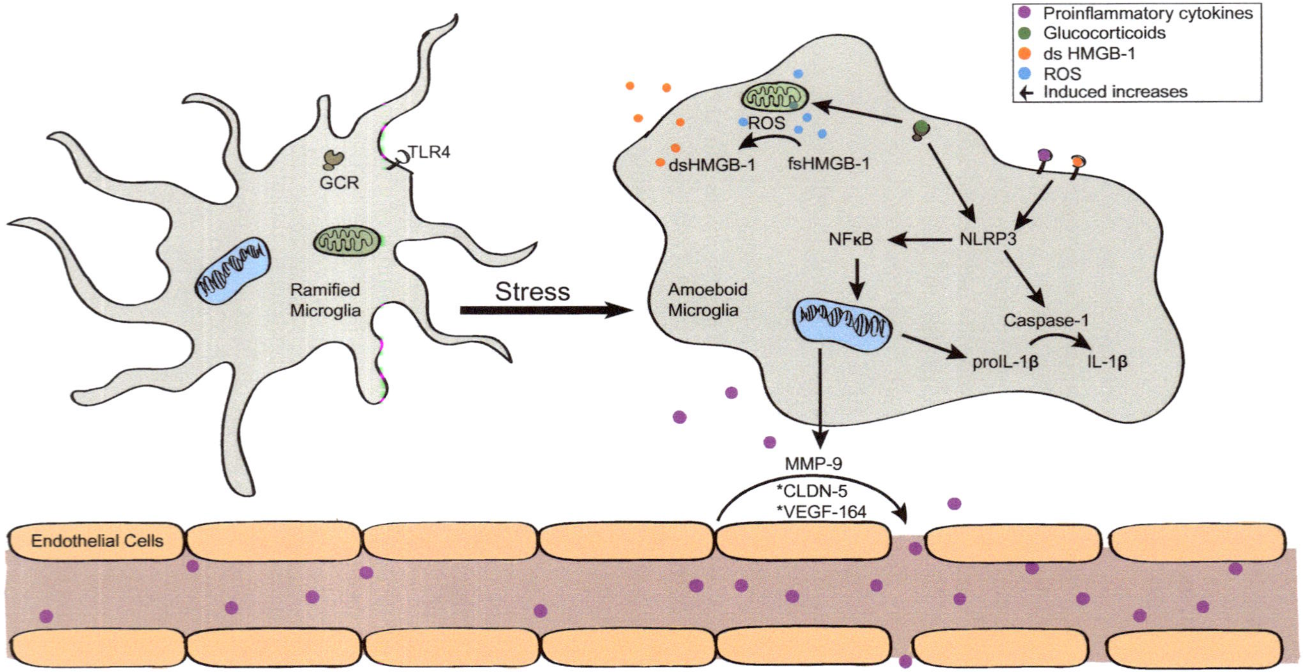

Fig. 8.1 ***Schematic highlighting key sources of stress-induced neuroinflammation.*** Stress exposure promotes a shift in microglial morphology from a highly ramified "resting" or "surveillant" state to an ameboid M1 proinflammatory state. One mechanism stimulating the release of cytokines from microglia is activation of microglial glucocorticoid receptors (GCRs). Moreover, GCR activation also results in priming of inflammatory responses. The process of inflammatory priming can occur directly through activation of the NLRP3 inflammasome. Inflammatory priming can also occur indirectly by reactive oxygen species (ROS) released from mitochondria within

(Continued)

(Zhang et al., 2011). In humans, administration of an anti-HMGB1 monoclonal antibody improves the integrity of the BBB in patients with Alzheimer's disease (Festoff et al., 2016). Furthermore, inflammatory factors, including HMGB-1 (Qiu et al., 2010), also stimulate leukocytes and microglia to release the zymogen matrix metallopeptidase-9 (MMP-9), thereby contributing to endothelial damage (Crocker et al., 2006) and BBB leakage (Seo et al., 2013). Taken together, stress initiates the induction of several proteins capable of disrupting the BBB, a likely mechanism that serves to enhance the flow of inflammatory factors from the circulation into the brain, thus exacerbating neuroinflammation.

Stress-induced activation and sensitization of microglia as a source of neuroinflammation

Microglia are highly adaptive innate immune cells that are capable of transitioning between pro-inflammatory (M1) and anti-inflammatory (M2) states. Under non-stress resting conditions, the morphology of microglia are highly ramified and associated with the M2 state of surveillance and maintenance of the neural cell microenvironment (Nimmerjahn et al., 2005; Kettenmann et al., 2011). Under stressful conditions, a greater number of microglia take on an M1 amoeboid morphology associated with a reactive inflammatory state (Gemma and Bachstetter, 2013; Brites and Fernandes, 2015; Tang et al., 2018). Moreover, activation of glucocorticoid receptors (Ros-Bernal et al., 2011; Liu et al., 2016) on the microglial cell

---

Fig. 8.1, cont'd microglia or within cells that then passively diffuse across the cell membrane or entering through plasma membrane aquaporin channels (not shown). ROS accumulation results in the oxidation of high mobility group box -1 (HMGB-l). Once released, HMGB-1 and proinflammatory cytokines such as interleukin (IL)-1β can act on toll like receptor 4 (TLR 4) on the surface of microglia to further stimulate the NLRP3 signaling cascade.
Another significant source of stress-induced neuroinflammation is the breakdown of the blood brain barrier (BBB). Prior to stress exposure, endothelial cells tightly adhere to one another, blocking the flow of circulating cytokines to the brain. However, following stress exposure, tight junctions between these endothelial cells break down permitting peripheral cytokines and inflammatory cells to pass into the brain. This process is facilitated by plasma vascular endothelial growth factor (VEFG)-164, endothelial claudin-5 (CLDN-5) and microglia released matrix metalloproteinase-9 (MMP-9). *(Adapted from Finnell, J.E., Wood, S.K., 2018. Putative inflammatory sensitive mechanisms underlying risk or resilience to social stress. Front. Behav. Neurosci. 12, 240. http://dx.doi.org/10.3389/fnbeh.2018.00240.)*

surface can initiate the shift in morphology towards the proinflammatory state. This suggests that stress-induced release of corticosterone (in rodents) and cortisol (in humans) could promote this shift to an M1 state, resulting in the release of a number of different effectors including cytokines and chemokines (Brites and Fernandes, 2015). Microglia are thus a conduit through which stress can propagate inflammatory signals throughout the brain (Fruhbeis et al., 2013).

In addition to stimulating the release of cytokines from microglia (Nair and Bonneau, 2006; Kreisel et al., 2014), stress is also capable of sensitizing microglia. Microglia that are sensitized respond to a subsequent stress or immune challenge with faster and more robust cytokine and chemokine release (Frank et al., 2012, 2018; Fonken et al., 2016). This stress-induced neuroinflammatory sensitization is termed microglial priming and is regulated in part through HMGB-1. For example, rats exposed to inescapable foot shock demonstrate increased HMGB-1 and increased gene expression of Interleukin-1 β (IL-1β), nuclear factor kappa (NFκ)B, and tumor necrosis factor (TNF)-α (Frank et al., 2018). Further investigation into the mechanism by which HMGB-1 increases the expression of proinflammatory cytokines revealed that this occurs through activation of the nucleotide-binding oligomerization domain-like receptor (NLRP3) inflammasome (Weber et al., 2015). First discovered in the early 2000s, the inflammasome is a protein complex that regulates the maturation and secretion of IL-1β (Martinon et al., 2002; Duewell et al., 2010). HMGB-1 activation of the NLRP3 inflammasome therefore potentiates proinflammatory signaling by enhancing the cleavage of proIL-1β to IL-1β via activation of caspase-1 (Yan et al., 2012). It is also important to note that the proinflammatory capacity of HMGB-1 is dependent on the redox state of the protein. In its fully reduced state, HMGB-1 lacks the ability to promote proinflammatory signaling. Alternatively, the oxidized state of HMGB-1 potentiates proinflammatory signaling (Yang et al., 2012). Evidence of HMGB-1 regulating microglial priming in the context of stress has been reported using various forms of stress exposure including inescapable foot shock (Yang et al., 2012; Weber et al., 2015), single prolonged stress (Lai et al., 2018), and chronic unpredictable stress (Franklin et al., 2018).

Pharmacological modulators of HMGB-1 induced inflammation have also been identified using cell-based screening in macrophages. Beyond the predicted efficacy of anti-inflammatory compounds, HMGB-1 induced inflammation was inhibited by glucocorticoids and adrenergic agonists (Gero et al., 2013). *In vivo* pharmacological studies have determined that norepinephrine (NE) can regulate immune function by acting on α- and β-adrenergic receptors (Johnson et al., 2005). Importantly, the major endogenous source of NE to the entire neuraxis is located within the locus coeruleus (LC) (Samuels and Szabadi, 2008) and studies support a crucial role for NE in neuroinflammatory processes (Johnson et al., 2005; Jardanhazi-Kurutz et al., 2011; Gyoneva and Traynelis, 2013; Bharani et al., 2017). For example, Jardanhazi-Kurutz et al. (2011) identified that lesioned LC-NE projections induced microgliosis in cortical and hippocampal structures, thus enhancing the inflammatory potential within these brain regions. More recently, a study used a social defeat stressor that is capable of increasing HMGB-1 (Finnell et al., 2017b), and identified a major role for endogenous NE on the inflammatory priming response in plasma and the brain (Finnell et al., 2019). While a history of repeated social defeat exposure resulted in neuroinflammatory priming in the central amygdala (a major projection region of the LC) as compared with rats with a history of control, rats treated with *N*-(2-chloroethyl)-*N*-ethyl-2-bromobenzylamine (DSP-4, a known LC selective noradrenergic neurotoxin) demonstrated a striking enhancement of stress-induced priming (Fig. 8.2A–C, adapted from (Finnell et al., 2019). These data indicate that the LC-NE system serves as a brake on neuroinflammatory priming, and when the LC-NE system is dampened using DSP-4, neuroinflammatory priming is exacerbated. It should also be noted that this brain region is also positioned to impact peripheral NE expression through descending projections to sympathetic spinal nerves (Bruinstroop et al., 2012) and studies support a role for LC-NE release in stress-related neuroinflammatory priming as discussed in "Inflammation in the pathogenesis of CVD" section of this chapter.

Stress-related dysfunction in the adaptive immune system

In addition to activating microglia, the resident cells of the innate immune system within the brain, stress is also capable of modifying T helper (Th) lymphocyte cells, components of the acquired (adaptive) immunity. It is important to note that two T cell subtypes exist and have opposing effects on inflammation; helper T cell type 1 (Th1) regulate release of proinflammatory cytokines while type 2 (Th2) regulates anti-inflammatory responses thereby counteracting the effects of Th1 (Chen, 2007). Following repeated restraint stress exposure, rats exhibit decreases in the total number

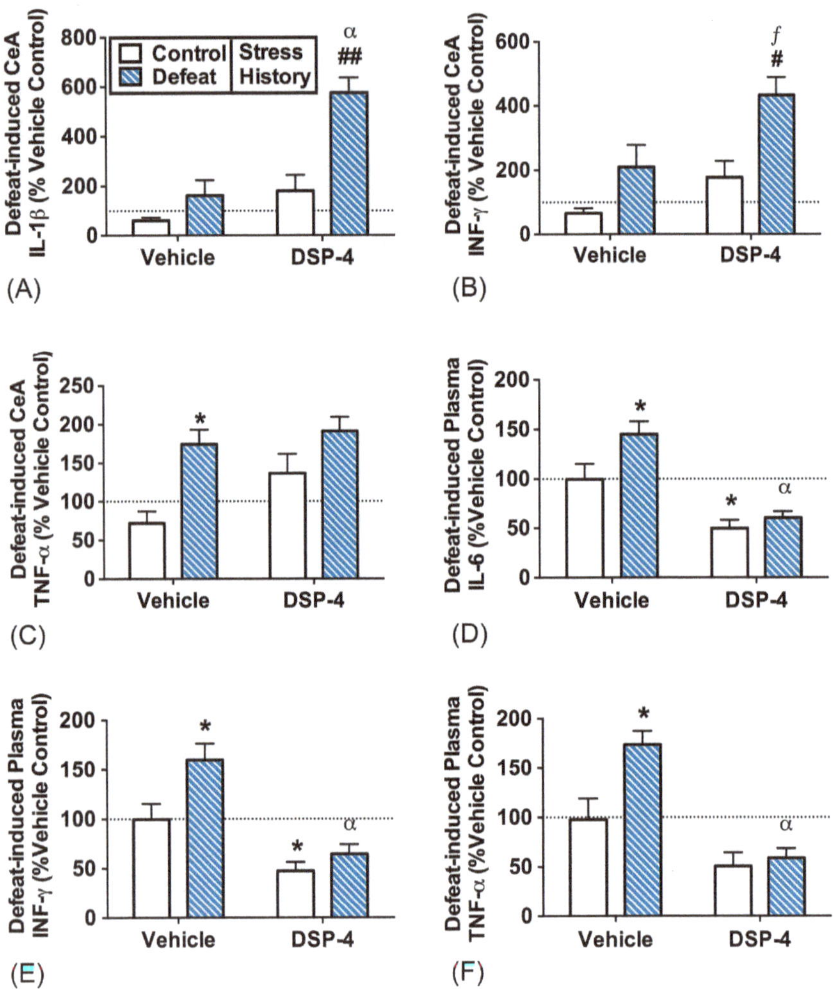

Fig. 8.2 ***Evidence of LC-NE regulation of cytokines in the brain and periphery.*** *Top panel:* Following the acute social defeat challenge, vehicle-treated rats with a history of social defeat exhibited moderate but consistent increases in pro-inflammatory cytokines (A) IL-1β, (B) INF-γ, and (C) TNF-α compared with vehicle treated rats with a history of control. Treatment with DSP-4 produced modest increases in stress-induced cytokines in rats with a history of control, while the LC-NE lesion robustly exacerbated the neuroinflammatory priming evident in rats with a history of social defeat exposure for all cytokines except TNF-α. *$p < 0.05$ vs. vehicle treated rats with a history of control; $^{f}p = 0.07$ vs. vehicle treated rats with a history of defeat; $^{\#}p < 0.05$, $^{\#\#}p < 0.01$, vs. DSP-4 treated rats with a history of control. Dotted horizontal line indicates 100% of vehicle control. *Bottom panel:* DSP-4 treatment suppresses circulating cytokine release in response to an acute social defeat challenge regardless of stress history. Following the acute social defeat challenge, vehicle treated animals with a history of social defeat exhibited

(Continued)

of suppressor/cytotoxic CD8 T cells in the blood and spleen (Batuman et al., 1990). Moreover, Batuman et al. (1990) identified that mitogenic activation of these T cells by concanavalin-A and phytohemagglutinin was diminished as evidenced by decreased IL-2 release following stimulation. Following repeated exposure to social defeat stress, rats exhibited an enhanced number of CD4+ splenocytes expressing pro-inflammatory cytokines following concanavalin-A treatment, but only in the susceptible subset of rats that developed depressive-like behaviors (Finnell et al., 2017b). Mature T cells from the periphery can also enter the brain via various routes including passage through leaky regions of the blood brain barrier such as the choroid plexus (Ransohoff et al., 2003). Following predator stress, T cells infiltrate the brain, especially in the region of the choroid plexus and are required as a defense mechanism against stress (Lewitus et al., 2008). Therefore, under stress conditions where the integrity of the BBB is impaired, T and B cell infiltration is likely increased. Taken together, the adaptive immune system is also critical for stress responses and modifications within this system could regulate stress-induced neuroinflammation.

Stress as a catalyst for reactive oxygen species (ROS) accumulation: A role for mitochondrial dysfunction

Stress exposure is commonly reported to elicit oxidative stress, a mechanism that is associated with many diseases such as atherosclerosis, CVD and depression. Oxidative stress occurs as a result of an imbalance between the production of free radicals, such as superoxide and the antioxidant neutralization of these free radicals. Neurons exhibit high rates of oxygen consumption and as a result, produce large amounts of ROS (Halliwell, 1992). While ROS play a role in several critical neuronal functions such as neuronal plasticity and learning and memory (Massaad and Klann, 2011), overproduction and accumulation of ROS can be detrimental. However, ROS levels are tightly regulated by an antioxidant system and when imbalance

Fig. 8.2, cont'd enhanced (D) IL-6, (E) INF-γ, and (F) TNF-α, compared with vehicle treated controls consistent with stress-induced inflammatory priming. Strikingly, treatment with DSP-4 suppressed stress-induced facilitation of IL-6, TNF-α, and INF-γ regardless of stress history. *$p < 0.05$ vs. vehicle control; $^{\alpha}p < 0.05$ vs. vehicle defeat. *(Adapted from Finnell, J.E., Moffitt, C.M., Hesser, L.A., Harrington, E., Melson, M.N., Wood, C.S., Wood, S.K., 2019. The contribution of the locus coeruleus-norepinephrine system in the emergence of defeat-induced inflammatory priming. Brain Behav. Immun. http://dx.doi.org/10.1016/j.bbi.2019.01.021.)*

occurs, a deleterious buildup of ROS ensues, as has been reported to occur in stress-sensitive brain regions following repeated stress exposure (De Oliveira et al., 2007; Salim et al., 2010, 2011; Lindqvist et al., 2017).

As the energy powerhouse of the cell, mitochondria represent the largest source of ROS production, above that of nitric oxide synthase and monoamine oxidase. As a result, there is a direct relationship between mitochondrial dysfunction and increased oxidative stress. It is important to note that repeated stress exposure has dramatic effects on mitochondria, and may represent the source of stress-induced ROS accumulation. For example, chronic immobilization stress and chronic mild stress inhibit the activity of the mitochondrial respiratory chain complexes within the rat brain cortex (Madrigal et al., 2001; Rezin et al., 2008) and was shown to reduce hippocampal Na^+, K^+-ATPase activity (Gamaro et al., 2003). Moreover, chronic mild stress reduced mitochondrial respiration rates within the mouse hippocampus, cortex, and hypothalamus (Gong et al., 2011). Taken together, these studies support the role of stress in decreased mitochondrial function and resulting ROS accumulation.

One mechanism through which stress could impact oxidative stress is by stimulating the release of proinflammatory cytokines, as ROS are dose dependently increased by treatment with either TNF-α or IL-6 (Rochfort et al., 2014). This may occur as a result of proinflammatory cytokine's ability to impair mitochondrial function. For example, mitochondrial dysfunction occurs in the presence of physiologically relevant levels of TNF-α; low (post-stroke) levels of TNF-α rapidly decrease mitochondrial function as indicated by decreased mitochondrial membrane potential and increased caspase 8 activity (Doll et al., 2015). This effect signaled selectively through TNF-receptor 1 (TNF-R1), highlighting the role that proinflammatory cytokines may play in mitochondrial dysfunction. The relationship between inflammation and mitochondrial function is also bidirectional such that decreasing the activity of mitochondria within microglia amplifies the NLRP3 inflammasome and IL-1β release (Sarkar et al., 2017). Taken together these studies demonstrate a striking relationship between neuroinflammation and mitochondrial dysfunction.

Inflammation in the pathophysiology of depression

While once thought only to represent the body's response to pathogens and injury, inflammation is now known to underlie a host of disease states including psychiatric disorders. In 1927 Julius Wagner-Jauregg won the Nobel

Prize for his reports that stimulating the immune system affected psychiatric functioning. Nearly 60 years later, the link between inflammatory cytokines and depression was first reported (Maes et al., 1990). More recently studies demonstrate that specific subpopulations of depressed patients exhibit enhanced circulating inflammation evident within plasma (Irwin and Miller, 2007) and increased inflammatory markers in cerebrospinal fluid (Sasayama et al., 2013; Devorak et al., 2015). In fact, there is considerable evidence to suggest that the level of neuroinflammation may be correlated to severity of depressive symptoms (Sasayama et al., 2013; Kern et al., 2014; Devorak et al., 2015; Setiawan et al., 2015). Importantly, the vast body of literature that is emerging on the interaction between inflammation and depression has also revealed that elevated cytokines can produce depressive symptomatology. For example, clinical findings have demonstrated that chronic administration of cytokines for use as chemotherapeutics induced depression in a large number of patients (Denicoff et al., 1987; Renault et al., 1987). Moreover, it should be noted that inflammatory diseases such as CVD (Anda et al., 1993; Riba et al., 2011; Huffman et al., 2013), irritable bowel disease, allergic rhinitis, and rheumatoid arthritis (Cuffel et al., 1999; Stauder and Kovacs, 2003; Katon et al., 2004; Marrie et al., 2017) increase one's risk of developing psychiatric comorbidities.

Inflammation in the pathogenesis of depression: Correlation or causation?

While it remains equivocal as to whether inflammation causes depression or if it is merely a consequence of a depressive episode, there is in fact considerable evidence to point towards a causative role for inflammation. Most convincing are the findings that compounds that promote inflammation also promote depression. For example, granulocyte macrophage colony stimulating factor (GM-CSF), interferon (IFN)-γ, and IFN-α have been administered as anti-cancer therapy (Capuron et al., 2001; Schmeler et al., 2009). Of the many side effects associated with these treatments is the development of depressive episodes in patients with no history of depression (Capuron et al., 2001; Schmeler et al., 2009). Moreover, epidemiological studies identified that elevated inflammatory markers can be predictive of developing depression (Gimeno et al., 2009; Felger et al., 2013; Eraly et al., 2014). Based on this evidence, researchers propose that inflammation could be causative rather than merely a result of depressive episodes. Furthermore, studies in animal models also support this relationship. For example, stimulation of the immune system through administration of pro-inflammatory cytokines, injection of LPS, or inducing a bacterial infection can produce neuroinflammation (O'connor et al., 2003; Johnson et al., 2004; Ji et al., 2014; Zhang et al., 2014; Biesmans et al., 2015) and depressive-like behavior as evidenced by anhedonia (Kaster et al., 2012; Ji et al., 2014; Zhang et al., 2014) or behavioral despair (O'connor et al., 2003; Ji et al., 2014; Zhang et al., 2014; Zhu et al., 2015). Importantly, cytokines have also been shown to play a causal role in social stress-induced depressive-like behavior; IL-1β inhibition in the brain using IL-1 receptor antagonist prior to social stress prevented the development of anhedonia in susceptible rats (Wood et al., 2015) as did treatment with neuroactive doses of resveratrol (Finnell et al., 2017a) a natural polyphenol with potent anti-inflammatory properties (Falchetti et al., 2001; Donnelly et al., 2004; Fordham et al., 2014) (Fig. 8.2). In sum, both clinical and preclinical studies support the role for proinflammatory cytokines to precipitate the pathogenesis of depression.

Microglial polarization in depression

Of particular importance to better understand the role of neuroinflammation in the pathophysiology of depression is evaluating how the activation state of microglia (the resident cells in the innate immune system) shifts during depression. A recent clinical study showed for the first time that depressed patients exhibit significant increases in translocator protein density, a marker for activated microglia (Setiawan et al., 2015). The involvement of M1 (proinflammatory) type microglia in stress-induced neuroinflammation is supported by studies utilizing the tetracycline analogue minocycline, an antibiotic that is well documented to inhibit M1 polarization of microglia (Kobayashi et al., 2013). For example, minocycline treatment in conjunction with chronic mild stress (Wang et al., 2018) and inescapable foot shock (Blandino et al., 2006) have shown that inhibition of the M1 microglial phenotype, and subsequent decrease in proinflammatory cytokine release, inhibits the development of stress-induced anxiety- and depressive-like responses in rodents. Notably, clinical trials identified that minocycline was capable of producing anti-depressant effects in 90% of study participants (Murrough et al., 2018). Immune dysfunction has also been reported in studies of patients with biopolar depression and several proof-of-concept trials support the use of anti-inflammatory agents as treatment (Rosenblat, 2019). Taken together, these studies provide clear evidence for the involvement of M1 microglia in the emergence of depressive symptomatology.

Adaptive immune system imbalance in depression

Considerable evidence suggests that the adaptive immune system is also dysregulated in depression. For example, lower numbers of regulatory T cells (Treg) and B cells have been identified in patients suffering from major depression (Li et al., 2010; Ahmetspahic et al., 2018). Because Treg cells inhibit T cell proliferation, it is not surprising that Th1/Th2 imbalance also occurs in patients with major depression (Pavon et al., 2006; Li et al., 2010). Similar to changes induced by repeated stress exposure, T cell function is also dysregulated in depressed patients, as their T cells have reduced ability to respond to stimulation (Zorrilla et al., 2001). In strong support of the role of T cell dysfunction in depression, are the findings that antidepressant therapy can restore the Th1/Th2 balance in depressed patients (Martino et al., 2012). T cell dysregulation is also associated with depressive-like behaviors in studies in mice, where depressive-like behavior is related to accumulations of brain Th1 and Th17 cells (Beurel et al., 2013). Moreover, Th17 administration was sufficient to elicit depressive-like behaviors in mice, suggesting a possible causal role for Th17 in a depressive-like behavioral phenotype (Beurel et al., 2013).

BBB permeability in depression

In addition to neuroinflammation occurring as a result of activated microglia, increased BBB permeability may also play a role in depressed patient populations. Quantitative measurement of BBB function in the clinical setting can be conducted using the analysis of cerebrospinal fluid for proteins only found in the serum (i.e., albumin) or alternatively the assessment of brain constituents within the plasma (i.e., S100B) (Marchi et al., 2003). While there is a severe lack of studies evaluating BBB function in depressed cohorts, the association between disturbances in BBB permeability and depression has been recognized for more than two decades. Correlational studies in humans with various subtypes of depression have associated increased BBB permeability with the symptomatology of depressive illness (Niklasson and Agren, 1984). Moreover, CSF-serum albumin ratio was increased in elderly women suffering from depression (Gudmundsson et al., 2007). More recently, in the same study that confirmed that social defeat stress suppressed expression of the tight junction protein claudin-5 in susceptible mice, leading to a leaky BBB, suppressed claudin-5 expression was also identified in post mortem tissue from the nucleus accumbens of depressed patients (Menard et al., 2017). Taken together, both M1 polarized microglia and disruption of the BBB are likely susceptibility mechanisms driving increased neuroinflammation and may contribute to depressive symptomatology in humans.

ROS in the pathogenesis of depression

There is a strong link between oxidative stress and major depressive disorder that is underscored by the involvement of catecholamine metabolism in the production of ROS. Importantly, overproduction of monoamine oxidase's (MAO) ROS byproducts (including hydrogen peroxide) have been shown to elicit mitochondrial dysfunction and neuronal apoptosis (Scapagnini et al., 2012). Strikingly, studies of patients with depression and postpartum depression identified elevated MAO in these subjects compared with non-depressed controls (Sacher et al., 2015). As would be expected, there is also considerable evidence that ROS production is upregulated in depressed patients compared with non-depressed controls (Scapagnini et al., 2012; Vavakova et al., 2015). For example, xanthine oxidase (XO) is an enzyme that catalyzes oxidation of xanthine to produce the free radicals superoxide and hydrogen peroxide. Studies in depressed patients identified elevated XO in serum and in the thalamus and putamen postmortem (Michel et al., 2010; Leonard and Maes, 2012). Several studies have also attempted to determine whether the antioxidant systems that serve to decrease ROS are also impaired in depressed patients (Bilici et al., 2001; Herken et al., 2007; Gawryluk et al., 2011; Maes et al., 2011). The findings have been equivocal, likely due to small sample sizes and variable patient populations, however several studies have reported decreased activity of various antioxidants. For example, glutathione peroxidase (GPX) is one of the main antioxidants that serve to protect the organism from excessive buildup of free radicals by catalyzing hydrogen peroxides. Several studies have identified lower activity of GPX in the brains of depressed patients as compared with controls (Gawryluk et al., 2011; Maes et al., 2011). Decreased activity of GPX results in ROS accumulation and in fact, is negatively correlated with the severity of depressive symptoms. Moreover, superoxide dismutase activity, another major antioxidant, is also significantly lower in erythrocytes of patients with depression and is correlated with symptom severity (Bilici et al., 2001; Herken et al., 2007). Preclinical studies have demonstrated that ROS play a permissive role in the anxiety-like behavior following social defeat in rats (Solanki et al., 2017). Finally, lending evidence to a role for ROS in the pathogenesis of psychiatric disorders in humans, patients with depression not only exhibited elevated markers of the oxidative stress marker F2-isoprostanes and inflammation, but compared to individuals who readily respond to antidepressants, non-responders had higher levels of both inflammation and oxidative stress markers (Strawbridge et al., 2015; Vavakova et al., 2015; Lindqvist et al., 2017).

It is clear from these findings that there is a link between ROS accumulation and depression. While elevated MAO levels and reduced antioxidant activity (as discussed above) may be likely sources of increased ROS byproducts, mitochondrial structure and function have also been studied by a variety of techniques in depressed patients, as mitochondria remain the largest source of ROS production (Cadenas and Davies, 2000; Turrens, 2003). In addition to findings that stress exposure is capable of producing mitochondrial dysfunction (see "Inflammation in the pathophysiology of depression" section), there is mounting evidence that depressed patients exhibit mitochondrial abnormalities. Levels of mitochondrial complex I subunits were significantly reduced in the cerebellum of depressed patients (Ben-Shachar and Karry, 2008). At the functional level, brain glucose metabolism measured by positron emission tomography identified reduced glucose utilization in the brains of depressed patients (Videbech, 2000). Moreover, mitochondrial ATP production, which is required for synaptic ion homeostasis and phosphorylation (Mattson et al., 2008), was also decreased in patients suffering from depression (Gardner et al., 2003). Furthermore, there is preclinical evidence from studies in rodents to suggest that differences in mitochondrial function regulate an anxiety-like phenotype. For example, rats that exhibited high anxiety-like behavioral tendencies also demonstrated reduced expression of mitochondrial complex I and II proteins and decreased respiratory capacity and ATP (Hollis et al., 2015). Interestingly, rats demonstrating a high anxiety-like phenotype also exhibit increased ROS production within the nucleus accumbens (Hollis et al., 2015). Taken together, it is clear that proinflammatory cytokines are capable of shifting the balance of ROS production/elimination from a healthy balance towards maladaptive.

Finally, while it is unclear exactly how mitochondrial damage or ROS accumulation contribute to depressive symptoms, the success of antioxidant therapy as adjuvants to traditional antidepressants supports the role of oxidative stress in the pathophysiology of depression. For example, *N*-Acetyl-cysteine (NAC) can act as a direct ROS scavenger and also acts as a precursor to glutathione, the major antioxidant in the brain (Holdiness, 1991). In rodent studies, NAC readily crosses the blood brain barrier and replenishes brain glutathione following depletion using 2-cyclohexene-1-one, a compound that depletes glutathione in the striatum and frontal cortex (Dean et al., 2011). In preclinical studies, NAC has demonstrated antidepressant-like activity (Ferreira et al., 2008) and clinical trials support this notion. Specifically, NAC was tested as adjuvant therapy for depressive episodes in bipolar patients and open-label data indicate a robust reduction in depression scores (Berk et al., 2011). Natural plant derived polyphenols have also been comprehensively evaluated in preclinical studies for their anti-inflammatory and antioxidant capacity. Curcumin, for example, is highly effective in quenching ROS and NO-based free radicals (Sreejayan and Rao, 1997; Priyadarsini et al., 1998). In animal models, daily administration of curcumin reversed immobility in the forced swim test in an animal model of chronic fatigue syndrome (Gupta et al., 2009) as well as in chronically stressed rats (Bhutani et al., 2009), suggesting an antidepressant-like effect. Similarly, the natural polyphenol resveratrol which has anti-inflammatory and antioxidant properties (Falchetti et al., 2001; Donnelly et al., 2004; De La Lastra and Villegas, 2005; Zhang et al., 2010; Fordham et al., 2014) demonstrates anxiolytic (Patki et al., 2013a) and antidepressant efficacy in various animal models including social defeat (Finnell et al., 2017a), forced swim test (Xu et al., 2010; Ahmed et al., 2014), single prolonged stress (Solanki et al., 2015), and Wistar-Kyoto depressive-like rat model (Hurley et al., 2014). Taken together, both clinical and preclinical studies support the validity of antioxidants as therapy for psychosocial disorders.

Inflammation in the pathogenesis of CVD

Innate and adaptive immunity in CVD

Proinflammatory systems are firmly established in the pathogenesis of CVD. Not only are markers of inflammation elevated in patients with CVD, but increases in acute phase inflammatory markers such as C-reactive protein (CRP) and serum amyloid A, are in fact predictive of CVD risk (Biasucci et al., 1999; Ridker, 2003; Valgimigli et al., 2005). However, utility of these biomarkers is limited due to the nonspecific nature of these measurements (Ruparelia et al., 2017). Importantly, just as the innate and adaptive immune systems exhibit dysfunction following stress ("Stress-related dysfunction in the adaptive immune system" section) and in major depression ("BBB permeability in depression" section), there is also evidence of a role for deregulation of immunity in CVD which is thought to contribute to its pathogenesis.

Preclinical studies have established the role of various aspects of the immune system in CVD progression as targeting features of each of these systems attenuates CVD progression. Mast cells represent critical cell types that regulate processes of both innate and adaptive immunity (Shelburne and Abraham, 2011). Mast cells, granulated cells derived from the myeloid stem cell, represent a major sensory arm of the immune system by recruiting monocytes and macrophages to areas of injury (St John and Abraham, 2013). In healthy cardiac and vascular tissue, mast cells are present in low numbers, however patients with CVD such as atherosclerosis, have increased levels of mast cells and enhanced activity within these mast cells (Shi et al., 2015). Moreover, activation of mast cells initiates the release of cytokines and chemokines (Conti et al., 2017). Considerable data suggests that mast cells can regulate atherosclerosis progression. For example, mast cell degranulation induces atherosclerotic lesion development while mast cell stabilizers inhibit atherosclerotic plaque formation (Tang et al., 2009; Heikkila et al., 2010; Wang et al., 2013).

Monocytes represent another major cell type recruited by mast cells and play an important role in adaptive immunity. Specifically, inhibition of monocyte recruitment also has a beneficial effect on CVD; For example, therapeutic targeting of pro-inflammatory monocytes using siRNA CCR2-targeted nanoparticles allowed for selective silencing of inflammatory monocyte subsets, while sparing the innate immune system's other essential functions (Leuschner et al., 2011). This study determined that degradation of CCR2 in monocytes prevented accumulation of inflammatory monocytes at sites of inflammation such as atherosclerotic plaques. Moreover, Leuschner et al., identified that treatment also reduced infarct size after coronary artery occlusion. In addition, T cells are also responsible for releasing proinflammatory cytokines and activating monocytes and tissue macrophages and are elevated in the blood of patients following an acute myocardial infarction (Zal et al., 2004). Moreover, increased levels of T effector cells positively correlated with risk of future acute coronary events (Liuzzo et al., 2007). Finally, in agreement with findings in depressed individuals, following myocardial infarction there is an imbalance in T-cell subtypes, specifically a reduction in Treg expression (Mor et al., 2006). Treg cells play a critically important role in modulating inflammation, and therefore dysregulation of Treg levels or function is related to cardiovascular pathology including atherosclerosis, hypertension, stroke, myocardial infarction and heart failure (Meng et al., 2016). Moreover, similar to that observed in depressed patients, Th-1 and Th-17 effectors play a direct role in hypertension (Agita and Alsagaff, 2017).

Identical to the innate immune mechanism of the NLRP3 inflammasome in the brain discussed in "Stress-induced activation and sensitization of microglia as a source of neuroinflammation" section, cardiac NLRP3 also enhances cleavage of pro-IL-1β to active IL-1β, stimulating release of several other proinflammatory cytokines (Keller et al., 2008). Importantly, the NLRP3 pathway is an important mediator of post-myocardial ischaemia injury (Mezzaroma et al., 2011). IL-1β is also produced predominately by macrophages and therefore strategies that target monocytes/macrophages (Leuschner et al., 2011), IL-1β (Abbate et al., 2013), and NLRP3 (Van Hout et al., 2017) have all demonstrated cardiovascular protection.

Putative neurogenic regulation of CVD

While there is no doubt that peripheral, circulating inflammatory mediators regulate CVD, neuroinflammation is also a likely inflammatory regulator of cardiovascular function. Preclinical studies reveal a role for central inflammation in cardiovascular responses as injection of IL-1β into the brain increases blood pressure (Ye et al., 2000; Wood et al., 2017). Furthermore, centrally administered IL-1 receptor antagonist blunts the pressor response to footshock and air jet stress. In addition, the role of brain IL-1β (and not peripheral circulating cytokines) in this stress-induced pressor response was confirmed by revealing that intravenous injection of IL-1ra was ineffective (Zou et al., 2001; Ufnal et al., 2008). In humans, cytokines are also linked to the autonomic nervous system, as reductions in heart rate variability (indicative of enhanced sympathetic and reduced vagal control) are associated with higher concentrations of IL-6 and TNF-α (Gonzalez-Clemente et al., 2007; Von Kanel et al., 2008).

Moreover, centrally derived NE from the LC can directly impact inflammatory processes in both the brain (see "Stress-induced activation and sensitization of microglia as a source of neuroinflammation" section) and in the periphery. In the periphery, NE signals through β-adrenergic receptors on the macrophage cell surface to stimulate the release of proinflammatory cytokines (Flierl et al., 2009; Li et al., 2015) and can augment the peripheral inflammatory response to an immune challenge (Spengler et al., 1990; Kavelaars et al., 1997; Grisanti et al., 2010). Importantly, stress-induced inflammatory priming, which is a known to occur following repeated stress exposure, is regulated by LC-NE release. Specifically, DSP-4 lesion of noradrenergic terminals from the LC resulted in decreased circulating NE in plasma following social stress exposure (Finnell et al., 2019). Consequently, rats with an LC-NE lesion exhibited attenuated stress-induced circulating cytokine expression, regardless of stress history (Fig. 8.2D–F). Moreover, findings in nonstressed rats indicate that IL-1β signals through the LC-NE system to increase blood pressure. Specifically, the cytokine-induced pressor response to an intracerebroventricular injection of IL-1β is inhibited following LC-NE lesion (Wood et al., 2017). Taken together these data indicate that neuroinflammatory processes in the brain as well as neurotransmitter regulation within stress sensitive brain regions are capable of regulating peripheral cytokine responses and thus CVD risk.

ROS in the pathogenesis of cardiovascular disease

Oxidative stress plays a vital role in the pathogenesis of various forms of CVD. Just as occurs in the brain, the imbalance between ROS production and antioxidant mechanisms leads to the accumulation of ROS and consequently pathological conditions (Heitzer et al., 2001). Most clearly defined is the role of ROS in endothelial dysfunction. Endothelial cells on the internal surface of blood and lymphatic vessels regulate vascular tone by releasing vasodilatory and vasoconstricting factors (Zuo et al., 2003). In endothelial dysfunction, occurring when endothelial cells are injured or in the presence of metabolic stress, there is an imbalance between constriction and dilation of blood vessels (Zuo et al., 2003), thus initiating endothelial dysfunction (Chatzizisis et al., 2007). Importantly, ROS are signaling molecules and contribute to endothelial dysfunction in clinical and preclinical models of atherosclerosis (Taniyama and Griendling, 2003; Bayir, 2005). Just as occurs in the brain, mitochondria represent a considerable source of ROS generation in cardiac and vascular tissue. In fact, nearly 1/3 of the volume of cardiomyocytes consists of mitochondria (Barth et al., 1992). Despite the convincing role that ROS plays in cardiovascular disorders, multiple antioxidants have been tested in clinical trials to treat CVD but have been generally ineffective (reviewed in: Kornfeld et al., 2015). One potential explanation for the poor efficacy of these compounds to treat CVD is that ROS scavengers are highly unstable and can undergo auto-oxidation, which renders them with oxidant properties and may explain how some trials have concluded that antioxidant therapy is not only ineffective but in some cases harmful (Sugamura and Keaney, 2011; Lee et al., 2014). Therefore one strategy to preserve physiological ROS production while inhibiting pathological mitochondrial ROS production is to selectively target damaged mitochondria as described in length in (Kornfeld et al., 2015). Finally, in addition to ROS accumulation in the periphery contributing to CVD, induction of oxidative stress in the brain is in fact associated with hypertension (Salim et al., 2010). As such, ROS accumulation in both the brain and the periphery are likely contributing factors to the progression of CVD.

Inflammation and ROS responses to stress as a mechanism through which stress leads to depression and cardiovascular disease comorbidity

It was noted by Maltzberg nearly 80 years ago that patients suffering from depression were at increased risk of mortality (Maltzberg, 1937). In recent years the unmistakable association between stress-related psychosocial disorders and CVD is a mechanism for the increased mortality in depressed patients (Seligman and Nemeroff, 2015). It is tempting to suggest that the link between diseases characterized by peripheral inflammation such as CVD, and the striking elevated risk of major depression in these patients (Anda et al., 1993; Huffman et al., 2013; Marrie et al., 2017) may be driven by exaggerated neuroinflammation and ROS. In support of this notion, social stress exposure, another risk factor for depression is recognized to increase circulating proinflammatory cytokines and ROS in animals and humans (Pace et al., 2006; Patki et al., 2013b; Hodes et al., 2014; Wood et al., 2015; Quinn et al., 2018). While circulating cytokine levels typically return to baseline following cessation of an acute stress (Cheng et al., 2015), preclinical models generating a chronic stress-induced depressive-like phenotype is dependent upon persistent enhancement in inflammatory sensitivity (Hodes et al., 2014; Finnell et al., 2017a). In fact, preclinical models of social stress have identified that centrally acting IL-1ra (Fig. 8.3A) and resveratrol (Fig. 8.3B) are both capable of eliciting protection over the development of stress-induced anhedonia.

Studies that focus on individual differences in the response to stress provide support for the concept that stress-induced inflammatory mediators ultimately regulate susceptibility to stress related psychiatric and cardiovascular dysfunction. Preclinical and clinical studies demonstrate individual variability in the behavioral and inflammatory consequences of stress resulting in the emergence of susceptible and resilient populations. Specifically, greater stress-induced inflammatory responses are related to greater negative affect in humans (Dickerson et al., 2009) and promote depressive-like behaviors in rodents (Wohleb et al., 2013, 2014a,b; Hodes et al., 2014; Wood et al., 2015; Finnell and Wood, 2016; Finnell et al., 2017a,b, 2018). Importantly, these studies note that social stress exposure in their resilient cohort of animals does not produce exaggerated inflammatory responses and as a result, does not produce a depressive-like phenotype. Moreover,

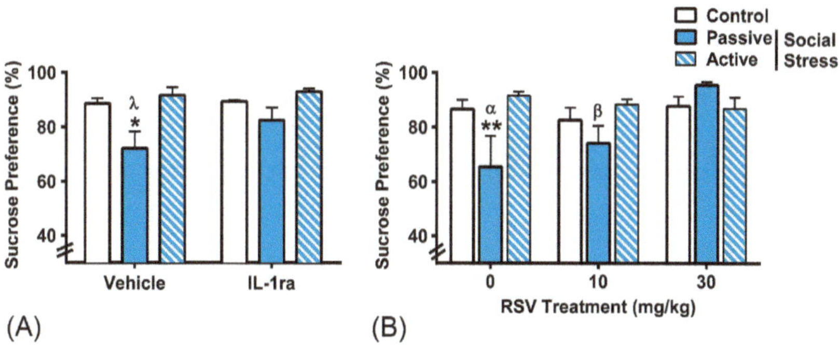

Fig. 8.3 ***Antidepressant-like effects of centrally acting anti-inflammatory compounds.*** Vehicle-treated rats demonstrating passive coping during social defeat (i.e., susceptible) develop anhedonia as measured by decreased preference for 1% sucrose over water as compared with vehicle-treated control rats (*$p < 0.05$) and active coping defeated rats ($\lambda p < 0.05$). Passive coping rats that were treated with IL-1ra (0.25 µg/rat, intracerebroventricular) before each social defeat episode did not develop anhedonia ($p < 0.05$, Tukey post hoc versus control). (B) Decreased sucrose preference in passive coping rats was also dose dependently blocked by resveratrol (RSV) treatment. Importantly, only neuroactive doses of RSV (30 mg/kg) exhibited antidepressant-like effects. (**$p < 0.01$ vs. 0 mg/kg control and active coping, $\alpha p < 0.001$ vs. 30 mg/kg passive coping; $\beta p < 0.01$ vs. 30 mg/kg passive coping). *((A) Reprinted by permission of Elsevier from Inflammatory Factors Mediate Vulnerability to a Social Stress-Induced Depressive-like Phenotype in Passive Coping Rats, by Wood, S.K., Wood, C.S., Lombard, C.M., Lee, C.S., Zhang, X.Y., Finnell, J.E., and Valentino, R.J. Biological Psychiatry, 78, 38–48, Copyright 2015 by the Society of Biological Psychiatry. (B) From Finnell, J.E., Lombard, C.M., Melson, M.N., Singh, N.P., Nagarkatti, M., Nagarkatti, P., Fadel, J.R., Wood, C.S., and Wood, S.K. (2017a). The protective effects of resveratrol on social stress-induced cytokine release and depressive-like behavior. Brain Behav Immun 59, 147–157.)*

individuals that cope with stress in a passive manner are at increased risk of hypertension (Harburg et al., 1964) and exposure to psychosocial stress alone is also capable of promoting atherosclerosis (Black and Garbutt, 2002), likely as a result of inflammatory leukocytes (Heidt et al., 2014). Because inflammation is central to the pathogenesis of both hypertension (Savoia and Schiffrin, 2006) and atherosclerosis (Black and Garbutt, 2002), inflammation lies at the intersection between depression and CVD and under stress exposure, may promote the comorbidity between these two disorders.

Conclusion

Hans Selye once wrote "It's not the stress that kills us, its our reaction to it". This seminal observation highlights the fact that many individuals are exposed to stress and remain resilient; yet another population copes poorly with the stress and develops psychiatric disorders as a result. While it has long been recognized that psychosocial stress leads to the pathogenesis of psychiatric disorders, it is striking that psychosocial factors account for significantly more of the modifiable risk of suffering from CVD above that of obesity and hypertension (Yusuf et al., 2004). It is clear that if the response to psychosocial stress elicits persistent inflammatory or mitochondrial/ROS dysfunction, a cascade is initiated promoting psychiatric disorders and CVD. At this point there is support for the notion that inflammation and/or ROS is causative to both of psychiatric disorders and CVD, however gaining a better understanding of the mechanisms of stress-related behavioral and cardiovascular dysfunction is critical as it will provide better targeted therapeutics.

References

Abbate, A., Van Tassell, B.W., Biondi-Zoccai, G., Kontos, M.C., Grizzard, J.D., Spillman, D.W., Oddi, C., Roberts, C.S., Melchior, R.D., Mueller, G.H., Abouzaki, N.A., Rengel, L.R., Varma, A., Gambill, M.L., Falcao, R.A., Voelkel, N.F., Dinarello, C.A., Vetrovec, G.W., 2013. Effects of interleukin-1 blockade with anakinra on adverse cardiac remodeling and heart failure after acute myocardial infarction [from the Virginia Commonwealth University-Anakinra Remodeling Trial (2) (VCU-ART2) pilot study]. Am. J. Cardiol. 111, 1394–1400. https://doi.org/10.1016/j.amjcard.2013.01.287.

Agita, A., Alsagaff, M.T., 2017. Inflammation, immunity, and hypertension. Acta Med. Indones. 49, 158–165.

Ahmed, R.F., Abdel-Rahman, R.F., Farid, O.A.H.A., El-Marasy, S.A., Hessin, A.F., 2014. Combined hepatoprotective and antidepressant effects of resveratrol in an acute model of depression. Bull. Fac. Pharm. Cairo Univ. 52, 191–197.

Ahmetspahic, D., Schwarte, K., Ambree, O., Burger, C., Falcone, V., Seiler, K., Kooybaran, M.R., Grosse, L., Roos, F., Scheffer, J., Jorgens, S., Koelkebeck, K., Dannlowski, U., Arolt, V., Scheu, S., Alferink, J., 2018. Altered B cell homeostasis in patients with major depressive disorder and normalization of CD5 surface expression on regulatory B cells in treatment responders. J. NeuroImmune Pharmacol. 13, 90–99. https://doi.org/10.1007/s11481-017-9763-4.

Anda, R., Williamson, D., Jones, D., Macera, C., Eaker, E., Glassman, A., Marks, J., 1993. Depressed affect, hopelessness, and the risk of ischemic heart disease in a cohort of U.S. adults. Epidemiology 4, 285–294.

Banks, W.A., 2005. Blood-brain barrier transport of cytokines: a mechanism for neuropathology. Curr. Pharm. Des. 11, 973–984.

Banks, W.A., Kastin, A.J., Durham, D.A., 1989. Bidirectional transport of interleukin-1 alpha across the blood-brain barrier. Brain Res. Bull. 23, 433–437.

Banks, W.A., Ortiz, L., Plotkin, S.R., Kastin, A.J., 1991. Human interleukin (IL) 1 alpha, murine IL-1 alpha and murine IL-1 beta are transported from blood to brain in the mouse by a shared saturable mechanism. J. Pharmacol. Exp. Ther. 259, 988–996.

Barth, E., Stammler, G., Speiser, B., Schaper, J., 1992. Ultrastructural quantitation of mitochondria and myofilaments in cardiac muscle from 10 different animal species including man. J. Mol. Cell. Cardiol. 24, 669–681.

Batuman, O.A., Sajewski, D., Ottenweller, J.E., Pitman, D.L., Natelson, B.H., 1990. Effects of repeated stress on T cell numbers and function in rats. Brain Behav. Immun. 4, 105–117.

Bayir, H., 2005. Reactive oxygen species. Crit. Care Med. 33, S498–S501.
Ben-Shachar, D., Karry, R., 2008. Neuroanatomical pattern of mitochondrial complex I pathology varies between schizophrenia, bipolar disorder and major depression. PLoS One 3, e3676. https://doi.org/10.1371/journal.pone.0003676.
Berk, M., Dean, O., Cotton, S.M., Gama, C.S., Kapczinski, F., Fernandes, B.S., Kohlmann, K., Jeavons, S., Hewitt, K., Allwang, C., Cobb, H., Bush, A.I., Schapkaitz, I., Dodd, S., Malhi, G.S., 2011. The efficacy of N-acetylcysteine as an adjunctive treatment in bipolar depression: an open label trial. J. Affect. Disord. 135, 389–394. https://doi.org/10.1016/j.jad.2011.06.005.
Beurel, E., Harrington, L.E., Jope, R.S., 2013. Inflammatory T helper 17 cells promote depression-like behavior in mice. Biol. Psychiatry 73, 622–630. https://doi.org/10.1016/j.biopsych.2012.09.021.
Bharani, K.L., Derex, R., Granholm, A.C., Ledreux, A., 2017. A noradrenergic lesion aggravates the effects of systemic inflammation on the hippocampus of aged rats. PLoS One 12, e0189821. https://doi.org/10.1371/journal.pone.0189821.
Bhutani, M.K., Bishnoi, M., Kulkarni, S.K., 2009. Anti-depressant like effect of curcumin and its combination with piperine in unpredictable chronic stress-induced behavioral, biochemical and neurochemical changes. Pharmacol. Biochem. Behav. 92, 39–43. https://doi.org/10.1016/j.pbb.2008.10.007.
Biasucci, L.M., Liuzzo, G., Fantuzzi, G., Caligiuri, G., Rebuzzi, A.G., Ginnetti, F., Dinarello, C.A., Maseri, A., 1999. Increasing levels of interleukin (IL)-1Ra and IL-6 during the first 2 days of hospitalization in unstable angina are associated with increased risk of in-hospital coronary events. Circulation 99, 2079–2084.
Biesmans, S., Bouwknecht, J.A., Ver Donck, L., Langlois, X., Acton, P.D., De Haes, P., Davoodi, N., Meert, T.F., Hellings, N., Nuydens, R., 2015. Peripheral administration of tumor necrosis factor-alpha induces neuroinflammation and sickness but not depressive-like behavior in mice. Biomed. Res. Int. 2015, 716920. https://doi.org/10.1155/2015/716920.
Bilici, M., Efe, H., Koroglu, M.A., Uydu, H.A., Bekaroglu, M., Deger, O., 2001. Antioxidative enzyme activities and lipid peroxidation in major depression: alterations by antidepressant treatments. J. Affect. Disord. 64, 43–51.
Black, P.H., Garbutt, L.D., 2002. Stress, inflammation and cardiovascular disease. J. Psychosom. Res. 52, 1–23.
Blandino Jr., P., Barnum, C.J., Deak, T., 2006. The involvement of norepinephrine and microglia in hypothalamic and splenic IL-1beta responses to stress. J. Neuroimmunol. 173, 87–95. https://doi.org/10.1016/j.jneuroim.2005.11.021.
Brites, D., Fernandes, A., 2015. Neuroinflammation and depression: microglia activation, extracellular microvesicles and microRNA dysregulation. Front. Cell. Neurosci. 9, 476. https://doi.org/10.3389/fncel.2015.00476.
Bruinstroop, E., Cano, G., Vanderhorst, V.G., Cavalcante, J.C., Wirth, J., Sena-Esteves, M., Saper, C.B., 2012. Spinal projections of the A5, A6 (locus coeruleus), and A7 noradrenergic cell groups in rats. J. Comp. Neurol. 520, 1985–2001. https://doi.org/10.1002/cne.23024.
Cadenas, E., Davies, K.J., 2000. Mitochondrial free radical generation, oxidative stress, and aging. Free Radic. Biol. Med. 29, 222–230.
Capuron, L., Ravaud, A., Gualde, N., Bosmans, E., Dantzer, R., Maes, M., Neveu, P.J., 2001. Association between immune activation and early depressive symptoms in cancer patients treated with interleukin-2-based therapy. Psychoneuroendocrinology 26, 797–808.
Carroll, J.E., Low, C.A., Prather, A.A., Cohen, S., Fury, J.M., Ross, D.C., Marsland, A.L., 2011. Negative affective responses to a speech task predict changes in interleukin (IL)-6. Brain Behav. Immun. 25, 232–238. https://doi.org/10.1016/j.bbi.2010.09.024.
Chatzizisis, Y.S., Coskun, A.U., Jonas, M., Edelman, E.R., Feldman, C.L., Stone, P.H., 2007. Role of endothelial shear stress in the natural history of coronary atherosclerosis and

vascular remodeling: molecular, cellular, and vascular behavior. J. Am. Coll. Cardiol. 49, 2379–2393. https://doi.org/10.1016/j.jacc.2007.02.059.
Chen, E., 2007. Psychological Stress and Its Relationship to Cytokines and Inflammatory Diseases. Taylor & Francis Group, Boca Raton, FL.
Cheng, Y., Jope, R.S., Beurel, E., 2015. A pre-conditioning stress accelerates increases in mouse plasma inflammatory cytokines induced by stress. BMC Neurosci. 16, 31. https://doi.org/10.1186/s12868-015-0169-z.
Christian, L.M., Glaser, R., Porter, K., Iams, J.D., 2013. Stress-induced inflammatory responses in women: effects of race and pregnancy. Psychosom. Med. 75, 658–669. https://doi.org/10.1097/PSY.0b013e31829bbc89.
Conti, P., Lessiani, G., Kritas, S.K., Ronconi, G., Caraffa, A., Theoharides, T.C., 2017. Mast cells emerge as mediators of atherosclerosis: special emphasis on IL-37 inhibition. Tissue Cell 49, 393–400. https://doi.org/10.1016/j.tice.2017.04.002.
Crocker, S.J., Milner, R., Pham-Mitchell, N., Campbell, I.L., 2006. Cell and agonist-specific regulation of genes for matrix metalloproteinases and their tissue inhibitors by primary glial cells. J. Neurochem. 98, 812–823. https://doi.org/10.1111/j.1471-4159.2006.03927.x.
Cuffel, B., Wamboldt, M., Borish, L., Kennedy, S., Crystal-Peters, J., 1999. Economic consequences of comorbid depression, anxiety, and allergic rhinitis. Psychosomatics 40, 491–496. https://doi.org/10.1016/S0033-3182(99)71187-4.
De La Lastra, C.A., Villegas, I., 2005. Resveratrol as an anti-inflammatory and anti-aging agent: mechanisms and clinical implications. Mol. Nutr. Food Res. 49, 405–430. https://doi.org/10.1002/mnfr.200500022.
De Oliveira, M.R., Silvestrin, R.B., Mello, E.S.T., Moreira, J.C., 2007. Oxidative stress in the hippocampus, anxiety-like behavior and decreased locomotory and exploratory activity of adult rats: effects of sub acute vitamin A supplementation at therapeutic doses. Neurotoxicology 28, 1191–1199. https://doi.org/10.1016/j.neuro.2007.07.008.
Dean, O.M., Van Den Buuse, M., Berk, M., Copolov, D.L., Mavros, C., Bush, A.I., 2011. N-acetyl cysteine restores brain glutathione loss in combined 2-cyclohexene-1-one and d-amphetamine-treated rats: relevance to schizophrenia and bipolar disorder. Neurosci. Lett. 499, 149–153. https://doi.org/10.1016/j.neulet.2011.05.027.
Den Haan, J.M., Arens, R., Van Zelm, M.C., 2014. The activation of the adaptive immune system: cross-talk between antigen-presenting cells, T cells and B cells. Immunol. Lett. 162, 103–112. https://doi.org/10.1016/j.imlet.2014.10.011.
Denicoff, K.D., Rubinow, D.R., Papa, M.Z., Simpson, C., Seipp, C.A., Lotze, M.T., Chang, A.E., Rosenstein, D., Rosenberg, S.A., 1987. The neuropsychiatric effects of treatment with interleukin-2 and lymphokine-activated killer cells. Ann. Intern. Med. 107, 293–300.
Devorak, J., Torres-Platas, S.G., Davoli, M.A., Prud'homme, J., Turecki, G., Mechawar, N., 2015. Cellular and molecular inflammatory profile of the choroid plexus in depression and suicide. Front. Psychiatry 6, 138. https://doi.org/10.3389/fpsyt.2015.00138.
Dickerson, S.S., Gable, S.L., Irwin, M.R., Aziz, N., Kemeny, M.E., 2009. Social-evaluative threat and proinflammatory cytokine regulation: an experimental laboratory investigation. Psychol. Sci. 20, 1237–1244. https://doi.org/10.1111/j.1467-9280.2009.02437.x.
Doll, D.N., Rellick, S.L., Barr, T.L., Ren, X., Simpkins, J.W., 2015. Rapid mitochondrial dysfunction mediates TNF-alpha-induced neurotoxicity. J. Neurochem. 132, 443–451. https://doi.org/10.1111/jnc.13008.
Donnelly, L.E., Newton, R., Kennedy, G.E., Fenwick, P.S., Leung, R.H., Ito, K., Russell, R.E., Barnes, P.J., 2004. Anti-inflammatory effects of resveratrol in lung epithelial cells: molecular mechanisms. Am. J. Phys. Lung Cell. Mol. Phys. 287, L774–L783. https://doi.org/10.1152/ajplung.00110.2004.
Duewell, P., Kono, H., Rayner, K.J., Sirois, C.M., Vladimer, G., Bauernfeind, F.G., Abela, G.S., Franchi, L., Nunez, G., Schnurr, M., Espevik, T., Lien, E., Fitzgerald, K.A.,

Rock, K.L., Moore, K.J., Wright, S.D., Hornung, V., Latz, E., 2010. NLRP3 inflammasomes are required for atherogenesis and activated by cholesterol crystals. Nature 464, 1357–1361. https://doi.org/10.1038/nature08938.
Elenkov, I.J., Chrousos, G.P., 1999. Stress hormones, Th1/Th2 patterns, pro/anti-inflammatory cytokines and susceptibility to disease. Trends Endocrinol. Metab. 10, 359–368.
Eraly, S.A., Nievergelt, C.M., Maihofer, A.X., Barkauskas, D.A., Biswas, N., Agorastos, A., O'connor, D.T., Baker, D.G., Marine Resiliency Study, T, 2014. Assessment of plasma C-reactive protein as a biomarker of posttraumatic stress disorder risk. JAMA Psychiatry 71, 423–431. https://doi.org/10.1001/jamapsychiatry.2013.4374.
Falchetti, R., Fuggetta, M.P., Lanzilli, G., Tricarico, M., Ravagnan, G., 2001. Effects of resveratrol on human immune cell function. Life Sci. 70, 81–96.
Felger, J.C., Li, L., Marvar, P.J., Woolwine, B.J., Harrison, D.G., Raison, C.L., Miller, A.H., 2013. Tyrosine metabolism during interferon-alpha administration: association with fatigue and CSF dopamine concentrations. Brain Behav. Immun. 31, 153–160. https://doi.org/10.1016/j.bbi.2012.10.010.
Ferreira, F.R., Biojone, C., Joca, S.R., Guimaraes, F.S., 2008. Antidepressant-like effects of N-acetyl-L-cysteine in rats. Behav. Pharmacol. 19, 747–750. https://doi.org/10.1097/FBP.0b013e3283123c98.
Festoff, B.W., Sajja, R.K., Van Dreden, P., Cucullo, L., 2016. HMGB1 and thrombin mediate the blood-brain barrier dysfunction acting as biomarkers of neuroinflammation and progression to neurodegeneration in Alzheimer's disease. J. Neuroinflammation 13, 194. https://doi.org/10.1186/s12974-016-0670-z.
Finnell, J.E., Wood, S.K., 2016. Neuroinflammation at the interface of depression and cardiovascular disease: evidence from rodent models of social stress. Neurobiol. Stress 4, 1–14. https://doi.org/10.1016/j.ynstr.2016.04.001.
Finnell, J.E., Lombard, C.M., Melson, M.N., Singh, N.P., Nagarkatti, M., Nagarkatti, P., Fadel, J.R., Wood, C.S., Wood, S.K., 2017a. The protective effects of resveratrol on social stress-induced cytokine release and depressive-like behavior. Brain Behav. Immun. 59, 147–157. https://doi.org/10.1016/j.bbi.2016.08.019.
Finnell, J.E., Lombard, C.M., Padi, A.R., Moffitt, C.M., Wilson, L.B., Wood, C.S., Wood, S.K., 2017b. Physical versus psychological social stress in male rats reveals distinct cardiovascular, inflammatory and behavioral consequences. PLoS One 12, e0172868. https://doi.org/10.1371/journal.pone.0172868.
Finnell, J.E., Muniz, B.L., Padi, A.R., Lombard, C.M., Moffitt, C.M., Wood, C.S., Wilson, L.B., Reagan, L.P., Wilson, M.A., Wood, S.K., 2018. Essential role of ovarian hormones in susceptibility to the consequences of witnessing social defeat in female rats. Biol. Psychiatry. https://doi.org/10.1016/j.biopsych.2018.01.013.
Finnell, J.E., Moffitt, C.M., Hesser, L.A., Harrington, E., Melson, M.N., Wood, C.S., Wood, S.K., 2019. The contribution of the locus coeruleus-norepinephrine system in the emergence of defeat-induced inflammatory priming. Brain Behav. Immun. https://doi.org/10.1016/j.bbi.2019.01.021.
Flierl, M.A., Rittirsch, D., Nadeau, B.A., Sarma, J.V., Day, D.E., Lentsch, A.B., Huber-Lang, M.S., Ward, P.A., 2009. Upregulation of phagocyte-derived catecholamines augments the acute inflammatory response. PLoS One 4, e4414. https://doi.org/10.1371/journal.pone.0004414.
Fonken, L.K., Weber, M.D., Daut, R.A., Kitt, M.M., Frank, M.G., Watkins, L.R., Maier, S.F., 2016. Stress-induced neuroinflammatory priming is time of day dependent. Psychoneuroendocrinology 66, 82–90. https://doi.org/10.1016/j.psyneuen.2016.01.006.
Fordham, J.B., Naqvi, A.R., Nares, S., 2014. Leukocyte production of inflammatory mediators is inhibited by the antioxidants phloretin, silymarin, hesperetin, and resveratrol. Mediat. Inflamm. 2014, 938712. https://doi.org/10.1155/2014/938712.

Frank, M.G., Thompson, B.M., Watkins, L.R., Maier, S.F., 2012. Glucocorticoids mediate stress-induced priming of microglial pro-inflammatory responses. Brain Behav. Immun. 26, 337–345. https://doi.org/10.1016/j.bbi.2011.10.005.

Frank, M.G., Fonken, L.K., Annis, J.L., Watkins, L.R., Maier, S.F., 2018. Stress disinhibits microglia via down-regulation of CD200R: a mechanism of neuroinflammatory priming. Brain Behav. Immun. 69, 62–73. https://doi.org/10.1016/j.bbi.2017.11.001.

Franklin, T.C., Wohleb, E.S., Zhang, Y., Fogaca, M., Hare, B., Duman, R.S., 2018. Persistent increase in microglial RAGE contributes to chronic stress-induced priming of depressive-like behavior. Biol. Psychiatry 83, 50–60. https://doi.org/10.1016/j.biopsych.2017.06.034.

Fruhbeis, C., Frohlich, D., Kuo, W.P., Kramer-Albers, E.M., 2013. Extracellular vesicles as mediators of neuron-glia communication. Front. Cell. Neurosci. 7, 182. https://doi.org/10.3389/fncel.2013.00182.

Gamaro, G.D., Streck, E.L., Matte, C., Prediger, M.E., Wyse, A.T., Dalmaz, C., 2003. Reduction of hippocampal Na+, K+-ATPase activity in rats subjected to an experimental model of depression. Neurochem. Res. 28, 1339–1344.

Gardner, A., Johansson, A., Wibom, R., Nennesmo, I., Von Dobeln, U., Hagenfeldt, L., Hallstrom, T., 2003. Alterations of mitochondrial function and correlations with personality traits in selected major depressive disorder patients. J. Affect. Disord. 76, 55–68.

Gawryluk, J.W., Wang, J.F., Andreazza, A.C., Shao, L., Young, L.T., 2011. Decreased levels of glutathione, the major brain antioxidant, in post-mortem prefrontal cortex from patients with psychiatric disorders. Int. J. Neuropsychopharmacol. 14, 123–130. https://doi.org/10.1017/S1461145710000805.

Gemma, C., Bachstetter, A.D., 2013. The role of microglia in adult hippocampal neurogenesis. Front. Cell. Neurosci. 7, 229. https://doi.org/10.3389/fncel.2013.00229.

Gero, D., Szoleczky, P., Modis, K., Pribis, J.P., Al-Abed, Y., Yang, H., Chevan, S., Billiar, T.R., Tracey, K.J., Szabo, C., 2013. Identification of pharmacological modulators of HMGB1-induced inflammatory response by cell-based screening. PLoS One 8, e65994. https://doi.org/10.1371/journal.pone.0065994.

Gimeno, D., Kivimaki, M., Brunner, E.J., Elovainio, M., De Vogli, R., Steptoe, A., Kumari, M., Lowe, G.D., Rumley, A., Marmot, M.G., Ferrie, J.E., 2009. Associations of C-reactive protein and interleukin-6 with cognitive symptoms of depression: 12-year follow-up of the Whitehall II study. Psychol. Med. 39, 413–423. https://doi.org/10.1017/S0033291708003723.

Gong, Y., Chai, Y., Ding, J.H., Sun, X.L., Hu, G., 2011. Chronic mild stress damages mitochondrial ultrastructure and function in mouse brain. Neurosci. Lett. 488, 76–80. https://doi.org/10.1016/j.neulet.2010.11.006.

Gonzalez-Clemente, J.M., Vilardell, C., Broch, M., Megia, A., Caixas, A., Gimenez-Palop, O., Richart, C., Simon, I., Martinez-Riquelme, A., Arroyo, J., Mauricio, D., Vendrell, J., 2007. Lower heart rate variability is associated with higher plasma concentrations of IL-6 in type 1 diabetes. Eur. J. Endocrinol. 157, 31–38. https://doi.org/10.1530/EJE-07-0090.

Grisanti, L.A., Evanson, J., Marchus, E., Jorissen, H., Woster, A.P., Dekrey, W., Sauter, E.R., Combs, C.K., Porter, J.E., 2010. Pro-inflammatory responses in human monocytes are beta1-adrenergic receptor subtype dependent. Mol. Immunol. 47, 1244–1254. https://doi.org/10.1016/j.molimm.2009.12.013.

Gudmundsson, P., Skoog, I., Waern, M., Blennow, K., Palsson, S., Rosengren, L., Gustafson, D., 2007. The relationship between cerebrospinal fluid biomarkers and depression in elderly women. Am. J. Geriatr. Psychiatry 15, 832–838. https://doi.org/10.1097/JGP.0b013e3180547091.

Gupta, A., Vij, G., Sharma, S., Tirkey, N., Rishi, P., Chopra, K., 2009. Curcumin, a polyphenolic antioxidant, attenuates chronic fatigue syndrome in murine water immersion stress model. Immunobiology 214, 33–39. https://doi.org/10.1016/j.imbio.2008.04.003.

Gyoneva, S., Traynelis, S.F., 2013. Norepinephrine modulates the motility of resting and activated microglia via different adrenergic receptors. J. Biol. Chem. 288, 15291–15302. https://doi.org/10.1074/jbc.M113.458901.
Haapakoski, R., Ebmeier, K.P., Alenius, H., Kivimaki, M., 2016. Innate and adaptive immunity in the development of depression: an update on current knowledge and technological advances. Prog. Neuro-Psychopharmacol. Biol. Psychiatry 66, 63–72. https://doi.org/10.1016/j.pnpbp.2015.11.012.
Halliwell, B., 1992. Reactive oxygen species and the central nervous system. J. Neurochem. 59, 1609–1623.
Harburg, E., Julius, S., Mcginn, N.F., Mcleod, J., Hoobler, S.W., 1964. Personality traits and behavioral patterns associated with systolic blood pressure levels in college males. J. Chronic Dis. 17, 405–414.
Heidt, T., Sager, H.B., Courties, G., Dutta, P., Iwamoto, Y., Zaltsman, A., Von Zur Muhlen, C., Bode, C., Fricchione, G.L., Denninger, J., Lin, C.P., Vinegoni, C., Libby, P., Swirski, F.K., Weissleder, R., Nahrendorf, M., 2014. Chronic variable stress activates hematopoietic stem cells. Nat. Med. 20, 754–758. https://doi.org/10.1038/nm.3589.
Heikkila, H.M., Trosien, J., Metso, J., Jauhiainen, M., Pentikainen, M.O., Kovanen, P.T., Lindstedt, K.A., 2010. Mast cells promote atherosclerosis by inducing both an atherogenic lipid profile and vascular inflammation. J. Cell. Biochem. 109, 615–623. https://doi.org/10.1002/jcb.22443.
Heitzer, T., Schlinzig, T., Krohn, K., Meinertz, T., Munzel, T., 2001. Endothelial dysfunction, oxidative stress, and risk of cardiovascular events in patients with coronary artery disease. Circulation 104, 2673–2678.
Herken, H., Gurel, A., Selek, S., Armutcu, F., Ozen, M.E., Bulut, M., Kap, O., Yumru, M., Savas, H.A., Akyol, O., 2007. Adenosine deaminase, nitric oxide, superoxide dismutase, and xanthine oxidase in patients with major depression: impact of antidepressant treatment. Arch. Med. Res. 38, 247–252. https://doi.org/10.1016/j.arcmed.2006.10.005.
Hodes, G.E., Pfau, M.L., Leboeuf, M., Golden, S.A., Christoffel, D.J., Bregman, D., Rebusi, N., Heshmati, M., Aleyasin, H., Warren, B.L., Lebonte, B., Horn, S., Lapidus, K.A., Stelzhammer, V., Wong, E.H., Bahn, S., Krishnan, V., Bolanos-Guzman, C.A., Murrough, J.W., Merad, M., Russo, S.J., 2014. Individual differences in the peripheral immune system promote resilience versus susceptibility to social stress. Proc. Natl. Acad. Sci. U. S. A. 111, 16136–16141. https://doi.org/10.1073/pnas.1415191111.
Holdiness, M.R., 1991. Clinical pharmacokinetics of N-acetylcysteine. Clin. Pharmacokinet. 20, 123–134. https://doi.org/10.2165/00003088-199120020-00004.
Hollis, F., Van Der Kooij, M.A., Zanoletti, O., Lozano, L., Canto, C., Sandi, C., 2015. Mitochondrial function in the brain links anxiety with social subordination. Proc. Natl. Acad. Sci. U. S. A. 112, 15486–15491. https://doi.org/10.1073/pnas.1512653112.
Huffman, J.C., Celano, C.M., Beach, S.R., Motiwala, S.R., Januzzi, J.L., 2013. Depression and cardiac disease: epidemiology, mechanisms, and diagnosis. Cardiovasc. Psychiatry Neurol. 2013, 695925. https://doi.org/10.1155/2013/695925.
Hurley, L.L., Akinfiresoye, L., Kalejaiye, O., Tizabi, Y., 2014. Antidepressant effects of resveratrol in an animal model of depression. Behav. Brain Res. 268, 1–7. https://doi.org/10.1016/j.bbr.2014.03.052.
Irwin, M.R., Miller, A.H., 2007. Depressive disorders and immunity: 20 years of progress and discovery. Brain Behav. Immun. 21, 374–383. https://doi.org/10.1016/j.bbi.2007.01.010.
Jardanhazi-Kurutz, D., Kummer, M.P., Terwel, D., Vogel, K., Thiele, A., Heneka, M.T., 2011. Distinct adrenergic system changes and neuroinflammation in response to induced locus ceruleus degeneration in APP/PS1 transgenic mice. Neuroscience 176, 396–407. https://doi.org/10.1016/j.neuroscience.2010.11.052.

Ji, W.W., Wang, S.Y., Ma, Z.Q., Li, R.P., Li, S.S., Xue, J.S., Li, W., Niu, X.X., Yan, L., Zhang, X., Fu, Q., Qu, R., Ma, S.P., 2014. Effects of perillaldehyde on alternations in serum cytokines and depressive-like behavior in mice after lipopolysaccharide administration. Pharmacol. Biochem. Behav. 116, 1–8. https://doi.org/10.1016/j.pbb.2013.10.026.

Johnson, J.D., O'connor, K.A., Watkins, L.R., Maier, S.F., 2004. The role of IL-1beta in stress-induced sensitization of proinflammatory cytokine and corticosterone responses. Neuroscience 127, 569–577. https://doi.org/10.1016/j.neuroscience.2004.05.046.

Johnson, J.D., Campisi, J., Sharkey, C.M., Kennedy, S.L., Nickerson, M., Greenwood, B.N., Fleshner, M., 2005. Catecholamines mediate stress-induced increases in peripheral and central inflammatory cytokines. Neuroscience 135, 1295–1307. https://doi.org/10.1016/j.neuroscience.2005.06.090.

Kaster, M.P., Gadotti, V.M., Calixto, J.B., Santos, A.R.S., Rodrigues, A.L.S., 2012. Depressive-like behavior induced by tumor necrosis factor-a in mice. Neuropharmacology 2 (62), 419–426.

Katon, W.J., Richardson, L., Lozano, P., Mccauley, E., 2004. The relationship of asthma and anxiety disorders. Psychosom. Med. 66, 349–355.

Kavelaars, A., Van De Pol, M., Zijlstra, J., Heijnen, C.J., 1997. Beta 2-adrenergic activation enhances interleukin-8 production by human monocytes. J. Neuroimmunol. 77, 211–216.

Keller, M., Ruegg, A., Werner, S., Beer, H.D., 2008. Active caspase-1 is a regulator of unconventional protein secretion. Cell 132, 818–831. https://doi.org/10.1016/j.cell.2007.12.040.

Kern, S., Skoog, I., Borjesson-Hanson, A., Blennow, K., Zetterberg, H., Ostling, S., Kern, J., Gudmundsson, P., Marlow, T., Rosengren, L., Waern, M., 2014. Higher CSF interleukin-6 and CSF interleukin-8 in current depression in older women. Results from a population-based sample. Brain Behav. Immun. 41, 55–58. https://doi.org/10.1016/j.bbi.2014.05.006.

Kettenmann, H., Hanisch, U.K., Noda, M., Verkhratsky, A., 2011. Physiology of microglia. Physiol. Rev. 91, 461–553. https://doi.org/10.1152/physrev.00011.2010.

Kobayashi, K., Imagama, S., Ohgomori, T., Hirano, K., Uchimura, K., Sakamoto, K., Hirakawa, A., Takeuchi, H., Suzumura, A., Ishiguro, N., Kadomatsu, K., 2013. Minocycline selectively inhibits M1 polarization of microglia. Cell Death Dis. 4, e525. https://doi.org/10.1038/cddis.2013.54.

Kornfeld, O.S., Hwang, S., Disatnik, M.H., Chen, C.H., Qvit, N., Mochly-Rosen, D., 2015. Mitochondrial reactive oxygen species at the heart of the matter: new therapeutic approaches for cardiovascular diseases. Circ. Res. 116, 1783–1799. https://doi.org/10.1161/CIRCRESAHA.116.305432.

Kreisel, T., Frank, M.G., Licht, T., Reshef, R., Ben-Menachem-Zidon, O., Baratta, M.V., Maier, S.F., Yirmiya, R., 2014. Dynamic microglial alterations underlie stress-induced depressive-like behavior and suppressed neurogenesis. Mol. Psychiatry 19, 699–709. https://doi.org/10.1038/mp.2013.155.

Lai, S., Wu, G., Jiang, Z., 2018. Glycyrrhizin treatment facilitates extinction of conditioned fear responses after a single prolonged stress exposure in rats. Cell. Physiol. Biochem. 45, 2529–2539. https://doi.org/10.1159/000488271.

Lee, B.J., Lin, J.S., Lin, Y.C., Lin, P.T., 2014. Effects of L-carnitine supplementation on oxidative stress and antioxidant enzymes activities in patients with coronary artery disease: a randomized, placebo-controlled trial. Nutr. J. 13, 79. https://doi.org/10.1186/1475-2891-13-79.

Leonard, B., Maes, M., 2012. Mechanistic explanations how cell-mediated immune activation, inflammation and oxidative and nitrosative stress pathways and their sequels and concomitants play a role in the pathophysiology of unipolar depression. Neurosci. Biobehav. Rev. 36, 764–785. https://doi.org/10.1016/j.neubiorev.2011.12.005.

Leuschner, F., Dutta, P., Gorbatov, R., Novobrantseva, T.I., Donahoe, J.S., Courties, G., Lee, K.M., Kim, J.I., Markmann, J.F., Marinelli, B., Panizzi, P., Lee, W.W., Iwamoto, Y., Milstein, S., Epstein-Barash, H., Cantley, W., Wong, J., Cortez-Retamozo, V., Newton, A., Love, K., Libby, P., Pittet, M.J., Swirski, F.K., Koteliansky, V., Langer, R., Weissleder, R., Anderson, D.G., Nahrendorf, M., 2011. Therapeutic siRNA silencing in inflammatory monocytes in mice. Nat. Biotechnol. 29, 1005–1010. https://doi.org/10.1038/nbt.1989.

Lewitus, G.M., Cohen, H., Schwartz, M., 2008. Reducing post-traumatic anxiety by immunization. Brain Behav. Immun. 22, 1108–1114. https://doi.org/10.1016/j.bbi.2008.05.002.

Li, Y., Xiao, B., Qiu, W., Yang, L., Hu, B., Tian, X., Yang, H., 2010. Altered expression of CD4(+) CD25(+) regulatory T cells and its 5-HT(1a) receptor in patients with major depression disorder. J. Affect. Disord. 124, 68–75. https://doi.org/10.1016/j.jad.2009.10.018.

Li, M., Yao, W., Li, S., Xi, J., 2015. Norepinephrine induces the expression of interleukin-6 via beta-adrenoreceptor-NAD(P)H oxidase system -NF-kappaB dependent signal pathway in U937 macrophages. Biochem. Biophys. Res. Commun. 460, 1029–1034. https://doi.org/10.1016/j.bbrc.2015.02.172.

Lindqvist, D., Dhabhar, F.S., James, S.J., Hough, C.M., Jain, F.A., Bersani, F.S., Reus, V.I., Verhoeven, J.E., Epel, E.S., Mahan, L., Rosser, R., Wolkowitz, O.M., Mellon, S.H., 2017. Oxidative stress, inflammation and treatment response in major depression. Psychoneuroendocrinology 76, 197–205. https://doi.org/10.1016/j.psyneuen.2016.11.031.

Liu, H.-S., Shi, H.-L., Huang, F., Peterson, K.E., Wu, H., Lan, Y.-Y., Zhang, B.-B., He, Y.-X., Woods, T., Du, M., Wu, X.-J., Wang, Z.-T., 2016. Astragaloside IV inhibits microglia activation via glucocorticoid receptor mediated signaling pathway. Sci. Rep. 6, 19137. https://doi.org/10.1038/srep19137.

Liuzzo, G., Biasucci, L.M., Trotta, G., Brugaletta, S., Pinnelli, M., Digianuario, G., Rizzello, V., Rebuzzi, A.G., Rumi, C., Maseri, A., Crea, F., 2007. Unusual CD4 + CD28null T lymphocytes and recurrence of acute coronary events. J. Am. Coll. Cardiol. 50, 1450–1458. https://doi.org/10.1016/j.jacc.2007.06.040.

Madrigal, J.L., Olivenza, R., Moro, M.A., Lizasoain, I., Lorenzo, P., Rodrigo, J., Leza, J.C., 2001. Glutathione depletion, lipid peroxidation and mitochondrial dysfunction are induced by chronic stress in rat brain. Neuropsychopharmacology 24, 420–429. https://doi.org/10.1016/S0893-133X(00)00208-6.

Maes, M., Bosmans, E., Suy, E., Vandervorst, C., De Jonckheere, C., Raus, J., 1990. Immune disturbances during major depression: upregulated expression of interleukin-2 receptors. Neuropsychobiology 24, 115–120. https://doi.org/10.1159/000119472.

Maes, M., Mihaylova, I., Kubera, M., Uytterhoeven, M., Vrydags, N., Bosmans, E., 2011. Lower whole blood glutathione peroxidase (GPX) activity in depression, but not in myalgic encephalomyelitis/chronic fatigue syndrome: another pathway that may be associated with coronary artery disease and neuroprogression in depression. Neuro. Endocrinol. Lett. 32, 133–140.

Maltzberg, B., 1937. Mortality among patients with involutional melancholia. Am. J. Psychiatry 93, 1231–1238.

Marchi, N., Rasmussen, P., Kapural, M., Fazio, V., Kight, K., Mayberg, M.R., Kanner, A., Ayumar, B., Albensi, B., Cavaglia, M., Janigro, D., 2003. Peripheral markers of brain damage and blood-brain barrier dysfunction. Restor. Neurol. Neurosci. 21, 109–121.

Marrie, R.A., Walld, R., Bolton, J.M., Sareen, J., Walker, J.R., Patten, S.B., Singer, A., Lix, L.M., Hitchon, C.A., El-Gabalawy, R., Katz, A., Fisk, J.D., Bernstein, C.N., Burden, C.T.I.D.T, Managing the Effects of Psychiatric Comorbidity in Chronic Immunoinflammatory, D, 2017. Increased incidence of psychiatric disorders in immune-mediated inflammatory disease. J. Psychosom. Res. 101, 17–23. https://doi.org/10.1016/j.jpsychores.2017.07.015.

Martino, M., Rocchi, G., Escelsior, A., Fornaro, M., 2012. Immunomodulation mechanism of antidepressants: interactions between serotonin/norepinephrine balance and Th1/Th2 balance. Curr. Neuropharmacol. 10, 97–123. https://doi.org/10.2174/157015912800604542.

Martinon, F., Burns, K., Tschopp, J., 2002. The inflammasome: a molecular platform triggering activation of inflammatory caspases and processing of proIL-beta. Mol. Cell 10, 417–426.

Massaad, C.A., Klann, E., 2011. Reactive oxygen species in the regulation of synaptic plasticity and memory. Antioxid. Redox Signal. 14, 2013–2054. https://doi.org/10.1089/ars.2010.3208.

Mattson, M.P., Gleichmann, M., Cheng, A., 2008. Mitochondria in neuroplasticity and neurological disorders. Neuron 60, 748–766. https://doi.org/10.1016/j.neuron.2008.10.010.

Medina-Rodriguez, E.M., Lowell, J.A., Worthen, R.J., Syed, S.A., Beurel, E., 2018. Involvement of innate and adaptive immune systems alterations in the pathophysiology and treatment of depression. Front. Neurosci. 12, 547. https://doi.org/10.3389/fnins.2018.00547.

Menard, C., Pfau, M.L., Hodes, G.E., Kana, V., Wang, V.X., Bouchard, S., Takahashi, A., Flanigan, M.E., Aleyasin, H., Leclair, K.B., Janssen, W.G., Labonte, B., Parise, E.M., Lorsch, Z.S., Golden, S.A., Heshmati, M., Tamminga, C., Turecki, G., Campbell, M., Fayad, Z.A., Tang, C.Y., Merad, M., Russo, S.J., 2017. Social stress induces neurovascular pathology promoting depression. Nat. Neurosci. 20, 1752–1760. https://doi.org/10.1038/s41593-017-0010-3.

Meng, X., Yang, J., Dong, M., Zhang, K., Tu, E., Gao, Q., Chen, W., Zhang, C., Zhang, Y., 2016. Regulatory T cells in cardiovascular diseases. Nat. Rev. Cardiol. 13, 167–179. https://doi.org/10.1038/nrcardio.2015.169.

Mezzaroma, E., Toldo, S., Farkas, D., Seropian, I.M., Van Tassell, B.W., Salloum, F.N., Kannan, H.R., Menna, A.C., Voelkel, N.F., Abbate, A., 2011. The inflammasome promotes adverse cardiac remodeling following acute myocardial infarction in the mouse. Proc. Natl. Acad. Sci. U. S. A. 108, 19725–19730. https://doi.org/10.1073/pnas.1108586108.

Michel, T.M., Camara, S., Tatschner, T., Frangou, S., Sheldrick, A.J., Riederer, P., Grunblatt, E., 2010. Increased xanthine oxidase in the thalamus and putamen in depression. World J. Biol. Psychiatry 11, 314–320. https://doi.org/10.3109/15622970802123695.

Mor, A., Luboshits, G., Planer, D., Keren, G., George, J., 2006. Altered status of CD4(+) CD25(+) regulatory T cells in patients with acute coronary syndromes. Eur. Heart J. 27, 2530–2537. https://doi.org/10.1093/eurheartj/ehl222.

Muramatsu, R., Takahashi, C., Miyake, S., Fujimura, H., Mochizuki, H., Yamashita, T., 2012. Angiogenesis induced by CNS inflammation promotes neuronal remodeling through vessel-derived prostacyclin. Nat. Med. 18, 1658–1664. https://doi.org/10.1038/nm.2943.

Murrough, J.W., Huryk, K.M., Mao, X., Iacoviello, B., Collins, K., Nierenberg, A.A., Kang, G., Shungu, D.C., Iosifescu, D.V., 2018. A pilot study of minocycline for the treatment of bipolar depression: effects on cortical glutathione and oxidative stress in vivo. J. Affect. Disord. 230, 56–64. https://doi.org/10.1016/j.jad.2017.12.067.

Nair, A., Bonneau, R.H., 2006. Stress-induced elevation of glucocorticoids increases microglia proliferation through NMDA receptor activation. J. Neuroimmunol. 171, 72–85. https://doi.org/10.1016/j.jneuroim.2005.09.012.

Niklasson, F., Agren, H., 1984. Brain energy metabolism and blood-brain barrier permeability in depressive patients: analyses of creatine, creatinine, urate, and albumin in CSF and blood. Biol. Psychiatry 19, 1183–1206.

Nimmerjahn, A., Kirchhoff, F., Helmchen, F., 2005. Resting microglial cells are highly dynamic surveillants of brain parenchyma in vivo. Science 308, 1314–1318. https://doi.org/10.1126/science.1110647.

O'connor, K.A., Johnson, J.D., Hansen, M.K., Wieseler Frank, J.L., Maksimova, E., Watkins, L.R., Maier, S.F., 2003. Peripheral and central proinflammatory cytokine response to a severe acute stressor. Brain Res. 991, 123–132.

Pace, T.W., Mletzko, T.C., Alagbe, O., Musselman, D.L., Nemeroff, C.B., Miller, A.H., Heim, C.M., 2006. Increased stress-induced inflammatory responses in male patients with major depression and increased early life stress. Am. J. Psychiatry 163, 1630–1633. https://doi.org/10.1176/appi.ajp.163.9.1630.

Patki, G., Allam, F.H., Atrooz, F., Dao, A.T., Solanki, N., Chugh, G., Asghar, M., Jafri, F., Bohat, R., Alkadhi, K.A., Salim, S., 2013a. Grape powder intake prevents ovariectomy-induced anxiety-like behavior, memory impairment and high blood pressure in female Wistar rats. PLoS One 8, e74522. https://doi.org/10.1371/journal.pone.0074522.

Patki, G., Solanki, N., Atrooz, F., Allam, F., Salim, S., 2013b. Depression, anxiety-like behavior and memory impairment are associated with increased oxidative stress and inflammation in a rat model of social stress. Brain Res. 1539, 73–86. https://doi.org/10.1016/j.brainres.2013.09.033.

Pavon, L., Sandoval-Lopez, G., Eugenia Hernandez, M., Loria, F., Estrada, I., Perez, M., Moreno, J., Avila, U., Leff, P., Anton, B., Heinze, G., 2006. Th2 cytokine response in major depressive disorder patients before treatment. J. Neuroimmunol. 172, 156–165. https://doi.org/10.1016/j.jneuroim.2005.08.014.

Pearson-Leary, J., Eacret, D., Chen, R., Takano, H., Nicholas, B., Bhatnagar, S., 2017. Inflammation and vascular remodeling in the ventral hippocampus contributes to vulnerability to stress. Transl. Psychiatry 7, e1160. https://doi.org/10.1038/tp.2017.122.

Powell, N.D., Sloan, E.K., Bailey, M.T., Arevalo, J.M., Miller, G.E., Chen, E., Kobor, M.S., Reader, B.F., Sheridan, J.F., Cole, S.W., 2013. Social stress up-regulates inflammatory gene expression in the leukocyte transcriptome via beta-adrenergic induction of myelopoiesis. Proc. Natl. Acad. Sci. U. S. A. 110, 16574–16579. https://doi.org/10.1073/pnas.1310655110.

Priyadarsini, K.I., Guha, S.N., Rao, M.N., 1998. Physico-chemical properties and antioxidant activities of methoxy phenols. Free Radic. Biol. Med. 24, 933–941.

Qiu, J., Xu, J., Zheng, Y., Wei, Y., Zhu, X., Lo, E.H., Moskowitz, M.A., Sims, J.R., 2010. High-mobility group box 1 promotes metalloproteinase-9 upregulation through toll-like receptor 4 after cerebral ischemia. Stroke 41, 2077–2082. https://doi.org/10.1161/STROKEAHA.110.590463.

Quinn, A.M., Williams, A.R., Sivilli, T.I., Raison, C.L., Pace, T.W.W., 2018. The plasma interleukin-6 response to acute psychosocial stress in humans is detected by a magnetic multiplex assay: comparison to high-sensitivity ELISA. Stress, 1–6. https://doi.org/10.1080/10253890.2018.1446518.

Ransohoff, R.M., Kivisakk, P., Kidd, G., 2003. Three or more routes for leukocyte migration into the central nervous system. Nat. Rev. Immunol. 3, 569–581. https://doi.org/10.1038/nri1130.

Renault, P.F., Hoofnagle, J.H., Park, Y., Et, A., 1987. Psychiatric complications of long-term interferon alfa therapy. Arch. Intern. Med. 147, 1577–1580. https://doi.org/10.1001/archinte.1987.00370090055011.

Rezin, G.T., Cardoso, M.R., Goncalves, C.L., Scaini, G., Fraga, D.B., Riegel, R.E., Comim, C.M., Quevedo, J., Streck, E.L., 2008. Inhibition of mitochondrial respiratory chain in brain of rats subjected to an experimental model of depression. Neurochem. Int. 53, 395–400. https://doi.org/10.1016/j.neuint.2008.09.012.

Riba, M., Wulsin, L., Rubenfire, M., 2011. Psychiatry and Heart Disease: The Mind, Brain, and Heart. Wiley-Blackwell.

Ridker, P.M., 2003. Clinical application of C-reactive protein for cardiovascular disease detection and prevention. Circulation 107, 363–369.

Rochfort, K.D., Collins, L.E., Murphy, R.P., Cummins, P.M., 2014. Downregulation of blood-brain barrier phenotype by proinflammatory cytokines involves NADPH oxidase-dependent ROS generation: consequences for interendothelial adherens and tight junctions. PLoS One 9, e101815. https://doi.org/10.1371/journal.pone.0101815.
Ros-Bernal, F., Hunot, S., Herrero, M.T., Parnadeau, S., Corvol, J.-C., Lu, L., Alvarez-Fischer, D., Carrillo-De Sauvage, M.A., Saurini, F., Coussieu, C., Kinugawa, K., Prigent, A., Höglinger, G., Hamon, M., Tronche, F., Hirsch, E.C., Vyas, S., 2011. Microglial glucocorticoid receptors play a pivotal role in regulating dopaminergic neurodegeneration in parkinsonism. Proc. Natl. Acad. Sci. 108, 6632.
Rosenblat, J.D., 2019. Targeting the immune system in the treatment of bipolar disorder. Psychopharmacology. https://doi.org/10.1007/s00213-019-5175-x.
Ruparelia, N., Chai, J.T., Fisher, E.A., Choudhury, R.P., 2017. Inflammatory processes in cardiovascular disease: a route to targeted therapies. Nat. Rev. Cardiol. 14, 133–144. https://doi.org/10.1038/nrcardio.2016.185.
Sacher, J., Rekkas, P.V., Wilson, A.A., Houle, S., Romano, L., Hamidi, J., Rusjan, P., Fan, I., Stewart, D.E., Meyer, J.H., 2015. Relationship of monoamine oxidase-A distribution volume to postpartum depression and postpartum crying. Neuropsychopharmacology 40, 429–435. https://doi.org/10.1038/npp.2014.190.
Salim, S., Asghar, M., Chugh, G., Taneja, M., Xia, Z., Saha, K., 2010. Oxidative stress: a potential recipe for anxiety, hypertension and insulin resistance. Brain Res. 1359, 178–185. https://doi.org/10.1016/j.brainres.2010.08.093.
Salim, S., Asghar, M., Taneja, M., Hovatta, I., Chugh, G., Vollert, C., Vu, A., 2011. Potential contribution of oxidative stress and inflammation to anxiety and hypertension. Brain Res. 1404, 63–71. https://doi.org/10.1016/j.brainres.2011.06.024.
Samuels, E.R., Szabadi, E., 2008. Functional neuroanatomy of the noradrenergic locus coeruleus: its roles in the regulation of arousal and autonomic function part I: principles of functional organisation. Curr. Neuropharmacol. 6, 235–253. https://doi.org/10.2174/157015908785777229.
Sarkar, S., Malovic, E., Harishchandra, D.S., Ghaisas, S., Panicker, N., Charli, A., Palanisamy, B.N., Rokad, D., Jin, H., Anantharam, V., Kanthasamy, A., Kanthasamy, A.G., 2017. Mitochondrial impairment in microglia amplifies NLRP3 inflammasome proinflammatory signaling in cell culture and animal models of Parkinson's disease. NPJ Parkinsons Dis. 3, 30. https://doi.org/10.1038/s41531-017-0032-2.
Sasayama, D., Hattori, K., Wakabayashi, C., Teraishi, T., Hori, H., Ota, M., Yoshida, S., Arima, K., Higuchi, T., Amano, N., Kunugi, H., 2013. Increased cerebrospinal fluid interleukin-6 levels in patients with schizophrenia and those with major depressive disorder. J. Psychiatr. Res. 47, 401–406. https://doi.org/10.1016/j.jpsychires.2012.12.001.
Savoia, C., Schiffrin, E.L., 2006. Inflammation in hypertension. Curr. Opin. Nephrol. Hypertens. 15, 152–158. https://doi.org/10.1097/01.mnh.0000203189.57513.76.
Scapagnini, G., Davinelli, S., Drago, F., De Lorenzo, A., Oriani, G., 2012. Antioxidants as antidepressants: fact or fiction? CNS Drugs 26, 477–490. https://doi.org/10.2165/11633190-000000000-00000.
Schmeler, K.M., Vadhan-Raj, S., Ramirez, P.T., Apte, S.M., Cohen, L., Bassett, R.L., Iyer, R.B., Wolf, J.K., Levenback, C.L., Gershenson, D.M., Freedman, R.S., 2009. A phase II study of GM-CSF and rIFN-gamma1b plus carboplatin for the treatment of recurrent, platinum-sensitive ovarian, fallopian tube and primary peritoneal cancer. Gynecol. Oncol. 113, 210–215. https://doi.org/10.1016/j.ygyno.2009.02.007.
Seligman, F., Nemeroff, C.B., 2015. The interface of depression and cardiovascular disease: therapeutic implications. Ann. N. Y. Acad. Sci. 1345, 25–35. https://doi.org/10.1111/nyas.12738.
Seo, J.H., Miyamoto, N., Hayakawa, K., Pham, L.D., Maki, T., Ayata, C., Kim, K.W., Lo, E.H., Arai, K., 2013. Oligodendrocyte precursors induce early blood-brain barrier opening after white matter injury. J. Clin. Invest. 123, 782–786. https://doi.org/10.1172/JCI65863.

Setiawan, E., Wilson, A.A., Mizrahi, R., Rusjan, P.M., Miler, L., Rajkowska, G., Suridjan, I., Kennedy, J.L., Rekkas, P.V., Houle, S., Meyer, J.H., 2015. Role of translocator protein density, a marker of neuroinflammation, in the brain during major depressive episodes. JAMA Psychiatry 72, 268–275. https://doi.org/10.1001/jamapsychiatry.2014.2427.

Shelburne, C.P., Abraham, S.N., 2011. The mast cell in innate and adaptive immunity. Adv. Exp. Med. Biol. 716, 162–185. https://doi.org/10.1007/978-1-4419-9533-9_10.

Shi, G.P., Bot, I., Kovanen, P.T., 2015. Mast cells in human and experimental cardiometabolic diseases. Nat. Rev. Cardiol. 12, 643–658. https://doi.org/10.1038/nrcardio.2015.117.

Solanki, N., Alkadhi, I., Atrooz, F., Patki, G., Salim, S., 2015. Grape powder prevents cognitive, behavioral, and biochemical impairments in a rat model of posttraumatic stress disorder. Nutr. Res. 35, 65–75. https://doi.org/10.1016/j.nutres.2014.11.008.

Solanki, N., Salvi, A., Patki, G., Salim, S., 2017. Modulating oxidative stress relieves stress-induced behavioral and cognitive impairments in rats. Int. J. Neuropsychopharmacol. 20, 550–561. https://doi.org/10.1093/ijnp/pyx017.

Spengler, R.N., Allen, R.M., Remick, D.G., Strieter, R.M., Kunkel, S.L., 1990. Stimulation of alpha-adrenergic receptor augments the production of macrophage-derived tumor necrosis factor. J. Immunol. 145, 1430–1434.

Sreejayan, Rao, M.N., 1997. Nitric oxide scavenging by curcuminoids. J. Pharm. Pharmacol. 49, 105–107.

St John, A.L., Abraham, S.N., 2013. Innate immunity and its regulation by mast cells. J. Immunol. 190, 4458–4463. https://doi.org/10.4049/jimmunol.1203420.

Stauder, A., Kovacs, M., 2003. Anxiety symptoms in allergic patients: identification and risk factors. Psychosom. Med. 65, 816–823.

Stewart, A.M., Roy, S., Wong, K., Gaikwad, S., Chung, K.M., Kalueff, A.V., 2015. Cytokine and endocrine parameters in mouse chronic social defeat: implications for translational 'cross-domain' modeling of stress-related brain disorders. Behav. Brain Res. 276, 84–91. https://doi.org/10.1016/j.bbr.2014.08.037.

Strawbridge, R., Arnone, D., Danese, A., Papadopoulos, A., Herane Vives, A., Cleare, A.J., 2015. Inflammation and clinical response to treatment in depression: a meta-analysis. Eur. Neuropsychopharmacol. 25, 1532–1543. https://doi.org/10.1016/j.euroneuro.2015.06.007.

Sugamura, K., Keaney Jr., J.F., 2011. Reactive oxygen species in cardiovascular disease. Free Radic. Biol. Med. 51, 978–992. https://doi.org/10.1016/j.freeradbiomed.2011.05.004.

Tang, Y.L., Yang, Y.Z., Wang, S., Huang, T., Tang, C.K., Xu, Z.X., Sun, Y.H., 2009. Mast cell degranulator compound 48-80 promotes atherosclerotic plaque in apolipoprotein E knockout mice with perivascular common carotid collar placement. Chin. Med. J. 122, 319–325.

Tang, J., Yu, W., Chen, S., Gao, Z., Xiao, B., 2018. Microglia polarization and endoplasmic reticulum stress in chronic social defeat stress induced depression mouse. Neurochem. Res. 43, 985–994. https://doi.org/10.1007/s11064-018-2504-0.

Taniyama, Y., Griendling, K.K., 2003. Reactive oxygen species in the vasculature: molecular and cellular mechanisms. Hypertension 42, 1075–1081. https://doi.org/10.1161/01.HYP.0000100443.09293.4F.

Turrens, J.F., 2003. Mitochondrial formation of reactive oxygen species. J. Physiol. 552, 335–344. https://doi.org/10.1113/jphysiol.2003.049478.

Ufnal, M., Sikora, M., Szczepanska-Sadowska, E., 2008. Interleukin-1 receptor antagonist reduces the magnitude of the pressor response to acute stress. Neurosci. Lett. 448, 47–51. https://doi.org/10.1016/j.neulet.2008.10.010.

Valgimigli, M., Ceconi, C., Malagutti, P., Merli, E., Soukhomovskaia, O., Francolini, G., Cicchitelli, G., Olivares, A., Parrinello, G., Percoco, G., Guardigli, G., Mele, D., Pirani, R., Ferrari, R., 2005. Tumor necrosis factor-alpha receptor 1 is a major predictor of mortality

and new-onset heart failure in patients with acute myocardial infarction: the Cytokine-Activation and Long-Term Prognosis in Myocardial Infarction (C-ALPHA) study. Circulation 111, 863–870. https://doi.org/10.1161/01.CIR.0000155614.35441.69.
Van Hout, G.P., Bosch, L., Ellenbroek, G.H., De Haan, J.J., Van Solinge, W.W., Cooper, M.A., Arslan, F., De Jager, S.C., Robertson, A.A., Pasterkamp, G., Hoefer, I.E., 2017. The selective NLRP3-inflammasome inhibitor MCC950 reduces infarct size and preserves cardiac function in a pig model of myocardial infarction. Eur. Heart J. 38, 828–836. https://doi.org/10.1093/eurheartj/ehw247.
Vavakova, M., Durackova, Z., Trebaticka, J., 2015. Markers of oxidative stress and neuroprogression in depression disorder. Oxidative Med. Cell. Longev. 2015, 898393. https://doi.org/10.1155/2015/898393.
Videbech, P., 2000. PET measurements of brain glucose metabolism and blood flow in major depressive disorder: a critical review. Acta Psychiatr. Scand. 101, 11–20.
Von Kanel, R., Nelesen, R.A., Mills, P.J., Ziegler, M.G., Dimsdale, J.E., 2008. Relationship between heart rate variability, interleukin-6, and soluble tissue factor in healthy subjects. Brain Behav. Immun. 22, 461–468. https://doi.org/10.1016/j.bbi.2007.09.009.
Wang, J., Sjoberg, S., Tia, V., Secco, B., Chen, H., Yang, M., Sukhova, G.K., Shi, G.P., 2013. Pharmaceutical stabilization of mast cells attenuates experimental atherogenesis in low-density lipoprotein receptor-deficient mice. Atherosclerosis 229, 304–309. https://doi.org/10.1016/j.atherosclerosis.2013.05.025.
Wang, Y.L., Han, Q.Q., Gong, W.Q., Pan, D.H., Wang, L.Z., Hu, W., Yang, M., Li, B., Yu, J., Liu, Q., 2018. Microglial activation mediates chronic mild stress-induced depressive- and anxiety-like behavior in adult rats. J. Neuroinflammation 15, 21. https://doi.org/10.1186/s12974-018-1054-3.
Weber, M.D., Frank, M.G., Tracey, K.J., Watkins, L.R., Maier, S.F., 2015. Stress induces the danger-associated molecular pattern HMGB-1 in the hippocampus of male Sprague Dawley rats: a priming stimulus of microglia and the NLRP3 inflammasome. J. Neurosci. 35, 316–324. https://doi.org/10.1523/JNEUROSCI.3561-14.2015.
Wohleb, E.S., Fenn, A.M., Pacenta, A.M., Powell, N.D., Sheridan, J.F., Godbout, J.P., 2012. Peripheral innate immune challenge exaggerated microglia activation, increased the number of inflammatory CNS macrophages, and prolonged social withdrawal in socially defeated mice. Psychoneuroendocrinology 37, 1491–1505. https://doi.org/10.1016/j.psyneuen.2012.02.003.
Wohleb, E.S., Powell, N.D., Godbout, J.P., Sheridan, J.F., 2013. Stress-induced recruitment of bone marrow-derived monocytes to the brain promotes anxiety-like behavior. J. Neurosci. 33, 13820–13833. https://doi.org/10.1523/JNEUROSCI.1671-13.2013.
Wohleb, E.S., Mckim, D.B., Sheridan, J.F., Godbout, J.P., 2014a. Monocyte trafficking to the brain with stress and inflammation: a novel axis of immune-to-brain communication that influences mood and behavior. Front. Neurosci. 8, 447. https://doi.org/10.3389/fnins.2014.00447.
Wohleb, E.S., Patterson, J.M., Sharma, V., Quan, N., Godbout, J.P., Sheridan, J.F., 2014b. Knockdown of interleukin-1 receptor type-1 on endothelial cells attenuated stress-induced neuroinflammation and prevented anxiety-like behavior. J. Neurosci. 34, 2583–2591. https://doi.org/10.1523/JNEUROSCI.3723-13.2014.
Wood, S.K., Wood, C.S., Lombard, C.M., Lee, C.S., Zhang, X.Y., Finnell, J.E., Valentino, R.J., 2015. Inflammatory factors mediate vulnerability to a social stress-induced depressive-like phenotype in passive coping rats. Biol. Psychiatry 78, 38–48. https://doi.org/10.1016/j.biopsych.2014.10.026.
Wood, C.S., Valentino, R.J., Wood, S.K., 2017. Individual differences in the locus coeruleus-norepinephrine system: relevance to stress-induced cardiovascular vulnerability. Physiol. Behav. 172, 40–48. https://doi.org/10.1016/j.physbeh.2016.07.008.

Xu, Y., Wang, Z., You, W., Zhang, X., Li, S., Barish, P.A., Vernon, M.M., Du, X., Li, G., Pan, J., Ogle, W.O., 2010. Antidepressant-like effect of trans-resveratrol: involvement of serotonin and noradrenaline system. Eur. Neuropsychopharmacol. 20, 405–413. https://doi.org/10.1016/j.euroneuro.2010.02.013.

Yan, W., Chang, Y., Liang, X., Cardinal, J.S., Huang, H., Thorne, S.H., Monga, S.P., Geller, D.A., Lotze, M.T., Tsung, A., 2012. High-mobility group box 1 activates caspase-1 and promotes hepatocellular carcinoma invasiveness and metastases. Hepatology 55, 1863–1875. https://doi.org/10.1002/hep.25572.

Yang, H., Lundback, P., Ottosson, L., Erlandsson-Harris, H., Venereau, E., Bianchi, M.E., Al-Abed, Y., Andersson, U., Tracey, K.J., Antoine, D.J., 2012. Redox modification of cysteine residues regulates the cytokine activity of high mobility group box-1 (HMGB1). Mol. Med. 18, 250–259. https://doi.org/10.2119/molmed.2011.00389.

Ye, S., Mozayeni, P., Gamburd, M., Zhong, H., Campese, V.M., 2000. Interleukin-1beta and neurogenic control of blood pressure in normal rats and rats with chronic renal failure. Am. J. Physiol. Heart Circ. Physiol. 279, H2786–H2796.

Yusuf, S., Hawken, S., Ounpuu, S., Dans, T., Avezum, A., Lanas, F., Mcqueen, M., Budaj, A., Pais, P., Varigos, J., Lisheng, L., Investigators, I.S., 2004. Effect of potentially modifiable risk factors associated with myocardial infarction in 52 countries (the INTERHEART study): case-control study. Lancet 364, 937–952. https://doi.org/10.1016/S0140-6736(04)17018-9.

Zal, B., Kaski, J.C., Arno, G., Akiyu, J.P., Xu, Q., Cole, D., Whelan, M., Russell, N., Madrigal, J.A., Dodi, I.A., Baboonian, C., 2004. Heat-shock protein 60-reactive CD4 + CD28null T cells in patients with acute coronary syndromes. Circulation 109, 1230–1235. https://doi.org/10.1161/01.CIR.0000118476.29352.2A.

Zhang, F., Liu, J., Shi, J.S., 2010. Anti-inflammatory activities of resveratrol in the brain: role of resveratrol in microglial activation. Eur. J. Pharmacol. 636, 1–7. https://doi.org/10.1016/j.ejphar.2010.03.043.

Zhang, J., Takahashi, H.K., Liu, K., Wake, H., Liu, R., Maruo, T., Date, I., Yoshino, T., Ohtsuka, A., Mori, S., Nishibori, M., 2011. Anti-high mobility group box-1 monoclonal antibody protects the blood-brain barrier from ischemia-induced disruption in rats. Stroke 42, 1420–1428. https://doi.org/10.1161/strokeaha.110.598334.

Zhang, Y., Liu, L., Peng, Y.L., Liu, Y.Z., Wu, T.Y., Shen, X.L., Zhou, J.R., Sun, D.Y., Huang, A.J., Wang, X., Wang, Y.X., Jiang, C.L., 2014. Involvement of inflammasome activation in lipopolysaccharide-induced mice depressive-like behaviors. CNS Neurosci. Ther. 20, 119–124. https://doi.org/10.1111/cns.12170.

Zhu, L., Wei, T., Gao, J., Chang, X., He, H., Miao, M., Yan, T., 2015. Salidroside attenuates lipopolysaccharide (LPS) induced serum cytokines and depressive-like behavior in mice. Neurosci. Lett. 606, 1–6. https://doi.org/10.1016/j.neulet.2015.08.025.

Zorrilla, E.P., Luborsky, L., Mckay, J.R., Rosenthal, R., Houldin, A., Tax, A., Mccorkle, R., Seligman, D.A., Schmidt, K., 2001. The relationship of depression and stressors to immunological assays: a meta-analytic review. Brain Behav. Immun. 15, 199–226. https://doi.org/10.1006/brbi.2000.0597.

Zou, C.J., Liu, J.D., Zhou, Y.C., 2001. Roles of central interleukin-1 on stress-induced-hypertension and footshock-induced-analgesia in rats. Neurosci. Lett. 311, 41–44.

Zuo, L., Pasniciuc, S., Wright, V.P., Merola, A.J., Clanton, T.L., 2003. Sources for superoxide release: lessons from blockade of electron transport, NADPH oxidase, and anion channels in diaphragm. Antioxid. Redox Signal. 5, 667–675. https://doi.org/10.1089/152308603770310347.

CHAPTER 9

Social support and stress, depression, and cardiovascular disease

Bert N. Uchino[a], Sierra Cronan[a], Emily Scott[a], Joshua Landvatter[a], Michael Papadakis[b]
[a]Department of Psychology and Health Psychology Program, University of Utah, Salt Lake City, UT, United States
[b]Department of Kinesiology, University of Utah, Salt Lake City, UT, United States

Contents

The quality of one's social relationships is one of the most reliable predictors of better health outcomes, including lower cardiovascular disease risk (Barth et al., 2010; Holt-Lunstad et al., 2010). It is important to note that most of the research in the area has been concerned with establishing links between social support and health outcomes. However, much less work has considered the possibility that social support might be an important mediator of links between other psychological factors and cardiovascular disease risk. Of these, stress and depression have reliably been related to higher cardiovascular disease morbidity and mortality (Richardson et al., 2012; Gan et al., 2014). This chapter will thus examine if social support mediates links between stress/depression and cardiovascular disease risk.

A conceptual model of such linkages is depicted in Fig. 9.1. According to the model, stress and depression have direct positive and negative (+/−) influences on social support processes (e.g., support mobilization/erosion, Path A). Social support can be separated along two dimensions: what support an individual perceives may be *available* to him/her, and the support an

Cardiovascular Implications of Stress and Depression
https://doi.org/10.1016/B978-0-12-815015-3.00009-X

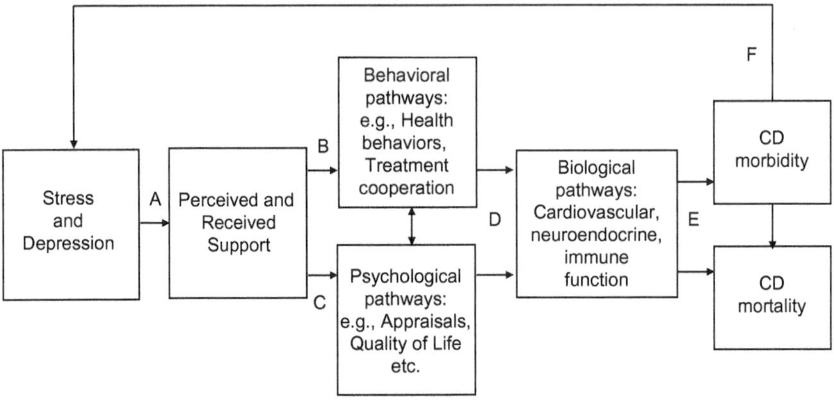

Fig. 9.1 Broad conceptual model linking stress/depression with cardiovascular disease.

individual actually *receives* (Wethington and Kessler, 1986). It is important to note that the perceived availability and receipt of support are not always related in a clear-cut manner and have often been associated with different effects on well-being (Wethington and Kessler, 1986; Dunkel-Schetter and Skokan, 1990; Wills and Shinar, 2000). For instance, perceived support appears to have stronger protective links to mortality than receiving support (e.g. Holt-Lunstad et al., 2010; Uchino, 2009). The links between received support and health have been more variable (either health-promoting or health-damaging) and it has been hypothesized that it is a contextual factor whose efficacy depends on provider, recipient, and stressor factors (e.g., relationship quality, type of stress, Uchino, 2009). For instance, receiving support from a low quality relationship is less beneficial, and in some cases, can exacerbate cardiovascular stress responses (Reblin et al., 2010).

The proposed links between measures of social support and health-relevant biological pathways (e.g., blood pressure, oxytocin, inflammation) are also depicted. The generally beneficial influence of perceived support is shown (+, Path B) along with the contextual nature of received support (+/−, Path C). These health-relevant biological processes are postulated to have a direct influence on the development and/or progression of cardiovascular disease as shown in Path D. Each of these major pathways is detailed below. Finally, although not an emphasis of this review, Path E indicates that the diagnosis of cardiovascular disease can feedback and further exacerbate the association between stress/depression and health (Nicassio and Smith, 1995; Berg and Upchurch, 2007).

The link between stress and social support

There is strong evidence for an association between stress and perceived/received social support as depicted in Fig. 9.1 (path A, Gore, 1981; Kaniasty and Norris, 2008). Conceptually, stress can lead to a mobilization of received social support which can then have beneficial influences if it is responsive to the recipient's needs (Uchino et al., 2011). Benefits may occur due to cognitive reappraisals of stress which facilitates better coping (Ozbay et al., 2007; Schwarzer and Knoll, 2007). In addition, receiving social support is thought to be directly involved in buffering against biological reactivity to stress (Heinrichs et al., 2003) as shown by a meta-analysis (Thorsteinsson and James, 1999). It should be noted that consistent with Fig. 9.1, the meta-analysis also found that receiving support was not always effective and moderated by factors such as the level of evaluation which might interfere with effective support (Thorsteinsson and James, 1999).

Although support might be mobilized early in the stress process, over time stress can lead to an erosion of received support. In one study, researchers directly tested this possibility by examining how support was influenced when couples attempted to cope with the diagnosis and treatment of breast cancer (Bolger et al., 1996). These researchers found that although support was initially mobilized in response to the diagnosis, the patients' distress was related to less enacted support from the spouse over time (also see Alferi et al., 2001). Such associations have also been found for perceived support as longitudinal studies show that chronic stress is related to declines in perceived support (Lepore et al., 1991; Prati and Pietrantoni, 2010). In other cases, stress can also lead to an erosion of support if one directly withdraws from their support network. For instance, both husbands and wives who experience workplace stressors tend to withdraw from their spouse at home (Story and Repetti, 2006). Besides depriving the individual of social resources to cope with stress, isolation can have deleterious effects on its own by disrupting biological homeostasis (Cacioppo et al., 2015).

The link between depression and social support

There is a relatively large literature linking depression to social support. Most research in this area has focused on whether perceived social support predicts depression over time. For instance, a nationally representative study found that people with low levels of social support had more than double the risk of developing depression over a 10-year period than those with high levels of social support (Teo et al., 2013). More generally, several meta-analyses show an association between depression and lower social support (Gariépy et al., 2016; Rueger et al., 2016).

The meta-analyses linking depression and social support are at least consistent with the possibility that depression predicts lower social support given many were cross-sectional studies (see path A, Fig. 9.1). As the case for stress, it is likely that depression is initially associated with a mobilization of received support (Griffiths et al., 2011). However, according to interpersonal models, depression is likely to interfere with receiving effective support (Coyne, 1976). Based on such perspectives, individuals interacting with a depressed person may begin to act in unsupportive ways (e.g., anger) in response to continued problems (Coyne, 1976). This may undermine the support process and inadvertently contribute to the maintenance of depression (Benazon, 2000). Consistent with this possibility, one prospective study found that depression was related to an erosion of perceived peer support over a 2 year period in adolescent girls (Stice et al., 2004). In addition, a longitudinal study of adolescents found that depression predicted decreases in perceived support over a 6 month period (Ren et al., 2018).

The link between social support, biological risk, and cardiovascular disease

The evidence linking perceived social support to biological risk factors for cardiovascular disease is strong (Uchino, 2006). In terms of the autonomic nervous system (ANS), besides the association between received support and lower cardiovascular reactivity covered above (e.g., Thorsteinsson and James, 1999), perceived support is related to lower ambulatory blood pressure during everyday life (Steptoe et al., 2000; Rodriguez et al., 2008). Perceived support is also associated with lower catecholamine levels (Grewen et al., 2005). Much work in this area has also focused on the association between social support and inflammation which is linked to future cardiovascular risk (Libby et al., 2016; Ruparelia et al., 2017). A recent meta-analysis found that social support was related to lower levels of inflammatory cytokines, and consistent with Fig. 9.1, the effect size appeared larger for perceived in contrast to received support (Uchino et al., 2018). It is likely that the association between social support and peripheral physiological measures of the ANS and inflammatory outcomes reflect central brain mechanisms (e.g., anterior cingulate cortex, Eisenberger et al., 2011), although more work is needed on the specificity of such links.

The oxytocin system is also emerging as an important integrative biological mechanism linking social support to health outcomes (Uchino and Way, 2017). It is ideally situated to be involved in social support influences as it is activated during social interactions with close others (Uchino and Way, 2017). For instance, variations in the oxytocin receptor gene are related to supportive behaviors and support quality (Mattson et al., 2019) and plasma oxytocin levels are associated with increased informal help-seeking in clinically depressed individuals (Thomas and Larkin, 2018). Individuals who were raised in unsupportive and abusive households show lower cerebrospinal fluid oxytocin concentrations in adulthood (Heim et al., 2009) and intranasal oxytocin has less positive social effects for such individuals, suggesting high levels of early life stress result in deficits in the oxytocin system (Riem et al., 2013). Although close, positive relationship processes such as support are related to activation of the oxytocin system, it should be noted that oxytocin can have different effects depending the context of the interaction (Bartz, 2016). For instance, administration of intranasal oxytocin increases in-group trust but has non-significant influences on out-group trust (see meta-analysis by Van IJzendoorn and Bakermans-Kranenburg, 2012).

The oxytocin system is also of interest because it has cascading links to other biological systems implicated in support processes (Uchino and Way, 2017). Oxytocin neurons reside in the hypothalamus and has projections to the pituitary as well as stress-related forebrain regions (e.g., ventral striatum, amygdala, Knobloch et al., 2012). Consistent with the beneficial biological influences of social support detailed above, oxytocin has been associated with decreased HPA responses (Neumann et al., 2000), blood pressure levels (Gutkowska and Jankowski, 2012), inflammation (Clodi et al., 2008), and increased respiratory sinus arrhythmia (Norman et al., 2011). As a result, oxytocin may be a key integrative biological mechanism given its association with both central and peripheral physiological processes (Radke et al., 2017; Uchino and Way, 2017). Although animal models show that oxytocin has protective effects on cardiovascular conditions such as myocardial infarction and hypertension (Moghimian et al., 2013; Jameson et al., 2016), there is currently a lack of human studies directly examining cardiovascular disease outcomes.

Consistent with the biological mechanisms reviewed above, perceived social support is associated with the initial development of coronary heart disease (CHD). The strongest test comes from prospective cohort studies of originally healthy individuals with unambiguous clinical endpoints such as myocardial infarction (MI; Barth et al., 2010). For CHD, there is a clear effect of social support on MI risk (André-Petersson et al., 2007, 2006; but see Piros et al., 2000). Studies including broader endpoints such as angina generally support the relationship between incident CHD and low social support (Orth-Gomér et al., 1993; Rosengren et al., 2004). In contrast, several large prospective cohort studies suggest that social support is likely unrelated to incident stroke (Ikeda et al., 2008; Ikehara et al., 2009; Nagayoshi et al., 2014). Data linking social support to the development of hypertension is mixed and requires more work (Räikkönen et al., 2001; Kamiya et al., 2010; Croezen et al., 2012).

A larger body of evidence connects perceived social support to prognosis and mortality for individuals diagnosed with cardiovascular disease. Barth et al. (2010) meta-analysis reported a hazard ratio of 1.59 for social support on all-cause mortality in patients with existing cardiovascular disease, after adjusting for other risk factors (see recent study by Weiss-Faratci et al., 2016). These data are consistent with research suggesting that social support is related to slower disease progression in cardiovascular patients as indexed by imaging techniques (Angerer et al., 2000; Wang et al., 2005). However, there is some heterogeneity among mortality studies both in terms of effect size and in the measurement of social support (Berkman et al., 1992; Brummett et al., 2005; Ikeda et al., 2008). The effect of perceived support on prognosis is likely smaller as cardiovascular disease severity increases (Luttik et al., 2005; Weiss-Faratci et al., 2016), and there is some evidence for a stronger relationship in women than in men (Leifheit-Limson et al., 2010; Low et al., 2010).

There is much less data on the link between received support and cardiovascular disease outcomes. As noted earlier, the receipt of support can be associated with variable outcomes, including higher mortality rates (Uchino, 2009). This variability is likely due to contextual and relationship factors such as whether the support actually meets an individual's needs or is perceived as responsive (Selcuk and Ong, 2013). Accordingly, a meta-analysis concluded that while perceived support predicts lower all-cause mortality, received support may not (Holt-Lunstad et al., 2010). However, some evidence is more positive. One early intervention found that patients who received volunteer support visits survived longer than individuals who did not (Herbst-Damm and Kulik, 2005). A recent longitudinal study also found that received support predicted mortality in patients with chronic disease (Becofsky et al., 2015). These data indicate that receiving support might be beneficial under some circumstances, but future studies focused on cardiovascular disease outcomes will be needed to make stronger inferences.

Social support as a mediator

There is a relatively large literature examining if depression or stress mediates the link between social support and health outcomes (Uchino et al., 2012). It is worth noting that these studies have generally failed to show mediation, although there were important conceptual and statistical issues not addressed by prior work (Uchino et al., 2012). But what about the possibility that social support is a mediator of the link between depression/stress and cardiovascular disease risk? Very few studies appear to have tested that possibility although this is plausible given the data reviewed above showing the necessary links (see Fig. 9.1, paths A–C). More formal tests, however, would require prospective designs and advanced statistical techniques such as structural equation modeling (SEM) or bootstrapping (Rucker et al., 2011).

Two studies appear to have partially tested social support as a mediator of depression/stress and an index of cardiovascular function. Using SEM, one study found the prospective association between left-ventricular ejection fraction and depression in 75 cardiovascular patients was partially mediated by the combination of received social support, identity illness, and self-efficacy (Greco et al., 2014). A second study (National Longitudinal Study of Adolescent to Adult Health) examined if maternal perceived social support mediated links between adolescent SES (a potential indicator of stress) and cardiovascular risk as indexed by the Framingham risk score (Doom et al., 2016). Although maternal support predicted lower cardiovascular risk, it was not shown to statistically mediate the link between adolescent SES and disease risk. Of course, the specificity of the links might have been improved by examining only social support in the first study or directly measuring stressors/depression (instead of using SES as a proxy) in the second study. Overall, studies in the area are sparse so additional work will be needed to test the support mediational model in Fig. 9.1.

Intervention implications

The model postulated in Fig. 9.1 has several important clinical implications. If the model is correct, it suggests that social support is a more proximal factor linking depression and stress to cardiovascular risk. This implies that social support interventions may be used to effectively reduce cardiovascular morbidity and mortality associated with stress and depression. There is a relatively large literature on social support interventions that can be leveraged to address this question (Hogan et al., 2002; Holt-Lunstad and Uchino, 2015). Although this literature is complex, several studies have shown that support interventions might be beneficial to cardiovascular risk given its impact on diabetes regulation and health behaviors more generally (Heisler et al., 2010; Patten et al., 2011; Sorkin et al., 2014). For instance, one intervention used a dyadic support approach (i.e., mother-daughter) in an attempt to reduce diabetes complications in high-risk Latinas (Sorkin et al., 2014). This research showed that intervention participants had increases in social support and social control, as well as decreases in undermining behaviors. Importantly, the dyadic intervention group had greater weight loss and healthier dietary changes than a control group (Sorkin et al., 2014).

More generally, the model suggests the importance of a comprehensive psychosocial approach that targets multiple links in the theoretical chain. This is consistent with the aims and outcomes of cognitive behavioral therapy for stress which seeks to also improve interpersonal functioning (Fredette et al., 2016). For instance, one study utilized a broad intervention focused on reducing stress, increasing positive social ties, and decreasing negative social ties among women who had recently had a heart attack (Orth-Gomér et al., 2009). Results showed that mortality due to coronary disease was reduced by 66% over a 9 year period (Orth-Gomér et al. 2009). Such studies suggest the promise of the model in Fig. 9.1 to guide both comprehensive and more specific intervention strategies. Of course, interventions can be very informative to theory building as well so future interventions in the area should measure stress, depression, and social support to test the mediation model.

Conclusions

This review established several important theoretical links and some pathways that require future work. More specifically, the links between stress/depression and social support appear well-established. Social support is also related to biological risk as well as aspects of cardiovascular disease. Of these biological mechanisms, the oxytocin system is emerging as a promising, integrative pathway given its links to close relationship processes and cascading influences across other health-relevant biological systems (e.g., HPA, ANS). However, less known is whether social support could be a direct mechanism through which stress and depression are associated with cardiovascular disease, as well as the more precise conditions under which received support might be harmful or beneficial. The proposed social support mediational model thus needs further work but intervention approaches that focus on different aspects of the model, as well as a more comprehensive approach, might prove fruitful.

References

Alferi, S.M., Carver, C.S., Antoni, M.H., Weiss, S., Durán, R.E., 2001. An exploratory study of social support, distress, and life disruption among low-income Hispanic women under treatment for early stage breast cancer. Health Psychol. 20, 41–46.

André-Petersson, L., Hedblad, B., Janzon, L., Ostergren, P.-O., 2006. Social support and behavior in a stressful situation in relation to myocardial infarction and mortality: Who is at risk? Results from prospective cohort study "men born in 1914," Malmö, Sweden. Int. J. Behav. Med. 13 (4), 340–347.

André-Petersson, L., Engström, G., Hedblad, B., Janzon, L., Rosvall, M., 2007. Social support at work and the risk of myocardial infarction and stroke in women and men. Soc. Sci. Med. 64 (4), 830–841.

Angerer, P., Siebert, U., Kothny, W., Mühlbauer, D., Mudra, H., Von Schacky, C., 2000. Impact of social support, cynical hostility and anger expression on progression of coronary atherosclerosis. J. Am. Coll. Cardiol. 36 (6), 1781–1788.

Barth, J., Schneider, S., von Känel, R., 2010. Lack of social support in the etiology and the prognosis of coronary heart disease: a systematic review and meta-analysis. Psychosom. Med. 72 (3), 229–238.

Bartz, J.A., 2016. Oxytocin and the pharmacological dissection of affiliation. Curr. Dir. Psychol. Sci. 25 (2), 104–110.

Becofsky, K.M., Shook, R.P., Sui, X., Wilcox, S., Lavie, C.J., Blair, S.N., 2015. Influence of the source of social support and size of social network on all-cause mortality. Mayo Clin. Proc. 90 (7), 895–902.

Benazon, N.R., 2000. Predicting negative spousal attitudes toward depressed persons: a test of Coyne's interpersonal model. J. Abnorm. Psychol. 109 (3), 550–554.

Berg, C.A., Upchurch, R., 2007. A developmental-contextual model of couples coping with chronic illness across the adult life span. Psychol. Bull. 133 (6), 920–954.

Berkman, L.F., Leo-Summers, L., Horwitz, R.I., 1992. Emotional support and survival after myocardial infarction. A prospective, population-based study of the elderly. Ann. Intern. Med. 117, 1003–1009.

Bolger, N., Foster, M., Vinokur, A.D., Ng, R., 1996. Close relationships and adjustment to a life crisis: the case of breast cancer. J. Pers. Soc. Psychol. 70, 283–294.

Brummett, B.H., Mark, D.B., Siegler, I.C., Williams, R.B., Babyak, M.A., Clapp-Channing, N.E., Barefoot, J.C., 2005. Perceived social support as a predictor of mortality in coronary patients: effects of smoking, sedentary behavior, and depressive symptoms. Psychosom. Med. 67 (1), 40–45.

Cacioppo, J.T., Cacioppo, S., Capitanio, J.P., Cole, S., 2015. The neuroendocrinology of social isolation. Annu. Rev. Psychol. 66, 733–767.

Clodi, M., Vila, G., Geyeregger, R., Riedl, M., Stulnig, T.M., Struck, J., Luger, T.A., Luger, A., 2008. Oxytocin alleviates the neuroendocrine and cytokine response to bacterial endotoxin in healthy men. Am. J. Physiol. Endocrinol. Metab. 295 (3), E686–E691.

Coyne, J.C., 1976. Depression and the response of others. J. Abnorm. Psychol. 85 (2), 186–193.

Croezen, S., Picavet, H., Haveman-Nies, A., Verschuren, W., de Groot, L., van't Veer, P., 2012. Do positive or negative experiences of social support relate to current and future health? Results from the Doetinchem cohort study. BMC Public Health 12 (1), 65.

Doom, J.R., Gunnar, M.R., Clark, C.J., 2016. Maternal relationship during adolescence predicts cardiovascular disease risk in adulthood. Health Psychol. 35 (4), 376–386.
Dunkel-Schetter, C., Skokan, L.A., 1990. Differentiating the cognitive and behavioral aspects of social support. In: Sarason, B.R., Sarason, I.G., Pierce, G.R. (Eds.), Social Support: An Interactional View. John Wiley and Sons, New York, pp. 267–296.
Eisenberger, N.I., Master, S.L., Inagaki, T.K., Taylor, S.E., Shirinyan, D., Lieberman, M.D., Naliboff, B.D., 2011. Attachment figures activate a safety signal-related neural region and reduce pain experience. Proc. Natl. Acad. Sci. 108 (28), 11721–11726.
Fredette, C., El-Baalbaki, G., Palardy, V., Rizkallah, E., Guay, S., 2016. Social support and cognitive–behavioral therapy for posttraumatic stress disorder: a systematic review. Traumatology 22 (2), 131–144.
Gan, Y., Gong, Y., Tong, X., Sun, H., Cong, Y., Dong, X., Wang, Y., Xu, X., Yin, X., Deng, J., Li, L., Cao, S., Lu, Z., 2014. Depression and the risk of coronary heart disease: a meta-analysis of prospective cohort studies. BMC Psychiatry 14 (1), 1–11.
Gariépy, G., Honkaniemi, H., Quesnel-Vallée, A., 2016. Social support and protection from depression: systematic review of current findings in western countries. Br. J. Psychiatry 209 (4), 284–293.
Gore, S., 1981. Stress-buffering functions of social supports: an appraisal and clarification of research models. In: Dohrenwend, B.S., Dohrenwend, B.P. (Eds.), Stressful Life Events and Their Context. Prodist, New York, pp. 202–222.
Greco, A., Steca, P., Pozzi, R., Monzani, D., D'Addario, M., Villani, A., Rella, V., Giglio, A., Malfatto, G., Parati, G., 2014. Predicting depression from illness severity in cardiovascular disease patients: self-efficacy beliefs, illness perception, and perceived social support as mediators. Int. J. Behav. Med. 21 (2), 221–229.
Grewen, K.M., Girdler, S.S., Amico, J., Light, K.C., 2005. Effects of partner support on resting oxytocin, cortisol, norepinephrine, and blood pressure before and after warm partner contact. Psychosom. Med. 67 (4), 531–538.
Griffiths, K.M., Crisp, D.A., Barney, L., Reid, R., 2011. Seeking help for depression from family and friends: a qualitative analysis of perceived advantages and disadvantages. BMC Psychiatry 11 (1), 196.
Gutkowska, J., Jankowski, M., 2012. Oxytocin revisited: its role in cardiovascular regulation. J. Neuroendocrinol. 24 (4), 599–608.
Heim, C., Young, L.J., Newport, D.J., Mletzko, T., Miller, A.H., Nemeroff, C.B., 2009. Lower CSF oxytocin concentrations in women with a history of childhood abuse. Mol. Psychiatry 14 (10), 954–958.
Heinrichs, M., Baumgartner, T., Kirschbaum, C., Ehlert, U., 2003. Social support and oxytocin interact to suppress cortisol and subjective responses to psychosocial stress. Biol. Psychiatry 54 (12), 1389–1398.
Heisler, M., Vijan, S., Makki, F., Piette, J.D., 2010. Diabetes control with reciprocal peer support versus nurse care management. Ann. Intern. Med. 153 (8), 507.
Herbst-Damm, K.L., Kulik, J.A., 2005. Volunteer support, marital status, and the survival times of terminally ill patients. Health Psychol. 24 (2), 225–229.
Hogan, B.E., Linden, W., Najarian, B., 2002. Social support interventions do they work? Clin. Psychol. Rev. 22, 381–440.
Holt-Lunstad, J., Uchino, B.N., Rimer, B.K., Viswanath, K., 2015. Social support and health. In: Glanz, K. (Ed.), Health Behavior: Theory, Research, and Practice. Jossey-Bass, San Francisco, pp. 183–204.
Holt-Lunstad, J., Smith, T.B., Layton, J.B., 2010. Social relationships and mortality risk: a meta-analytic review. PLoS Med. 7 (7), e1000316.
Ikeda, A., Iso, H., Kawachi, I., Yamagishi, K., Inoue, M., Tsugane, S., 2008. Social support and stroke and coronary heart disease: the JPHC study cohorts II. Stroke 39 (3), 768–775.

Ikehara, S., Iso, H., Yamagishi, K., Yamamoto, S., Inoue, M., Tsugane, S., 2009. Alcohol consumption, social support, and risk of stroke and coronary heart disease among Japanese men: the JPHC study. Alcohol. Clin. Exp. Res. 33 (6), 1025–1032.
Jameson, H., Bateman, R., Byrne, P., Dyavanapalli, J., Wang, X., Jain, V., Mendelowitz, D., 2016. Oxytocin neuron activation prevents hypertension that occurs with chronic intermittent hypoxia/hypercapnia in rats. Am. J. Physiol. Heart Circ. Physiol., H1549–H1557.
Kamiya, Y., Whelan, B., Timonen, V., Kenny, R.A., 2010. The differential impact of subjective and objective aspects of social engagement on cardiovascular risk factors. BMC Geriatr. 10 (1), 81.
Kaniasty, K., Norris, F.H., 2008. Longitudinal linkages between perceived social support and posttraumatic stress symptoms: sequential roles of social causation and social selection. J. Trauma. Stress. 21 (3), 274–281.
Knobloch, H.S., Charlet, A., Hoffmann, L.C., Eliava, M., Khrulev, S., Cetin, A.H., Osten, P., Schwarz, M.K., Seeburg, P.H., Stoop, R., Grinevich, V., 2012. Evoked axonal oxytocin release in the central amygdala attenuates fear response. Neuron 73 (3), 553–566.
Leifheit-Limson, E.C., Reid, K.J., Kasl, S.V., Lin, H., Jones, P.G., Buchanan, D.M., Parashar, S., Peterson, P.N., Spertus, J.A., Lichtman, J.H., 2010. The role of social support in health status and depressive symptoms after acute myocardial infarction evidence for a stronger relationship among women. Circ. Cardiovasc. Qual. Outcomes 3 (2), 143–150.
Lepore, S.J., Evans, G.W., Schneider, M.L., 1991. Dynamic role of social support in the link between chronic stress and psychological distress. J. Pers. Soc. Psychol. 61 (6), 899–909.
Libby, P., Nahrendorf, M., Swirski, F.K., 2016. Leukocytes link local and systemic inflammation in ischemic cardiovascular disease an expanded cardiovascular continuum. J. Am. Coll. Cardiol. 67 (9), 1091–1103.
Low, C., Thurston, R.C., Matthews, K., 2010. Psychosocial factors in the development of heart disease in women: current research and future directions. Psychosom. Med. 72 (9), 842–854.
Luttik, M.L., Jaarsma, T., Sanderman, R., van Veldhuisen, D.J., 2005. The importance and impact of social support on outcomes in patients with heart failure: An overview of the literature. J. Cardiovasc. Nurs. 20 (3), 162–169.
Mattson, R.E., Cameron, N., Middleton, F.A., Starr, L.R., Davila, J., Johnson, M.D., 2019. Oxytocin receptor gene (OXTR) links to marital quality via social support behavior and perceived partner responsiveness. J. Fam. Psychol. 33, 44–53.
Moghimian, M., Faghihi, M., Karimian, S.M., Imani, A., Houshmand, F., Azizi, Y., 2013. The role of central oxytocin in stress-induced cardioprotection in ischemic-reperfused heart model. J. Cardiol. 61 (1), 79–86.
Nagayoshi, M., Everson-Rose, S.A., Iso, H., Mosley Jr., T.H., Rose, K.M., Lutsey, P.L., 2014. Social network, social support, and risk of incident stroke: atherosclerosis risk in communities study. Stroke 45, 2868–2873.
Neumann, I.D., Wigger, A., Torner, L., Holsboer, F., Landgraf, R., 2000. Brain oxytocin inhibits basal and stress-induced activity of the hypothalamo-pituitary-adrenal axis in male and female rats: partial action within the paraventricular nucleus. J. Neuroendocrinol. 12 (3), 235–243.
Nicassio, P.C., Smith, T.W., 1995. Psychosocial Management of Chronic Illness. American Psychological Association, Washington, DC.
Norman, G.J., Cacioppo, J.T., Morris, J.S., Malarkey, W.B., Berntson, G.G., DeVries, A.C., 2011. Oxytocin increases autonomic cardiac control: moderation by loneliness. Biol. Psychol. 86 (3), 174–180.
Orth-Gomér, K., Rosengren, A., Wilhelmsen, L., 1993. Lack of social support and incidence of coronary heart disease in middle-aged Swedish men. Psychosom. Med. 55, 37–43.
Orth-Gomér, K., Schneiderman, N., Wang, H.X., Walldin, C., Blom, M., Jernberg, T., 2009. Stress reduction prolongs life in women with coronary disease: the Stockholm women's intervention trial for coronary heart disease (SWITCHD). Circ. Cardiovasc. Qual. Outcomes 2 (1), 25–32.

Ozbay, F., Johnson, D.C., Dimoulas, E., Morgan, C.A., Charney, D., Southwick, S., 2007. Social support and resilience to stress: from neurobiology to clinical practice. Psychiatry 4 (5), 35–40.
Patten, C.A., Smith, C.M., Brockman, T.A., Decker, P.A., Hughes, C.A., Nadeau, A.M., Sinicrope, P.S., Offord, K.P., Lichtenstein, E., Zhu, S.-H., 2011. Support-person promotion of a smoking quitline. Am. J. Prev. Med. 41 (1), 17–23.
Piros, S., Karlehagen, S., Lappas, G., Wilhelmsen, L., 2000. Risk factors for myocardial infarction among Swedish railway engine drivers during 10 years follow-up. Eur. J. Cardiovasc. Risk 7 (5), 395–400.
Prati, G., Pietrantoni, L., 2010. An application of the social support deterioration deterrence model to rescue workers. J. Community Psychol. 38 (7), 901–917.
Radke, S., Volman, I., Kokal, I., Roelofs, K., de Bruijn, E.R.A., Toni, I., 2017. Oxytocin reduces amygdala responses during threat approach. Psychoneuroendocrinology 79, 160–166.
Räikkönen, K., Matthews, K.A., Kuller, L.H., 2001. Trajectory of psychological risk and incident hypertension in middle-aged women. Hypertension 38 (4), 798–802.
Reblin, M., Uchino, B.N., Smith, T.W., 2010. Provider and recipient factors that may moderate the effectiveness of received support: examining the effects of relationship quality and expectations for support on behavioral and cardiovascular reactions. J. Behav. Med. 33 (6), 423–431.
Ren, P., Qin, X., Zhang, Y., Zhang, R., 2018. Is social support a cause or consequence of depression? A longitudinal study of adolescents. Front. Psychol. 9, 1–8.
Richardson, S., Shaffer, J.A., Falzon, L., Krupka, D., Davidson, K.W., Edmondson, D., 2012. Meta-analysis of perceived stress and its association with incident coronary heart disease. Am. J. Cardiol. 110 (12), 1711–1716.
Riem, M.M.E., Bakermans-Kranenburg, M.J., Huffmeijer, R., van Ijzendoorn, M.H., 2013. Does intranasal oxytocin promote prosocial behavior to an excluded fellow player? A randomized-controlled trial with Cyberball. Psychoneuroendocrinology 38 (8), 1418–1425.
Rodriguez, C.J., Burg, M.M., Meng, J., Pickering, T.G., Jin, Z., Sacco, R.L., Boden-Albala, B., Homma, S., Di Tullio, M.R., 2008. Effect of social support on nocturnal blood pressure dipping. Psychosom. Med. 70 (1), 7–12.
Rosengren, A., Wilhelmsen, L., Orth-Gomér, K., 2004. Coronary disease in relation to social support and social class in Swedish men: A 15 year follow-up in the study of men born in 1933. Eur. Heart J. 25 (1), 56–63.
Rucker, D.D., Preacher, K.J., Tormala, Z.L., Petty, R.E., 2011. Mediation analysis in social psychology: current practices and new recommendations. Soc. Personal. Psychol. Compass 5 (6), 359–371.
Rueger, S.Y., Malecki, C.K., Pyun, Y., Aycock, C., Coyle, S., 2016. A meta-analytic review of the association between perceived social support and depression in childhood and adolescence. Psychol. Bull. 142 (10), 1017–1067.
Ruparelia, N., Chai, J.T., Fisher, E.A., Choudhury, R.P., 2017. Inflammatory processes in cardiovascular disease: a route to targeted therapies. Nat. Rev. Cardiol. 14 (3), 133–144.
Schwarzer, R., Knoll, N., 2007. Functional roles of social support within the stress and coping process: a theoretical and empirical overview. Int. J. Psychol. 42 (4), 243–252.
Selcuk, E., Ong, A.D., 2013. Perceived partner responsiveness moderates the association between received emotional support and all-cause mortality. Health Psychol. 32 (2), 231–235.
Sorkin, D.H., Mavandadi, S., Rook, K.S., Biegler, K.A., Kilgore, D., Dow, E., Ngo-Metzger, Q., 2014. Dyadic collaboration in shared health behavior change: the effects of a randomized trial to test a lifestyle intervention for high-risk Latinas. Health Psychol. 33 (6), 566–575.

Steptoe, A., Lundwall, K., Cropley, M., 2000. Gender, family structure and cardiovascular activity during the working day and evening. Soc. Sci. Med. 50 (4), 531–539.
Stice, E., Ragan, J., Randall, P., 2004. Prospective relations between social support and depression: differential direction of effects for parent and peer support? J. Abnorm. Psychol. 113 (1), 155–159.
Story, L.B., Repetti, R., 2006. Daily occupational stressors and marital behavior. J. Fam. Psychol. 20 (4), 690–700.
Teo, A.R., Choi, H.J., Valenstein, M., 2013. Social relationships and depression: ten-year follow-up from a nationally representative study. PLoS One 8 (4), 1–8.
Thomas, S., Larkin, T., 2018. Plasma cortisol and oxytocin levels predict help-seeking intentions for depressive symptoms. Psychoneuroendocrinology 87, 159–165.
Thorsteinsson, E.B., James, J.E., 1999. A meta-analysis of the effects of experimental manipulations of social support during laboratory stress. Psychol. Health 14 (5), 869–886.
Uchino, B.N., 2006. Social support and health: a review of physiological processes potentially underlying links to disease outcomes. J. Behav. Med. 29 (4), 377–387.
Uchino, B.N., 2009. Understanding the links between social support and physical health: a life-span perspective with emphasis on the separability of perceived and received support. Perspect. Psychol. Sci. 4 (3), 236–255.
Uchino, B.N., Way, B.M., 2017. Integrative pathways linking close family ties to health: A neurochemical perspective. Am. Psychol. 72 (6), 590–600.
Uchino, B.N., Carlisle, M., Birmingham, W., Vaughn, A.A., 2011. Social support and the reactivity hypothesis: conceptual issues in examining the efficacy of received support during acute psychological stress. Biol. Psychol. 86 (2), 137–142.
Uchino, B.N., Bowen, K., Carlisle, M., Birmingham, W., 2012. Psychological pathways linking social support to health outcomes: a visit with the "ghosts" of research past, present, and future. Soc. Sci. Med. 74 (7), 949–957.
Uchino, B.N., Trettevik, R., De Grey, R.G.K., Cronan, S., Hogan, J., Baucom, B.R.W., 2018. Social support, social integration, and inflammatory cytokines: a meta-analysis. Health Psychol. 37 (5), 462–471.
Van IJzendoorn, M.H., Bakermans-Kranenburg, M.J., 2012. A sniff of trust: meta-analysis of the effects of intranasal oxytocin administration on face recognition, trust to in-group, and trust to out-group. Psychoneuroendocrinology 37 (3), 438–443.
Wang, H.-X., Mittleman, M.A., Orth-Gomer, K., 2005. Influence of social support on progression of coronary artery disease in women. Soc. Sci. Med. 60 (3), 599–607.
Weiss-Faratci, N., Lurie, I., Neumark, Y., Malowany, M., Cohen, G., Benyamini, Y., Goldbourt, U., Gerber, Y., 2016. Perceived social support at different times after myocardial infarction and long-term mortality risk: a prospective cohort study. Ann. Epidemiol. 26 (6), 424–428.
Wethington, E., Kessler, R.C., 1986. Perceived support, received support, and adjustment to stressful life events. J. Health Soc. Behav. 27 (1), 78.
Wills, T.A., Shinar, O., 2000. Measuring perceived and received social support. In: Cohen, S., Gordon, L., Gottlieb, B. (Eds.), Social Support Measurement and Intervention: A Guide for Health and Social Scientists. Oxford University Press, New York, pp. 86–135.

CHAPTER 10

The influence of health behaviors upon the association between stress and depression and cardiovascular disease☆

Jeffrey L. Goodie, Kevin Wilfong, Phillip Kroke
Department of Medical & Clinical Psychology, Uniformed Services University, Bethesda, MD, United States

Contents

☆Disclaimer: The opinions and assertions expressed herein are those of the authors and do not necessarily reflect the official policy or position of the Uniformed Services University, Department of Defense, Department of Health and Human Services, or their agencies.

Cardiovascular Implications of Stress and Depression
https://doi.org/10.1016/B978-0-12-815015-3.00010-6

Multiple health behaviors not only contribute to mortality but are directly related to the development or exacerbation of cardiovascular disease (CVD) and depression. Adhering to a healthy lifestyle more than triples the likelihood of living to the oldest age (Loef and Walach, 2012; Willcox et al., 2006); however, fewer than 3% of the United States population engages in all four of the most important healthy lifestyle behaviors (i.e., being sufficiently active, eating a healthy diet, being a nonsmoker, and having the recommended body fat percentage; Loprinzi et al., 2016). Stress is commonly described as a reason why individuals engage in a range of unhealthy behaviors. Sometimes the behavior, such as tobacco use and unhealthy eating, is described as reducing perceived stress. In other situations, stress is cited as the cause for the unhealthy behavior, such as physical inactivity and decreased sleep. Regardless of the causal direction, health behaviors are important to consider in the relations between stress, CVD, and depression.

The researchers conducting the study Heart and Soul (Whooley et al., 2008) used a prospective cohort design to examine the association of depressive symptoms with the onset of cardiovascular events among those with coronary heart disease. Although those with baseline depressive symptoms had a 50% greater rate of subsequent cardiovascular events than those with no depressive symptoms, the relation between depressive symptoms and cardiovascular events was no longer significant when the researchers controlled for health behaviors, including smoking, medication nonadherence, and self-reported physical activity (Whooley et al., 2008). Adjusting for physical activity alone resulted in a 31.7% reduction in the strength of the association. A subsequent study based on these data found that while depressive symptoms were associated with subsequent higher levels of inflammation, after controlling for health behaviors (i.e., tobacco use, body mass index, and physical inactivity), the relation between depression and inflammation was no longer significant (Duivis et al., 2011).

Data from studies similar to Heart and Soul demonstrate the important relations between stress, CVD, depression, and health behaviors. This chapter focuses on the most recent research regarding the known relations between health behaviors (i.e., tobacco use, alcohol use, dietary and physical activity behaviors, and medication adherence) and stress, and then how those health behaviors impact the onset and exacerbation of CVD and depression. Unlike previous reviews (e.g., Carney et al., 1995; Lett et al., 2004) we also discuss sleep as an important health behavior for researchers to consider when examining relations between stress, CVD, and depression.

Tobacco use

Tobacco use takes multiple forms (e.g., smoking, cigarettes, cigars), but most research focuses on the impact of smoking. Tobacco use is the number one cause of preventable deaths (U.S. Department of Health and Human Services [DHHS], 2014). Although tobacco use rates have decreased over the last 50 years, 17.6% of the adult population continues to smoke (Phillips et al., 2017). There are well-established relations between tobacco use with stress, CVD, and depression; Thorndike and Rigotti (2009) refer to the established relations between coronary artery disease, nicotine addiction, and depression as the "tragic triad."

Unlike the decline in tobacco products, the use of e-cigarettes (i.e., also called electronic nicotine delivery systems [ENDS] and vaping) has increased in the United States, particularly among youth between 2011 and 2015; between 2015 and 2017 there was a decrease in e-cigarettes use (Wang et al., 2018). Those who used e-cigarettes were found to be young, non-Hispanic, white, non-married, college educated, current or former tobacco smokers (Wilson and Wang, 2017). Endorsing the use of e-cigarettes is controversial among health professionals. Although still an avenue for ingesting nicotine, some health agencies have encouraged e-cigarette use as a healthier alternative for active smokers (McNeil et al., 2015) and some data suggest that they may be as effective as nicotine replacement therapy for helping smokers quit or reduce their smoking (Glasser et al., 2017).

Relations with stress

Stress and tobacco use have well established bidirectional relations. One primary reason individuals report using tobacco, and returning to tobacco use after quitting, is to decrease stress (Kassel et al., 2003). Among a cohort of 4983 adults in the United States who were followed for 9 to 10 years as part of the Midlife in the US (MIDUS) study, a wide range of stressors (e.g., financial, work, perceived inequality, past-year family problems, cumulative) were associated with increased odds of being a persistent smoker compared to a non-smoker (Slopen et al., 2013). The stimulant properties of nicotine elevate mood, while at the same time, help individuals avoid withdrawal symptoms (Torres and O'Dell, 2016). Although the understanding of how tobacco use and stress interact continues to be refined (e.g., Torres and O'Dell, 2016) the fundamental, bidirectional relation continues to be replicated. Interestingly, some recent evidence suggests that those who are least physiologically (e.g., ACTH, beta-endorphin, cortisol and blood pressure) reactive to stress may be more at risk of relapse after quitting tobacco (Al'Absi, 2018). Relations between e-cigarettes and stress have not been widely studied.

Relations with CVD

It will not surprise anyone reading this chapter that smoking continues to demonstrate strong relations with the risk of dying from CVD. Quitting smoking remains the most important change most smokers can make to improve their health and reduce their risk of CVD. Aune et al.'s (2018) review of the literature concluded that current smokers (i.e., adjusted relative risk = 3.06) and former smokers (i.e., adjusted relative risk = 1.38) were more likely to die from sudden cardiac death compared to never smokers. In another meta-analysis (Mons et al., 2015), which examined cohorts of individuals 60 years and older from all over the world, not only was current smoking associated with a twofold increase in cardiovascular mortality compared to never smokers, those who were former smokers were 37% more likely to die from a CVD-related disease. The increased risk of death among former smokers decreased in a dose-response manner with time. Women may be more susceptible to the impact of smoking on mortality and morbidity risks associated with smoking. In a meta-analysis of studies, Huxley and Woodward (2011) found that current women smokers had a 25% increased risk for developing coronary heart disease compared to their male counterparts.

The findings regarding relations between e-cigarette use and CVD risk has been mixed. There is some concern that e-cigarette use changes risk factors (e.g., heart rate, blood pressure, platelet function, vascular damage) associated with increased incidence of CVD compared to non-smokers, but longitudinal data are needed to understand these relations (Schweitzer et al., 2017). Alzahrani et al. (2018) found that e-cigarette use was associated with increased odds (OR = 1.79) of having a myocardial infarction compared to those who never used e-cigarettes and never smoked tobacco.

Relations with depression

Smokers have a higher incidence of depression and individuals who demonstrate depressive symptoms are less likely to quit smoking. There are multiple hypotheses regarding why this relation exists. As reviewed by Fluharty et al. (2017), individuals may smoke to help manage depressive symptoms, smoking may lead to physiological changes that increase one's susceptibility to depressive symptoms, there may be a bidirectional relation, or there may be some confounding factor contributing to the smoking and depressive symptoms (e.g., a common genetic predisposition). In their systematic review of the current literature, findings were inconsistent regarding the nature of these relations (Fluharty et al., 2017). When individuals quit tobacco use, studies find that depressive symptoms significantly decrease (Taylor et al., 2014). Individuals who were smoking when they had a cardiac event have been found to be more likely to be diagnosed with Major Depressive Disorder and other depressive diagnoses compared those who were not smokers. Although smoking was not related to depression among those with coronary artery disease 6 and 9 months after the cardiac event, individuals were more likely to be diagnosed with depressive symptoms at some point during the study if they smoked, compared to non-smokers (Stafford et al., 2013).

There are some recent data to suggest that depressive symptoms are positively associated with e-cigarette use. Bandiera et al. (2016) found that, among 5438 Texas college students, self-reported depressive symptoms were related to current e-cigarette use, after controlling for socio-demographic characteristics and current cigarette use. Similarly, depressive symptoms and e-cigarette use has been positively associated among a population of French adults (Wiernik et al., 2019). A broad sample of US adults reporting depression, anxiety, and/or an emotional problem were more likely to try an e-cigarette and e-cigarette users were more likely to report a depressive, anxiety, and/or emotional problem (Bianco, 2019). There are not enough data to know whether the relations between e-cigarettes and depression will be the same as those with tobacco, but these early data suggest that findings are similar.

The relations between smoking, stress, and depression continue to be replicated; however, some findings suggest that the construct of distress tolerance may be an important moderator for these relations. Leventhal and Zvolensky (2015) propose that transdiagnostic vulnerabilities of anhedonia, distress tolerance, and anxiety sensitivity may account for relations between smoking and emotion-related psychopathology (e.g., depression, anxiety, post-traumatic stress). As an example, Trujillo et al. (2017) found that among 212 current smokers, distress tolerance was predictive of dependence, craving, and negative reinforcement smoking motivation, above anxiety and depressive symptoms. Research examining models that relate tobacco use, stress, CVD, and depression should consider the potential moderating effects of such transdiagnostic vulnerabilities.

Alcohol use

Alcohol consumption in the U.S. is commonplace, with 71.7% of Americans consuming alcohol in the last 12 months and 90.8% consuming alcohol at some point in their lifetime. The effects of healthy alcohol consumption complicate the detrimental impact of alcohol use; still, the use of alcohol resulted in approximately 3 million global deaths (5.3% of all deaths) in 2016, when accounting for both the detrimental and beneficial health effects of alcohol (World Health Organization, 2018b). Problematic alcohol use (i.e., alcohol dependence or non-dependent alcohol abuse) has been reported in 7.4% of Americans between 2002 and 2014, with few changes in prevalence (Cheng et al., 2018). Additionally, problematic alcohol use is associated with the development of a range of health problems including CVD and depression (Rehm et al., 2017; WHO, 2018b).

Relations with stress

The relations between stress and alcohol use are well established. The nature, developmental period, duration, and severity of the stress experienced can impact the relation between stress and alcohol (Keyes et al., 2012; Moustafa et al., 2018; Virtanen et al., 2015). General life stressors, childhood maltreatment, and minority stress (objective and perceived) were predictive of increased risk for problematic alcohol use (Keyes et al., 2012; Moustafa et al., 2018; Virtanen et al., 2015). Additionally, childhood maltreatment predicted an earlier onset in drinking age. Catastrophic stress was related to an increase in population alcohol consumption, but not incidence of problematic alcohol use (Keyes et al., 2012). The relation between stress and alcohol use has been attributed to abnormalities in the hypothalamic-pituitary-adrenal (HPA) axis responsivity and cortisol secretion (Stephens and Wand, 2012). For example, those at risk for developing an alcohol use disorder demonstrate altered HPA axis functioning compared to those who are at low risk and deficiencies in cortisol response following abstinence in those who demonstrated alcohol dependent behaviors are more likely to relapse (Stephens and Wand, 2012). Studies have also found shared epigenetic-related pathways between stress-related and alcohol disorders (Palmisano and Pandey, 2017), suggesting genetic predispositions may contribute to the relations between alcohol use and stress.

Relations with CVD

The relation between CVD and alcohol use is dependent on the pattern of alcohol use. In 2016, alcohol was linked to 593,000 CVD deaths (3.3% of all CVD deaths) and 13 million CVD disability-adjusted life years (3.2% of all CVD DALYs; WHO, 2018a). At the same time, healthy alcohol use has been identified as a protective factor, particularly in ischemic diseases. Often alcohol consumption, particularly red wine, is described as having a J-shaped relation with CVD. That is, no alcohol consumption is associated with more risk of CVD compared to moderate alcohol use (i.e., ≤1 drink/day for women; ≤2 drinks/day for men), but once alcohol consumption exceeds these levels, risk for CVD mortality quickly rises. The beneficial effects of alcohol include higher levels of high-density lipoprotein cholesterol and adiponectin, as well as lower levels of fibrinogen (Rehm et al., 2017). However, a recent study (Wood et al., 2018) looking at almost 600,000 current drinkers challenged this J-shaped relation and found that the relation between alcohol consumption and disease risk may be more linear for certain subpopulations and forms of CVD.

Studies are consistent that chronic and episodic heavy drinking are both associated with increased risk of negative cardiovascular health outcomes (Rehm et al., 2017). Chronic heavy drinking contributes to the development of hypertension, ischemic heart disease, cardiomyopathy, atrial fibrillation and flutter, and all types of stroke (Briasoulis et al., 2012; George and Figueredo, 2011; Kodama et al., 2011; Patra et al., 2010; Roerecke and Rehm, 2014). Episodic heavy drinking is associated with the development of CVD through biological pathways such as greater risk of higher levels of low-density lipoprotein, thrombosis, arrhythmias, and acute or chronic hypertension (McKee and Britton, 1998; Rehm et al., 2017).

Relations with depression

The comorbidity of alcohol use disorders and depression present with increasing prevalence from adolescence (below 5%) into early adulthood (10–21%) and rests between 5% and 18% in adulthood. Unlike alcohol use disorders and depression, which present disproportionately in men and women respectively, the comorbid occurrence of these disorders does not differ significantly at any point across the lifetime (Brière et al., 2014).

One intersection between alcohol and depression is the substance-induced depressive disorder identified in the DSM-5 (American Psychiatric Association [APA], 2013). A review of the literature suggests both unidirectional and reciprocal causal pathways are plausible in the interaction of alcohol and depression, that is, alcohol could lead to depressive symptoms, depressive symptoms could lead to increased alcohol use, and the two may cyclically maintain one another (Rehm et al., 2017). Additionally, the course of depressive etiology is impacted by alcohol use such that individuals with depressive symptoms are at higher risk for injury due to self-harm and suicidal behaviors (Borges et al., 2017).

A meta-analysis of treatment outcome studies looking at depression in patients with problematic alcohol use suggests these co-occurring disorders still show improvement in treatment, particularly in the first 3 months and in individuals with high levels of depression at baseline. In cases determined to be substance-induced depression, the depressive symptoms ameliorated with treatment but the mechanism of change (reduction in alcohol versus depressive symptoms) was unclear; further, approximately 25% of these individuals better fit criteria for independent depression than substance-induced depression within 12 months (Foulds et al., 2015).

Physical inactivity

Estimates suggest that 23–31.1% of the world's population is physically inactive, with higher rates in richer, more developed countries (Hallal et al., 2012; Sallis et al., 2016). Estimates for the United States are similar to global rates, with a third of the population not meeting recommended levels of aerobic physical activity (Carison et al., 2009). Demographic characteristics including being female, older, having a lower education, and ethnicity are associated with low physical activity (Hawkins et al., 2009; Stubbs et al., 2017). Physical inactivity has a significant impact on mortality with a physically inactive lifestyle explaining 9% of total premature mortality (Lee et al., 2012).

Lett et al. (2004) highlighted the potential value of exercise as a treatment for depression among those with CVD. Physical activity and sedentarism have a bidirectional relation with stress, depression, and CVD. Physical activity reduces symptoms or risk factors related to stress, depression, and CVD. However, it is likely that sedentarism, regardless of whether an individual is meeting physical activity guidelines, has an independent, negative association on stress, CVD, and depression. Further, psychological stress, complications from cardiovascular illness, and depressive symptoms are negatively associated with physical activity.

Relations with stress

Recent literature has reported a reciprocal, bidirectional relation between physical activity and stress (Hiles et al., 2017; Teychenne et al., 2015). Low levels of physical activity are consistently associated with increased stress and anxiety (Stubbs et al., 2017; Stults-Kolehmainen and Sinha, 2014). In a population with diagnosed behavioral health disorders, higher overall stress scores were associated with higher prevalence of physical inactivity (Beutel et al., 2018). A similar relation exists between sedentary behaviors and anxiety (Stults-Kolehmainen and Sinha, 2014; Teychenne et al., 2015).

Historically, most research in this area has focused on the anxiolytic effects of exercise and physical activity on stress. The anxiolytic effects of physical activity are supported after one session of exercise (Guszkowska, 2004), with 78% endorsing the primary aim of their exercise regimen was to "reduce stress" (Firth et al., 2016). Those who have a comorbid condition or endorse greater anxiety symptom severity find the most improvement from physical activity (Hiles et al., 2017). The benefits of physical activity are present at low levels and show diminishing returns at moderate to high levels of physical activity (Hallal et al., 2012).

The effect of stress on physical activity has been a topic of recent research interest (Hiles et al., 2017; Teychenne et al., 2015; Stubbs et al., 2017; Stults-Kolehmainen and Sinha, 2014). One thorough review found that psychological stress was predictive of less physical activity and increased sedentarism (Stults-Kolehmainen and Sinha, 2014). Sixty one percent of respondents endorsed stress, as well as depression, as barriers to exercise (Firth et al., 2016). However, there is some evidence that stress positively influences physical activity and routine physical activity promotes resilience from future stress (Gerber and Pühse, 2009; Stults-Kolehmainen and Sinha, 2014). According to Gerber and Pühse (2009), the resilience associated with physical activity for stress is likely related to using exercise as a coping strategy during stressful events.

Relations with CVD

Low levels of physical activity and increased sedentarism has been linked to rates of all-cause mortality and CVD (Aggio et al., 2017; Lear et al., 2017; Li and Siegrist, 2012). It is estimated that 6% of the disease burden associated with coronary heart disease can be explained by physical inactivity (Lee et al., 2012). Decades of research supports that a high level of physical activity is protective for CVD and related risk factors.

The data relating physical activity and CVD are highly consistent with the Bradford Hill criteria for a causal relation (Fedak et al., 2015): the relation between low levels of physical activity and CVD is strong (Pate et al., 1995; Warburton et al., 2006); the relationship is maintained across diverse populations (Carlsson et al., 2013; Lear et al., 2017; Pate et al., 1995);. there is a temporal sequence of physical inactivity preceding CVD (Aggio et al., 2017; Li and Siegrist, 2012); there is a dose-dependent relation between physical activity levels and CVD risk (Li and Siegrist, 2012; Warburton and Bredin, 2018); and there are plausible physiological mechanisms to understand the association (Crisafulli et al., 2015).

Relations with depression

A meta-analysis of 25 randomized controlled trials completed over the past 30 years examining the relations between exercise and depression found a large, statistically and clinically significant effect for increased exercise reducing depressive symptoms (Schuch et al., 2016). The largest effects were found with moderate intensity, aerobic exercise, and interventions supervised by an exercise professional. Similarly, a meta-analysis of 14 studies examining patients diagnosed with ischemic heart disease, found only three studies that specifically measured depression at baseline and as an outcome. Although the evidence in this population was more limited given the small number of studies, these authors found significant effects for increased exercise reducing depressive symptoms (Verschueren et al., 2018). Conversely, increased depressive symptoms are associated with decreased physical activity, suggesting a bidirectional relation between these constructs. In the 8-year British Whitehall II prospective cohort (9309 participants), individuals reporting elevated levels of depressive symptoms during phase 1 were less likely to engage in recommended physical activity levels at phases 2 and 3 (Da Silva et al., 2012). In sum, increasing physical activity appears to be related to decreasing depressive symptoms; however, those experiencing depressive symptoms are more likely to be physically inactive.

Dietary habits

The impact of dietary habits on lifetime health is a growing cause for concern. A healthy diet can act as a protective factor against physical and mental illness and an unhealthy diet can increase the risk for poor physical and behavioral health (Jacka et al., 2017; Rodríguez-Monforte et al., 2015). Unhealthy dietary patterns may be attributed to increased production of processed foods, the socio-economic impact of urbanization, and shifting lifestyles (WHO, 2018a). The association between dietary patterns and health has been investigated for over a century, but has only recently yielded results that can be operationalized to target the growing risk created by poor dietary habits (Kaplan et al., 2015). Some suggest attention to dietary habits should be promoted as mainstream medicine, while others are more skeptical regarding the pragmatism of current research for clinical practice (Sarris et al., 2015b; Schumacher et al., 2017).

Important to note are the common factors that constitute a healthy diet. Although the precise nature of a healthy diet depends on several individual characteristics, the WHO suggests a well-balanced diet includes fruits, vegetables, legumes, nuts, and whole grains; additionally, individuals should be sure <10% of their total calories are from sugars, <30% are from fats, and that their diet includes <5 g of salt per day (WHO, 2018a).

Relations with stress

The association between stress and unhealthy eating is a well-founded phenomenon. Terms like "stress eating" have even entered modern vernacular (Ducharme, 2018). Children as young as 8 or 9 years old are likely to show increases in unhealthy eating related to stress (Hill et al., 2018). This relation has been linked to psychological mechanisms such as attention bias and habit formation around unhealthy eating behaviors. Additionally, biological mechanisms have been implicated that increase risk for obesogenic eating patterns (Tryon et al., 2013).

Relations with CVD

The relation between CVD and dietary patterns has been demonstrated across epidemiological studies. A meta-analysis revealed that unhealthy dietary patterns were related to a 14% increased risk of CVD, whereas healthy dietary patterns acted as a protective factor with a 31% decreased risk of CVD (Rodríguez-Monforte et al., 2015). Common items in the protective dietary patterns included vegetables, fruits, legumes, whole grains, fish, and poultry. Conversely, common items included in dietary patterns associated with increased risk were red and processed meat, refined grains, French fries, sugar-dense sweets, high-fat dairy products, and alcohol (Rodríguez-Monforte et al., 2015).

In addition to primary dietary patterns, supplements such as folic acid have shown moderate to low quality evidence for protective effects in CVD. Specifically, folic acid supplements are associated with a 22% reduction in risk for CVD with a Number-Needed-to-Treat of 111. However, the impact of dietary supplements on CVD is largely unclear with some yielding no evident effect (multivitamins, vitamins C, D, b-carotene, calcium, and selenium) and others showing some evidence of increased risk for all-cause mortality (antioxidant mixtures and niacin [with a statin] for all-cause mortality; Jenkins et al., 2018).

Of note, dietary habits can also be impacted by fiscal accessibility of healthy choices and public knowledge of dietary evidence. One prospective analysis using the US IMPACT Food Policy Model estimated a national 10% subsidy for producers of fruits and vegetables could prevent or postpone as many as 150,500 cardiovascular deaths (95% CI: 141,400–158,500) by 2030 in the United States (Pearson-Stuttard et al., 2017). A formal review of the literature focused on knowledge translation of dietary evidence pertaining to CVD suggests there are no identifiable studies published with the primary aim to translate knowledge into practice. Although empirical evidence exists for the relation between dietary change and CVD, there is insufficient evidence for detailed, pragmatic behavior change strategies (Schumacher et al., 2017). Two diets that show the greatest promise of reducing not only CVD, but also all-cause mortality, are the Mediterranean Diet and Dietary Approaches to Stop Hypertension (DASH; Anand et al., 2015). These diets generally adhere to similar dietary recommendations (e.g., high levels of fruits, vegetables, grains, and legumes; moderate to low levels of fish and poultry, and limited red meat), with the DASH diet particularly focusing on lower levels of sodium consumption (i.e., up to 2300 mg/day). Both of these diet plans are consistently associated with improved cardiovascular health.

Relations with depression

The link between dietary quality and behavioral health is another well-established relation. High-quality dietary patterns are associated with better behavioral health, whereas low-quality diets often co-occur with the presence of depressive symptoms (Li et al., 2017). However, given the focus on cross-sectional data, a causal relation is not well understood, researched, or replicated. This means that although several theories outline the potential mechanisms underlying the interaction of diet and depression, current data cannot conclusively support the extent to which dietary patterns are a risk factor for, symptom of, or merely concurrent with depression (Molendijk et al., 2018).

Regardless of the underlying mechanisms, studies consistently show that a healthy diet can act as a protective factor for depressive symptoms. One cohort study suggested moderate to high adherence to a quality diet as defined by the Mediterranean Diet Score, the Pro-vegetarian Dietary Pattern, or the Alternative Healthy Eating Index-2010 created an effective protective factor for depressive symptoms; further, this study found a threshold effect, with substantial risk reduction at moderate adherence and diminishing returns from moderate to maximum adherence (Sánchez-Villegas et al., 2015). Another randomized control trial experimented with dietary change as an isolated intervention for depression. In this intervention a clinical dietitian utilized dietary and nutritional counseling, motivational interviewing, goal setting, and mindful eating to promote dietary patterns similar to those suggested by the WHO above (Jacka et al., 2017). The outcomes of this intervention revealed symptoms ameliorated independent of weight change, and targeting dietary patterns indirectly impacted behavioral changes related to food such as meal preparation, shopping, and eating patterns (Jacka et al., 2017). In combination, these studies illustrate the potential for improved mental health outcomes through attention to dietary patterns despite insufficient evidence to understand the mechanisms of interaction.

Overweight and obesity

The primary cause of overweight and obesity is the consumption of too many calories relative to the number of calories burned by an individual; however, the factors determining why excessive calories are consumed is complex. The relation between weight and health is largely related to other health behaviors covered in this chapter such as dietary habits and physical activity, but weight provides a unique area of study with nuanced relations to stress, CVD, and depression. Although weight impacts health in a relatively bimodal distribution, for the purpose of simplicity this section will focus on overweight and obesity. For adults, obesity is typically defined as a body mass index (BMI) of 30 or higher, overweight as 25 or higher, and healthy weight between 18.5 and 25; additionally, obesity is stratified into Class 1 (BMI 30–35), Class 2 (BMI 35–40), and Class 3 (BMI 40 or higher) obesity (Centers for Disease Control and Prevention [CDC], 2016).

For all-cause mortality, classes 2 and 3 obesity have consistently been associated with significantly higher risk, with inconsistent directionality of risk for class 1 obesity (Flegal et al., 2013). Despite the well-established association with increased risk, obesity affects approximately 39.8% of adults and 18.5% of children and adolescents in the United States and is continuing to rise on a global scale (CDC, 2018a,b; NCD Risk Factor Collaboration, 2017).

Relations with stress

A systematic review and meta-analysis of 17 articles examining relations between stress and weight found significant associations between weight and waist circumference with self-reported stress levels (Tenk et al., 2018). The impact of stress on weight is of particular importance given the correlation with increased BMI and waist circumference (Tenk et al., 2018). Taken in combination, the distribution of weight around the abdomen is indicative of visceral adiposity, which poses a uniquely increased risk for metabolic syndromes such as hypertension, hyperglycemia, and abnormal serum lipids (Després and Lemieux, 2006; Tenk et al., 2018). The mechanisms by which stress is considered to impact obesity include cognitive processes (executive function and self-regulation), behavioral processes (overeating or unhealthy dietary patterns; decreased physical activity; reduced sleep), and physiological changes (neurocognitive reward processing via the HPA axis; increased production of biochemical hormones and peptides). Conversely, obesity can contribute to experienced stress through weight stigma and the obesogenic processes (Tomiyama, 2019). Obesity in children and adolescents is also impacted indirectly by maternal stress with the greatest impact on toddlers, which is likely attributable to factors including the increasingly diverse food and social environments of older children (Tate et al., 2015).

To illustrate this complex interaction between stress and weight, the mechanisms around a common stressor has been studied repeatedly. Job stress is suggested as a strong mediator for the development of visceral adiposity and increases in BMI (Tenk et al., 2018). However, longitudinal studies do not support a direct relation between job stress and BMI as stress decreases; this means interventions targeting solely reductions in a conditional stressor (i.e., do not address dietary habits, physical activity, and other health behaviors) are not likely to reduce BMI (Kivimäki et al., 2015; Tenk et al., 2018). Further, the bidirectional relation between stress and weight creates a self-sustaining cycle (Tomiyama, 2019).

Relations with CVD

The relation between weight and CVD is complicated by years of consistently replicated epidemiological studies supporting a phenomenon coined the "obesity paradox," which suggests individuals with overweight and class

1 obesity are at reduced risk for cardiovascular mortality compared to their healthy weight peers (Horwich et al., 2018; Sharma et al., 2015). The obesity paradox sustains significant results in chronic and acute heart failure, when controlling for non-cardiovascular mortality, and across heterogeneous biochemical and clinical profiles (Sharma et al., 2015).

Reviewers of the obesity paradox literature suggest methodological considerations including selection, survival, and treatment biases may explain some of the findings supporting this epidemiological paradox (Standl et al., 2013). Another methodological consideration is the use of BMI as an independent anthropomorphic measure, given individuals with high percentage lean mass may present as overweight on measures of BMI alone (Horwich et al., 2018). Additionally, several complicating factors that increase risk for CVD such as other related health behaviors (e.g., smoking), medical conditions that mimic healthy weight (e.g., anabolic deficiencies), and environmental factors (e.g., socio-economic status) reduce the homogeneity of the healthy-weight comparison (Horwich et al., 2018; Standl et al., 2013). Finally, cardiac cachexia following a cardiac event may result in unintentional weight loss and poor cardiovascular outcomes (Horwich et al., 2018; Standl et al., 2013).

Recent studies looking at lifetime risk for incident CVD revealed higher risk in overweight individuals compared to their healthy-weight peers, bringing the obesity paradox into question. Childhood obesity is associated with increased risk for CVD in adulthood such as hypertension and abnormal serum lipids (Umer et al., 2017). For middle-aged men and women, the risk of developing CVD compared to healthy peers increased by 21% and 32% for overweight and 214% and 153% for severe obesity, respectively. Overall, overweight and obesity were associated with earlier onset of CVD and greater years lived with disability. Additionally, obesity was associated with increased cardiovascular mortality and reduced longevity (Khan et al., 2018).

Although overweight and obesity are clear risk factors for CVD, current cardiovascular literature suggests weight loss is not as important as cardiorespiratory fitness (CRF; Elagizi et al., 2018). CRF nearly doubles the predictive value of BMI in cardiovascular mortality risk. Further, healthy CRF can nearly eliminate the risk associated with obesity, and those with low CRF and obesity have a cardiovascular mortality risk nearly triple that of fit, healthy-weight individuals (Barry et al., 2018). In total, cardiovascular intervention should focus on CRF as markers of health over weight loss (Barry et al., 2018; Elagizi et al., 2018; Horwich et al., 2018).

Relations with depression

The association between depression and weight has consistently shown a positive relation from adolescence to adulthood, across cross-sectional and longitudinal meta-analyses, with the strongest relation in females and in obese BMI ranges (Amiri et al., 2018; Marmorstein et al., 2014; Pereira-Miranda et al., 2017; Quek et al., 2017; Sutaria et al., 2019). In fact, the relation is so well founded that a diagnostic criterion for major depressive disorder is a significant and unintentional change in weight or appetite (APA, 2013). Longitudinal studies of obesity and depression suggest a bidirectional relation in the maintenance of these health concerns (Marmorstein et al., 2014). However, current research commonly calls for further study as to the directionality in the development of obesity and depression (Pereira-Miranda et al., 2017; Quek et al., 2017). One question that approaches causality is assessing the longitudinal risk associated with obesity. Obese children and adolescents are at 32% increased risk for current or future depression compared to their healthy-weight peers, with a 44% increased risk in adolescent females compared to males (Sutaria et al., 2019). Among 18 prospective-cohort studies, those who were obese were 15% more likely to demonstrate depressive symptoms than normal weight counterparts (Amiri et al., 2018).

The relation between depression and weight is explained by many of the same mechanisms as stress and obesity. Stressors associated with the onset of depressive symptoms may result in increased production of biochemical hormones and peptides leading to greater appetite (Amiri et al., 2018; Pereira-Miranda et al., 2017; Quek et al., 2017). Increases in weight may negatively impact body image, reduce interest or perceived ability to engage in physical activity, and promote maintenance of a sedentary lifestyle (Pereira-Miranda et al., 2017; Sutaria et al., 2019). Depressive emotional states may be accompanied by emotionally driven binge-eating patterns, seeking calorie-dense foods, and exacerbating weight-gain, creating a self-sustaining cycle (Pereira-Miranda et al., 2017).

Clinically, this suggests overweight and obese individuals may be at particularly high risk for developing depressive symptoms. Screening for depression may be particularly important in children and adolescents with obesity (Quek et al., 2017; Sutaria et al., 2019). Additionally, treatment strategies targeting depression (e.g., antidepressants, cognitive-behavior therapy) have been shown to ameliorate weight-related health concerns through reductions in negative body image, binge-eating habits, and appetitive or emotional drives to overeat (Amiri et al., 2018; Quek et al., 2017).

Medication and medical regimen adherence

Adherence to medication and medical recommendations is associated with a 26% increase in positive health outcomes (Dimatteo et al., 2002). Despite the importance of adherence in treating chronic disease, 30–60% of patients adhere poorly to treatment (Dunbar-Jacob and Mortimer-Stephens, 2001; Marcum et al., 2013). Poor adherence alone accounts for 3–10% of total US health care costs (Iuga and McGuire, 2014). Haynes et al. (2002) suggest that increasing adherence would be more impactful than improving any specific medical or behavioral intervention.

Depression and CVD are chronic diseases requiring long-term prevention and management with intensive adherence to lifestyle changes, medications, or a combination thereof (Brown and Bussell, 2011; Chowdhury et al., 2013; Dimatteo et al., 2002). Medical regimen adherence includes many of the health behaviors discussed in this chapter. As a result, this section will primarily focus on medication adherence in relation to stress, CVD, and depression.

Relations with stress

Stress is associated with decreased adherence to health behaviors for a number of chronic health conditions. Stress' influence on adherence can be classified into provider, patient, and systems-level factors (Brown and Bussell, 2011).

The cumulative effect of occupational stress on providers has been termed professional burn-out (Maslach, 1993). Professional burn-out is a provider-level factor which likely influences patient adherence. A provider plays a critical role in medical adherence through their competency in prescribing, monitoring, or intervening appropriately in any health condition. Failure in any of the aforementioned competencies may result in reduced patient adherence. Specifically, stressed providers have more negative patient outcomes, poorer physical and mental health, and interpersonal difficulties that may influence patient adherence (Morse et al., 2012; Salyers et al., 2017; Van Der Linden et al., 2005).

Self-reported symptoms of anxiety and perceived psychological distress are associated with an overall reduced adherence to treatment. (Warren et al., 2013; Kalichman and Grebler, 2010; Novak et al., 2013; Alcántara et al. 2014). This effect is particularly consistent for medication adherence. Other health behaviors evidence mixed results in their association with stress depending on the specific health behavior and the presence of comorbid medical conditions (Osborn et al., 2014; Roohafza et al., 2016).

Systems-level stressors such as discrimination and socioeconomic stressors are also associated with reduced adherence. Poor adherence was associated with food insecurity (Kalichman and Grebler, 2010), financial stress (Mcallister et al., 2013), and ethnic minority status (Kang et al., 2018).

Relations with CVD

Approximately 44% of the reduction in CVD mortality for the last 20 years can be attributed to reducing risk factors (e.g., cholesterol, smoking, physical inactivity) that are amenable to health behavior intervention (Ford et al., 2007). Medication adherence is one of the simplest and most effective health behaviors to reduce risk factors for CVD (Warren et al., 2013), with adherence yielding a 20% lower risk of CVD (Chowdhury et al., 2013). Additionally, the odds of blood pressure control for adherent patients, compared with those who were not adherent, was 3.44 (Dimatteo et al., 2002). Despite this demonstrated efficacy and over hundreds of developed blood pressure medications, 40–50% of patients with CVD or related risk factors are poorly adherent to their medication regimen (Chowdhury et al., 2013; Kronish and Ye, 2013).

Relations with depression

Depressed patients, when compared to nondepressed patients, are 1.76–3 times more likely to be non-adherent to provider recommendations (DiMatteo et al., 2000; Grenard et al., 2011). Reduced adherence is particularly pronounced for those with comorbid medical conditions or a history of significant health problems (DiMatteo et al., 2000). Kronish et al. (2006) found that, among those diagnosed with acute coronary syndromes, those who self-reported persistent depression were less likely to adhere to medication regimens compared to their persistently non-depressed peers (OR = 0.50).

Hopelessness, a limited social network, and impaired cognitive function have all been posited as depressive symptoms that reduce adherence (DiMatteo et al., 2000). It was found that those who are the least adherent were most disrupted by these depressive symptoms as opposed to specific stressors (Osborn et al., 2014). Another possibility is that depression leaves individuals more sensitive to the effects of stress. Bottonari et al. (2010) found the incidence of stress more effectively predicted decreased adherence in depressed patients than in nondepressed patients. Despite the consistent associations between depression and poor health outcomes (Kalichman and Grebler, 2010; Sin and DiMatteo, 2014), it is not clear whether adherence mediates the relation. Wing et al. (2002) suggest a model that specific health behaviors related to depression are associated with other medical conditions (e.g., CVD). That is to say, it may be that the behaviors of depression (e.g., hyperphagia, physical inactivity) are directly tied to the etiology of a co-morbid medical condition.

Sleep

In previous reviews of the relations between health behaviors with stress, CVD, and depression, sleep behaviors were not considered (Carney et al., 1995; Lett et al., 2004). However, there is growing evidence that sleep affects and is affected by stress, CVD, and depression. The trend in the United States is towards shorter sleep durations with 29.2% of adults reporting 6 or fewer hours of sleep (Ford et al., 2015). These trends along with the prevalence of sleep disorders (e.g., insomnia and sleep apnea) may be important for understanding the relations between stress, CVD, and depression.

Relations with stress

Experimental studies have found that stress results in decreased slow-wave and REM sleep, decreased sleep efficiency (i.e., time asleep while in bed), and increases in nighttime awakenings (Kim and Dimsdale, 2007). Poor sleep may also impact the stress response. Bassett et al. (2015) found that poor perceived sleep quality among men was related to increased cortisol levels following a standardized stress task; the same relation was not found among women and sleep length was not associated with changes in cortisol responses.

Relations with CVD

The risk for CVD morbidity and mortality has a U-shaped relation with sleep duration; shorter and longer sleepers demonstrate increased risk (Covassin and Singh, 2016; Hall et al., 2018; Sabanayagam and Shankar, 2010). Those reporting insomnia or poor sleep with short sleep were associated with a 29% higher risk of developing CVD compared to a reference group (Bertisch et al., 2018). This U-shaped relation pattern with sleep length is also found for obesity, diabetes, and high blood pressure (Buxton and Marcelli, 2010). Hall et al. (2018) provide a detailed summary of what is known about the relations between sleep and CVD that goes beyond the limits of this chapter. One critical finding is that short sleep duration and insomnia are associated with increased risk of CVD morbidity and mortality, independent of other risk factors (Hall et al., 2018).

Relations with depression

A systematic review of nine studies examining relations between sleep disturbances and depression, as well as anxiety, determined that most studies found a bidirectional relation between sleep disturbances and depression (Alvaro et al., 2013). These conclusions were based in part on four longitudinal studies examining this relation. One of these studies (Buysse et al., 2008) found that this bidirectional relation existed over 20-years, but that baseline insomnia was a more consistent predictor of depressive symptoms compared to baseline depression predicting insomnia.

Summary

Similar to the conclusions by Carney et al. (1995) and Lett et al. (2004), health behaviors continue to demonstrate strong, often bidirectional, relations with stress, CVD, and depression. Health behaviors may not only cause, but also may be influenced by stress and depression. Determining causal relations remains a challenge, but relations between health behaviors, stress, and depression often fulfill the Bradford Hill criteria for causal relations (Fedak et al., 2015). Some recent developments and directions in research may be reason to expand and modify the research questions examining the mechanisms and roles of health behaviors. Examining broad transdiagnostic constructs, such as distress tolerance, rather than categorizing responses as stress, depression, or any other emotional constructs (e.g., anxiety) may provide a clearer understanding of the relations between health behaviors, emotional responses, and CVD. Sleep behaviors are an important health behavior to consider in models relating stress, depression, and CVD. Despite the vast cumulative knowledge about the relations between health behaviors, stress, CVD, and depression, much less is known about the physiological mechanisms responsible for these relations and it remains largely unclear how to most effectively help individuals change these behaviors. Continued research and future reviews should focus on answering these important questions.

References

Aggio, D., Papacosta, O., Lennon, L., Whincup, P., Wannamethee, G., Jefferis, B.J., 2017. Association between physical activity levels in mid-life with physical activity in old age: a 20-year tracking study in a prospective cohort. BMJ Open 7.

Al'Absi, M., 2018. Stress and addiction: When a robust stress response indicates resiliency. Psychosom. Med. 80, 2–16.

Alcántara, C., Edmondson, D., Moise, N., Oyola, D., Hiti, D., Kronish, I.M., 2014. Anxiety sensitivity and medication nonadherence in patients with uncontrolled hypertension. J. Psychosom. Res. 77, 283–286.
Alvaro, P.K., Roberts, R.M., Harris, J.K., 2013. A systematic review assessing bidirectionality between sleep disturbances, anxiety, and depression. Sleep 36, 1059–1068.
Alzahrani, T., Pena, I., Temesgen, N., Glantz, S.A., 2018. Association between electronic cigarette use and myocardial infarction. Am. J. Prev. Med. 55, 455–461.
American Psychiatric Association, 2013. The Diagnostic and Statistical Manual of Mental Disorders, fifth ed. American Psychiatric Publishing, Arlington, VA.
Amiri, S., Behnezhad, S., Nadinlui, K.B., 2018. Body mass index (BMI) and risk of depression in adults: a systematic review and meta-analysis of longitudinal studies. Obes. Med. 12, 1–12.
Anand, S.S., Hawkes, C., de Souza, R.J., Mente, A., Dehghan, M., Nugent, R., Zulyniak, M.A., Weis, T., Bernstein, A.M., Krauss, R.M., Kromhout, D., Jenkins, D.J.A., Malik, V., Martinez-Gonzalez, M.A., Mozaffarian, D., Yusuf, S., Willett, W.C., Popkin, B.M., 2015. Food consumption and its impact on cardiovascular disease: importance of solutions focused on the globalized food system: a report from the workshop convened by the world heart federation. J. Am. Coll. Cardiol. 66, 1590–1614.
Aune, D., Schlesinger, S., Norat, T., Riboli, E., 2018. Tobacco smoking and the risk of heart failure: a systematic review and meta-analysis of prospective studies. Eur. J. Epidemiol. 33, 509–521.
Bandiera, F.C., Loukas, A., Wilkinson, A.V., Perry, C.L., 2016. Associations between tobacco and nicotine product use and depressive symptoms among college students in Texas. Addict. Behav. 63, 19–22.
Barry, V.W., Caputo, J.L., Kang, M., 2018. The joint association of fitness and fatness on cardiovascular disease mortality: a meta-analysis. Prog. Cardiovasc. Dis. 61, 136–141.
Bassett, S.M., Lupis, S.B., Gianferante, D., Rohleder, N., Wolf, J.M., 2015. Sleep quality but not sleep quantity effects on cortisol responses to acute psychosocial stress. Stress 18, 638–644.
Bertisch, S.M., Pollock, B.D., Mittleman, M.A., Buysse, D.J., Bazzano, L.A., Gottlieb, D.J., Redline, S., 2018. Insomnia with objective short sleep duration and risk of incident cardiovascular disease and all-cause mortality: Sleep heart health study. Sleep 41.
Beutel, T.F., Zwerenz, R., Michal, M., 2018. Psychosocial stress impairs health behavior in patients with mental disorders. BMC Psychiatry 18 (375).
Bianco, C.L., 2019. Rates of electronic cigarette use among adults with a chronic mental illness. Addict. Behav. 89, 1–4.
Borges, G., Bagge, C.L., Cherpitel, C.J., Conner, K.R., Orozco, R., Rossow, I., 2017. A meta-analysis of acute use of alcohol and the risk of suicide attempt. Psychol. Med. 47, 949–957.
Bottonari, K.A., Safren, S.A., McQuaid, J.R., Hsiao, C.B., Roberts, J.E., 2010. A longitudinal investigation of the impact of life stress on HIV treatment adherence. J. Behav. Med. 33, 486–495.
Briasoulis, A., Agarwal, V., Messerli, F.H., 2012. Alcohol consumption and the risk of hypertension in men and women: a systematic review and meta-analysis. J. Clin. Hypertens. 14, 792–798.
Brière, F.N., Rohde, P., Seeley, J.R., Klein, D., Lewinsohn, P.M., 2014. Comorbidity between major depression and alcohol use disorder from adolescence to adulthood. Compr. Psychiatry 55, 526–533.
Brown, M.T., Bussell, J.K., 2011. Medication adherence: WHO cares? Mayo Clin. Proc. 86, 304–314.
Buxton, O.M., Marcelli, E., 2010. Short and long sleep are positively associated with obesity, diabetes, hypertension, and cardiovascular disease among adults in the United States. Soc. Sci. Med. 71, 1027–1036.

Buysse, D.J., Angst, J., Gamma, A., Ajdacic, V., Eich, D., Rössler, W., 2008. Prevalence, course, and comorbidity of insomnia and depression in young adults. Sleep 31, 473–480.

Carison, S.A., Fulton, J.E., Galuska, D.A., Kruger, J., Lobelo, I.F., Loustalot, F.V., 2009. Prevalence of self-reported physically active adults United States, 2007. J. Am. Med. Assoc. 301, 926–927.

Carlsson, A.C., Wändell, P.E., Gigante, B., Leander, K., Hellenius, M.L., De Faire, U., 2013. Seven modifiable lifestyle factors predict reduced risk for ischemic cardiovascular disease and all-cause mortality regardless of body mass index: a cohort study. Int. J. Cardiol. 168, 946–952.

Carney, R.M., Freedland, K.E., Rich, M.W., Jaffe, A.S., 1995. Depression as a risk factor for cardiac events in established coronary heart disease: a review of possible mechanisms. Ann. Behav. Med. 17 (2), 142–149.

Centers for Disease Control and Prevention, 2016. Defining Adult Overweight and Obesity. Centers for Disease Control Website. Available from: https://www.cdc.gov/obesity/adult/defining.html. (Accessed 1 February 2019).

Centers for Disease Control and Prevention, 2018a. Childhood Obesity Facts. Centers for Disease Control Website. Available from: https://www.cdc.gov/obesity/data/childhood.html. (Accessed 1 February 2019).

Centers for Disease Control and Prevention, 2018b. Adult Obesity Facts. Centers for Disease Control Website. Available from: https://www.cdc.gov/obesity/data/adult.html. (Accessed 1 February 2019).

Cheng, H.G., Kaakarli, H., Breslau, J., Anthony, J.C., 2018. Assessing changes in alcohol use and alcohol use disorder prevalence in the United States: evidence from national surveys from 2002 through 2014. J. Am. Med. Assoc. Psychiatry 75, 211–213.

Chowdhury, R., Khan, H., Heydon, E., Shroufi, A., Fahimi, S., Moore, C., Stricker, B., Mendis, S., Hofman, A., Mant, J., Franco, O.H., 2013. Adherence to cardiovascular therapy: a meta-analysis of prevalence and clinical consequences. Eur. Heart J. 34, 2940–2948.

Covassin, N., Singh, P., 2016. Sleep duration and cardiovascular disease risk: epidemiologic and experimental evidence. Sleep Med. Clin. 11, 81–89.

Crisafulli, A., Pagliaro, P., Cohen-Solal, A., Coats, A.J., 2015. Effects of physical exercise on cardiovascular diseases: biochemical, cellular, and organ effects. Biomed Res. Int. 2015.

Da Silva, M.A., Singh-Manoux, A., Brunner, E.J., Kaffashian, S., Shipley, M.J., Kivimäki, M., Nabi, H., 2012. Bidirectional association between physical activity and symptoms of anxiety and depression: the Whitehall II study. Eur. J. Epidemiol. 27, 537–546.

Després, J.P., Lemieux, I., 2006. Abdominal obesity and metabolic syndrome. Nature 444, 881–887.

DiMatteo, M.R., Lepper, H.S., Croghan, T.W., 2000. Depression is a risk factor for noncompliance with medical treatment meta-analysis of the effects of anxiety and depression on patient adherence. Arch. Intern. Med. 160, 2101–2107.

Dimatteo, M.R., Giordani, P.J., Lepper, H.S., Thomas, W., 2002. Patient adherence and medical treatment outcomes a meta-analysis. Med. Care 40, 794–811.

Ducharme, J., 2018. Here's Why You Stress Eat: And How to Stop Doing it. *Time*, Available from: http://time.com/5347612/how-to-stop-stress-eating/. (Accessed 1 February 2019).

Duivis, H.E., De Jonge, P., Penninx, B.W., Na, B.Y., Cohen, B.E., Whooley, M.A., 2011. Depressive symptoms, health behaviors, and subsequent inflammation in patients with coronary heart disease: prospective findings from the heart and soul study. Am. J. Psychiatry 168, 913–920.

Dunbar-Jacob, J., Mortimer-Stephens, M.K., 2001. Treatment adherence in chronic disease. J. Clin. Epidemiol. 54.

Elagizi, A., Kachur, S., Lavie, C.J., Carbone, S., Pandey, A., Ortega, F.B., Milani, R.V., 2018. An overview and update on obesity and the obesity paradox in cardiovascular diseases. Prog. Cardiovasc. Dis. 61, 142–150.

Fedak, K.M., Bernal, A., Capshaw, Z.A., Gross, S., 2015. Applying the Bradford Hill criteria in the 21st century: how data integration has changed causal inference in molecular epidemiology. Emerg. Themes Epidemiol. 12, 14.

Firth, J., Rosenbaum, S., Stubbs, B., Gorczynski, P., Yung, A.R., Vancampfort, D., 2016. Motivating factors and barriers towards exercise in severe mental illness: a systematic review and meta-analysis. Psychol. Med. 46, 2869–2881.

Flegal, K.M., Kit, B.K., Orpana, H., Graubard, B.I., 2013. Association of all-cause mortality with overweight and obesity using standard body mass index categories a systematic review and meta-analysis. J. Am. Med. Assoc. 309, 71–82.

Fluharty, M., Taylor, A.E., Grabski, M., Munafò, M.R., 2017. The association of cigarette smoking with depression and anxiety: a systematic review. Nicotine Tob. Res. 19, 3–13.

Ford, E.S., Ajani, U.A., Croft, J.B., Critchley, J.A., Labarthe, D.R., Kottke, T.E., Giles, W.H., Capewell, S., 2007. Explaining the decrease in U.S. deaths from coronary disease, 1980–2000. N. Engl. J. Med. 356, 2388–2398.

Ford, E.S., Cunningham, T.J., Croft, J.B., 2015. Trends in self-reported sleep duration among US adults from 1985 to 2012. Sleep 38, 829–832.

Foulds, J.A., Adamson, S.J., Boden, J.M., Williman, J.A., Mulder, R.T., 2015. Depression in patients with alcohol use disorders: systematic review and meta-analysis of outcomes for independent and substance-induced disorders. J. Affect. Disord. 185, 47–59.

George, A., Figueredo, V.M., 2011. Alcoholic cardiomyopathy: a review. J. Card. Fail. 17, 844–849.

Gerber, M., Pühse, U., 2009. Do exercise and fitness protect against stress-induced health complaints? A review of the literature. Scand. J. Public Health 37, 801–819.

Glasser, A.M., Collins, L., Pearson, J.L., Abudayyeh, H., Niaura, R.S., Abrams, D.B., Villanti, A.C., 2017. Overview of electronic nicotine delivery systems: a systematic review. Am. J. Prev. Med. 52, e33–e66.

Grenard, J.L., Munjas, B.A., Adams, J.L., Suttorp, M., Maglione, M., McGlynn, E.A., Gellad, W.F., 2011. Depression and medication adherence in the treatment of chronic diseases in the United States: a meta-analysis. J. Gen. Intern. Med. 26, 1175–1182.

Guszkowska, M., 2004. Effects of exercise on anxiety, depression and mood. Psychiatr. Pol. 38, 611–620.

Hall, M.H., Brindle, R.C., Buysse, D.J., 2018. Sleep and cardiovascular disease: emerging opportunities for psychology. Am. Psychol. 73, 994–1006.

Hallal, P.C., Andersen, L.B., Bull, F.C., Guthold, R., Haskell, W., Ekelund, U., Alkandari, J.R., Bauman, A.E., Blair, S.N., Brownson, R.C., Craig, C.L., Goenka, S., Heath, G.W., Inoue, S., Kahlmeier, S., Katzmarzyk, P.T., Kohl, H.W., Lambert, E.V., Lee, I.M., Leetongin, G., Lobelo, F., Loos, R.J.F., Marcus, B., Martin, B.W., Owen, N., Parra, D.C., Pratt, M., Puska, P., Ogilvie, D., Reis, R.S., Sallis, J.F., Sarmiento, O.L., Wells, J.C., 2012. Global physical activity levels: surveillance progress, pitfalls, and prospects. Lancet 380, 247–257.

Hawkins, M.S., Storti, K.L., Richardson, C.R., King, W.C., Strath, S.J., Holleman, R.G., Kriska, A.M., 2009. Objectively measured physical activity of USA adults by sex, age, and racial/ethnic groups: a cross-sectional study. Int. J. Behav. Nutr. Phys. Act. 6 (1).

Haynes, R., McDonald, H., Garg, A., Montague, P., 2002. Interventions for helping patients to follow prescriptions for medications. Cochrane Database Syst. Rev. 2.

Hiles, S.A., Lamers, F., Milaneschi, Y., Penninx, B.W.J.H., 2017. Sit, step, sweat: longitudinal associations between physical activity patterns, anxiety and depression. Psychol. Med. 47, 1466–1477.

Hill, D.C., Moss, R.H., Sykes-Muskett, B., Conner, M., O'Connor, D.B., 2018. Stress and eating behaviors in children and adolescents: systematic review and meta-analysis. Appetite 123, 14–22.

Horwich, T.B., Fonarow, G.C., Clark, A.L., 2018. Obesity and the obesity paradox in heart failure. Prog. Cardiovasc. Dis. 61, 151–156.

Huxley, R.R., Woodward, M., 2011. Cigarette smoking as a risk factor for coronary heart disease in women compared with men: a systematic review and meta-analysis of prospective cohort studies. Lancet 378, 1297–1305.

Iuga, A.O., McGuire, M.J., 2014. Adherence and health care costs. Risk Manag. Healthc. Policy 7, 35–44.

Jacka, F.N., O'Neil, A., Opie, R., Itsiopoulos, C., Cotton, S., Mohebbi, M., Castle, D., Dash, S., Mihalopoulos, C., Chatterton, M.L., Brazionis, L., Dean, O.M., Hodge, A.M., Berk, M., 2017. A randomised controlled trial of dietary improvement for adults with major depression (the "SMILES" trial). BMC Med. 15, 23.

Jenkins, D.J.A., Spence, J.D., Giovannucci, E.L., Kim, Y.I., Josse, R., Vieth, R., Blanco Mejia, S., Viguiliouk, E., Nishi, S., Sahye-Pudaruth, S., Paquette, M., Patel, D., Mitchell, S., Kavanagh, M., Tsirakis, T., Bachiri, L., Maran, A., Umatheva, N., McKay, T., Trinidad, G., Bernstein, D., Chowdhury, A., Correa-Betanzo, J., Del Principe, G., Hajizadeh, A., Jayaraman, R., Jenkins, A., Jenkins, W., Kalaichandran, R., Kirupaharan, G., Manisekaran, P., Qutta, T., Shahid, R., Silver, A., Villegas, C., White, J., Kendall, C.W.C., Pichika, S.C., Sievenpiper, J.L., 2018. Supplemental vitamins and minerals for CVD prevention and treatment. J. Am. Coll. Cardiol. 71, 2570–2584.

Kalichman, S.C., Grebler, T., 2010. Stress and poverty predictors of treatment adherence among people with low-literacy living with HIV/AIDS. Psychosom. Med. 72, 810–816.

Kang, A.W., Dulin, A., Nadimpalli, S., Risica, P.M., 2018. Stress, adherence, and blood pressure control: a baseline examination of black women with hypertension participating in the SisterTalk II intervention. Prev. Med. Rep. 12, 25–32.

Kaplan, B.J., Rucklidge, J.J., Romijn, A., McLeod, K., 2015. The emerging field of nutritional mental health. Clin. Psychol. Sci. 3, 964–980.

Kassel, J.D., Stroud, L.R., Paronis, C.A., 2003. Smoking, stress, and negative affect: correlation, causation, and context across stages of smoking. Psychol. Bull. 129, 270–304.

Keyes, K.M., Hatzenbuehler, M.L., Grant, B.F., Hasin, D.S., 2012. Stress and alcohol: epidemiologic evidence. Alcohol Res. Curr. Rev. 34, 391–400.

Khan, S.S., Ning, H., Wilkins, J.T., Allen, N., Carnethon, M., Berry, J.D., Carnerthon, M., Berry, J.D., Sweis, R.N., Lloyd-Jones, D.M., 2018. Association of body mass index with lifetime risk of cardiovascular disease and compression of morbidity. J. Am. Med. Assoc. Cardiol. 3 (4), 280–287.

Kim, E.J., Dimsdale, J.E., 2007. The effect of psychosocial stress on sleep: a review of polysomnographic evidence. Behav. Sleep Med. 5, 256–278.

Kivimäki, M., Singh-Manoux, A., Nyberg, S., Jokela, M., Virtanen, M., 2015. Job strain and risk of obesity: systematic review and meta-analysis of cohort studies. Int. J. Obes. 39 (11).

Kodama, S., Saito, K., Tanaka, S., Horikawa, C., Saito, A., Heianza, Y., Anasako, Y., Nishigaki, Y., Yachi, Y., Iida, K.T., Ohashi, Y., Yamada, N., Sone, H., 2011. Alcohol consumption and risk of atrial fibrillation: a meta-analysis. J. Am. Coll. Cardiol. 57, 427–436.

Kronish, I.M., Ye, S., 2013. Adherence to cardiovascular medications: lessons learned and future directions. Prog. Cardiovasc. Dis. 55, 590–600.

Kronish, I.M., Rieckmann, N., Halm, E.A., Shimbo, D., Vorchheimer, D., Haas, D.C., Davidson, K.W., 2006. Persistent depression affects adherence to secondary prevention behaviors after acute coronary syndromes. J. Gen. Intern. Med. 21, 1178–1183.

Lear, S.A., Hu, W., Rangarajan, S., Gasevic, D., Leong, D., Iqbal, R., Casanova, A., Swaminathan, S., Anjana, R.M., Kumar, R., Rosengren, A., Wei, L., Yang, W., Chuangshi, W., Huaxing, L., Nair, S., Diaz, R., Swidon, H., Gupta, R., Mohammadifard, N., Lopez-Jaramillo, P., Oguz, A., Zatonska, K., Seron, P., Avezum, A., Poirier, P., Teo, K., Yusuf, S., 2017. The effect of physical activity on mortality and cardiovascular disease in 130,000 people from 17 high-income, middle-income, and low-income countries: the PURE study. Lancet 390, 2643–2654.

Lee, I.M., Shiroma, E.J., Lobelo, F., Puska, P., Blair, S.N., Katzmarzyk, P.T., Alkandari, J.R., Andersen, L.B., Bauman, A.E., Brownson, R.C., Bull, F.C., Craig, C.L., Ekelund, U., Goenka, S., Guthold, R., Hallal, P.C., Haskell, W.L., Heath, G.W., Inoue, S., Kahlmeier, S., Kohl, H.W., Lambert, E.V., Leetongin, G., Loos, R.J.F., Marcus, B., Martin, B.W., Owen, N., Parra, D.C., Pratt, M., Ogilvie, D., Reis, R.S., Sallis, J.F., Sarmiento, O.L., Wells, J.C., 2012. Effect of physical inactivity on major non-communicable diseases worldwide: an analysis of burden of disease and life expectancy. Lancet 380, 219–229.

Lett, H.S., Blumenthal, J.A., Babyak, M.A., Sherwoob, A., Strauman, T., Robins, C., Newman, M.F., 2004. Depression as a risk factor for coronary artery disease: evidence, mechanisms, and treatment. Psychosom. Med. 66, 305–315.

Leventhal, A.M., Zvolensky, M.J., 2015. Anxiety, depression, and cigarette smoking: A transdiagnostic vulnerability framework to understanding emotion-smoking comorbidity. Psychol. Bull. 141, 176–212.

Li, J., Siegrist, J., 2012. Physical activity and risk of cardiovascular disease-a meta-analysis of prospective cohort studies. Int. J. Environ. Res. Public Health 9, 391–407.

Li, Y., Lv, M., Wei, Y., Sun, L., Zhang, J., Zhang, H., Li, B., 2017. Dietary patterns and depression risk: a meta-analysis. Psychiatry Res. 253, 373–382.

Loef, M., Walach, H., 2012. The combined effects of healthy lifestyle behaviors on all cause mortality: a systematic review and meta-analysis. Prev. Med. (Baltim). https://doi.org/10.1016/j.ypmed.2012.06.017.

Loprinzi, P.D., Branscum, A., Hanks, J., Smit, E., 2016. Healthy lifestyle characteristics and their joint association with cardiovascular disease biomarkers in US adults. Mayo Clin. Proc. 91 (4), 432–442.

Marcum, Z.A., Sevick, M.A., Handler, S.M., 2013. Medication nonadherence: a diagnosable and treatable medical condition. J. Am. Med. Assoc. 309, 2105–2106.

Marmorstein, N.R., Iacono, W.G., Legrand, L., 2014. Obesity and depression in adolescence and beyond: Reciprocal risks. Int. J. Obes. 38, 906–911.

Maslach, C., 1993. Burnout: a multidimensional perspective. In: Schaufeli, W.B., Maslach, C., Marek, T. (Eds.), Series in Applied Psychology: Social Issues and Questions. Professional Burnout: Recent Developments in Theory and Research. Taylor & Francis, Philadelphia, PA, pp. 19–32.

Mcallister, J., Beardsworth, G., Lavie, E., Macrae, K., Carr, A., 2013. Financial stress is associated with reduced treatment adherence in HIV-infected adults in a resource-rich setting. HIV Med. 14, 120–124.

McKee, M., Britton, A., 1998. The positive relationship between alcohol and heart disease in eastern Europe: potential physiological mechanisms. J. R. Soc. Med. 91, 402–407.

McNeil, A., Brose, L.S., Calder, R., Hitchman, S.C., Hajek, P., McRobbie, H., 2015. E-cigarettes: an evidence update. . Public Health England, 111.

Molendijk, M., Molero, P., Ortuño Sánchez-Pedreño, F., Van der Does, W., Angel Martínez-González, M., 2018. Diet quality and depression risk: a systematic review and dose-response meta-analysis of prospective studies. J. Affect. Disord. 226, 346–354.

Mons, U., Müezzinler, A., Gellert, C., Schöttker, B., Abnet, C.C., Bobak, M., De Groot, L., Freedman, N.D., Jansen, E., Kee, F., Kromhout, D., Kuulasmaa, K., Tiina, L., O'Doherty, M.G., Bas, B.D.M., Orfanos, P., Peters, A., Van Der Schouw, Y.T., Wilsgaard, T., Wolk, A., Trichopoulou, A., Boffetta, P., Brenner, H., Baceviciene, M., Boer, J.M.A., Drygas, W., Eriksson, S., Feskens, E., Gafarov, V., Gardiner, J., Håkansson, N., Jansson, J.H., Jousilahti, P., Kampman, E., Kontto, J., Kubinova, R., Leenders, M., Linneberg, A., Løchen, M.L., Lorbeer, R., Malyutina, S., Mathiesen, E.B., Melhus, H., Michaëlsson, K., Njølstad, I., Orsini, N., Pająk, A., Pikhart, H., Pisinger, C., Salomaa, V., Sánchez, M.J., Sans, S., Schaan, B., Schneider, A., Siganos, G., Söderberg, S., Streppel, M., Tamošiunas, A., Veronesi, G., Waterham, E., Wennberg, P., 2015. Impact of smoking and smoking cessation on cardiovascular events and mortality among older adults: meta-analysis of individual participant data from prospective cohort studies of the CHANCES consortium. BMJ 350, 1–12.

Morse, G., Salyers, M.P., Rollins, A.L., Monroe-DeVita, M., Pfahler, C., 2012. Burnout in mental health services: a review of the problem and its remediation. Adm. Policy Ment. Health Ment. Health Serv. Res. 39, 341–352.

Moustafa, A.A., Parkes, D., Fitzgerald, L., Underhill, D., Garami, J., Levy-Gigi, E., Stramecki, F., Valikhani, A., Frydecka, D., Misiak, B., 2018. The relationship between childhood trauma, early-life stress, and alcohol and drug use, abuse, and addiction: an integrative review. Curr. Psychol. 1–6.

NCD Risk Factor Collaboration, 2017. Worldwide trends in body-mass index, underweight, overweight, and obesity from 1975 to 2016: a pooled analysis of 2416 population-based measurement studies in 128·9 million children, adolescents, and adults. Lancet 390, 2627–2642.

Novak, M., Björck, L., Giang, K.W., Heden-Ståhl, C., Wilhelmsen, L., Rosengren, A., 2013. Perceived stress and incidence of type 2 diabetes: a 35-year follow-up study of middle-aged Swedish men. Diabet. Med. 30, 8–16.

Osborn, C.Y., Mayberry, L.S., Wagner, J.A., Welch, G.W., 2014. Stressors may compromise medication adherence among adults with diabetes and low socioeconomic status. West. J. Nurs. Res. 36, 1091–1110.

Palmisano, M., Pandey, S.C., 2017. Epigenetic mechanisms of alcoholism and stress-related disorders. Alcohol 60, 7–18.

Pate, R.R., Macera, C.A., Pratt, M., Heath, G.W., Blair, S.N., Bouchard, C., Haskell, W.L., King, A.C., Buchner, D., Ettinger, W., Kriska, A., Leon, A.S., Marcus, B.H., Morris, J., Paffenbarger, R.S., Patrick, K., Pollock, M.L., Rippe, J.M., Sallis, J., Wilmore, J.H., 1995. Physical activity and public health: a recommendation from the Centers for Disease Control and Prevention and the American College of Sports Medicine. J. Am. Med. Assoc. 273, 402–407.

Patra, J., Taylor, B., Irving, H., Roerecke, M., Baliunas, D., Mohapatra, S., Rehm, J., 2010. Alcohol consumption and the risk of morbidity and mortality for different stroke types: a systematic review and meta-analysis. BMC Public Health 10, 258.

Pearson-Stuttard, J., Bandosz, P., Rehm, C.D., Penalvo, J., Whitsel, L., Gaziano, T., Conrad, Z., Wilde, P., Micha, R., Lloyd-Williams, F., Capewell, S., Mozaffarian, D., O'Flaherty, M., 2017. Reducing US cardiovascular disease burden and disparities through national and targeted dietary policies: a modelling study. PLoS Med. 14.

Pereira-Miranda, E., Costa, P.R.F., Queiroz, V.A.O., Pereira-Santos, M., Santana, M.L.P., 2017. Overweight and obesity associated with higher depression prevalence in adults: a systematic review and meta-analysis. J. Am. Coll. Nutr. 36, 223–233.

Phillips, E., Wang, T.W., Husten, C.G., Corey, C.G., Apelberg, B.J., Jamal, A., Homa, D.M., King, B.A., 2017. Tobacco product use among adults—United States, 2015. Morb. Mortal. Wkly Rep. 66, 1209–1215.

Quek, Y.H., Tam, W.W.S., Zhang, M.W.B., Ho, R.C.M., 2017. Exploring the association between childhood and adolescent obesity and depression: a meta-analysis. Obes. Rev. 18, 742–754.

Rehm, J., Gmel, G.E., Gmel, G., Hasan, O.S.M., Imtiaz, S., Popova, S., Probst, C., Roerecke, M., Room, R., Samokhvalov, A.V., Shield, K.D., Shuper, P.A., 2017. The relationship between different dimensions of alcohol use and the burden of disease—an update. Addiction 112, 968–1001.

Rodríguez-Monforte, M., Flores-Mateo, G., Sánchez, E., 2015. Dietary patterns and CVD: a systematic review and meta-analysis of observational studies. Br. J. Nutr. 114, 1341–1359.

Roerecke, M., Rehm, J., 2014. Chronic heavy drinking and ischaemic heart disease: a systematic review and meta-analysis. Open Heart 1.

Roohafza, H., Kabir, A., Sadeghi, M., Shokouh, P., Ahmadzad-Asl, M., Khadem-Maboudi, A.A., Sarrafzadegan, N., 2016. Stress as a risk factor for noncompliance with treatment regimens in patients with diabetes and hypertension. ARYA Atheroscler. 12, 166–171.

Sabanayagam, C., Shankar, A., 2010. Sleep duration and cardiovascular disease: results from the National Health Interview Survey. Sleep 33, 1037–1042.
Sallis, J.F., Bull, F., Guthold, R., Heath, G.W., Inoue, S., Kelly, P., Oyeyemi, A.L., Perez, L.G., Richards, J., Hallal, P.C., 2016. Progress in physical activity over the Olympic quadrennium. Lancet 388, 1325–1336.
Salyers, M.P., Bonfils, K.A., Luther, L., Firmin, R.L., White, D.A., Adams, E.L., Rollins, A.L., 2017. The relationship between professional burnout and quality and safety in healthcare: a meta-analysis. J. Gen. Intern. Med. 32, 475–482.
Sánchez-Villegas, A., Henríquez-Sánchez, P., Ruiz-Canela, M., Lahortiga, F., Molero, P., Toledo, E., Martínez-González, M.A., 2015. A longitudinal analysis of diet quality scores and the risk of incident depression in the SUN project. BMC Med. 13.
Sarris, J., Logan, A.C., Akbaraly, T.N., Paul Amminger, G., Balanzá-Martínez, V., Freeman, M.P., Hibbeln, J., Matsuoka, Y., Mischoulon, D., Mizoue, T., Nanri, A., Nishi, D., Parletta, N., Ramsey, D., Rucklidge, J.J., Sanchez-Villegas, A., Scholey, A., Su, K.P., Jacka, F.N., 2015b. International Society for Nutritional Psychiatry Research consensus position statement: nutritional medicine in modern psychiatry. World Psychiatry 14, 370–371.
Schuch, F.B., Vancampfort, D., Richards, J., Rosenbaum, S., Ward, P.B., Stubbs, B., 2016. Exercise as a treatment for depression: a meta-analysis adjusting for publication bias. J. Psychiatr. Res. 77, 42–51.
Schumacher, T.L., Burrows, T.L., Neubeck, L., Redfern, J., Callister, R., Collins, C.E., 2017. How dietary evidence for the prevention and treatment of CVD is translated into practice in those with or at high risk of CVD: a systematic review. Public Health Nutr. 20, 30–45.
Schweitzer, R.J., Wills, T.A., Behner, J.D., 2017. E-cigarette use and indicators of cardiovascular disease risk. Curr. Epidemiol. Rep. 4, 248–257.
Sharma, A., Lavie, C.J., Borer, J.S., Vallakati, A., Goel, S., Lopez-Jimenez, F., Arbab-Zadeh, A., Mukherjee, D., Lazar, J.M., 2015. Meta-analysis of the relation of body mass index to all-cause and cardiovascular mortality and hospitalization in patients with chronic heart failure. Am. J. Cardiol. 115, 1428–1434.
Sin, N.L., DiMatteo, M.R., 2014. Depression treatment enhances adherence to antiretroviral therapy: a meta-analysis. Ann. Behav. Med. 47, 259–269. https://doi.org/10.1007/s12160-013-9559-6.
Slopen, N., Kontos, E.Z., Ryff, C.D., Ayanian, J.Z., Albert, M.A., Williams, D.R., 2013. Psychosocial stress and cigarette smoking persistence, cessation, and relapse over 9–10 years: a prospective study of middle-aged adults in the United States. Cancer Causes Control 24, 1849–1863.
Stafford, L., Berk, M., Jackson, H.J., 2013. Tobacco smoking predicts depression and poorer quality of life in heart disease. BMC Cardiovasc. Disord. 13.
Standl, E., Erbach, M., Schnell, O., 2013. Defending the con side: Obesity paradox does not exist. Diabetes Care 36, S282–S286.
Stephens, M.A.C., Wand, G., 2012. Stress and the HPA axis: role of glucocorticoid in alcohol dependence. Alcohol Res 34, 468–483.
Stubbs, B., Vancampfort, D., Rosenbaum, S., Firth, J., Cosco, T., Veronese, N., Salum, G.A., Schuch, F.B., 2017. An examination of the anxiolytic effects of exercise for people with anxiety and stress-related disorders: a meta-analysis. Psychiatry Res. 249, 102–108.
Stults-Kolehmainen, M.A., Sinha, R., 2014. The effects of stress on physical activity and exercise. Sports Med. 44, 81–121.
Sutaria, S., Devakumar, D., Yasuda, S.S., Das, S., Saxena, S., 2019. Is obesity associated with depression in children? Systematic review and meta-analysis. Arch. Dis. Child. 104, 64–74.
Tate, E.B., Wood, W., Liao, Y., Dunton, G.F., 2015. Do stressed mothers have heavier children? A meta-analysis on the relationship between maternal stress and child body mass index. Obes. Rev. 16, 351–361.

Taylor, G., McNeill, A., Girling, A., Farley, A., Lindson-Hawley, N., Aveyard, P., 2014. Change in mental health after smoking cessation: systematic review and meta-analysis. BMJ 348, 1–22.

Tenk, J., Mátrai, P., Hegyi, P., Rostás, I., Garami, A., Szabó, I., Hartmann, P., Pétervári, E., Czopf, L., Hussain, A., Simon, M., Szujó, S., Balaskó, M., 2018. Perceived stress correlates with visceral obesity and lipid parameters of the metabolic syndrome: a systematic review and meta-analysis. Psychoneuroendocrinology 95, 63–73.

Teychenne, M., Costigan, S.A., Parker, K., 2015. The association between sedentary behaviour and risk of anxiety: a systematic review. BMC Public Health 15.

Thorndike, A.N., Rigotti, N.A., 2009. A tragic triad: coronary artery disease, nicotine addiction, and depression. Curr. Opin. Cardiol. 24, 447–453.

Tomiyama, A.J., 2019. Stress and obesity. Annu. Rev. Psychol. 70, 703–718.

Torres, O.V., O'Dell, L.E., 2016. Stress is a principal factor that promotes tobacco use in females. Prog. Neuropsychopharmacol. Biol. Psychiatry. https://doi.org/10.1016/j.pnpbp.2015.04.005.

Trujillo, M.A., Khoddam, R., Greenberg, J.B., Dyal, S.R., Ameringer, K.J., Zvolensky, M.J., Leventhal, A.M., 2017. Distress tolerance as a correlate of tobacco dependence and motivation: incremental relations over and above anxiety and depressive symptoms. Behav. Med. 43, 120–128.

Tryon, M.S., Carter, C.S., DeCant, R., Laugero, K.D., 2013. Chronic stress exposure may affect the brain's response to high calorie food cues and predispose to obesogenic eating habits. Physiol. Behav. 120, 233–242.

U.S. Department of Health and Human Services (DHHS), 2014. The health consequences of smoking: 50 years of progress. A report of the surgeon general. U.S. Department of Health and Human Services, Centers for Disease Control and Prevention, National Center for Chronic Disease Prevention and Health Promotion, Office on Smoking and Health, Atlanta, GA.

Umer, A., Kelley, G.A., Cottrell, L.E., Giacobbi, P., Innes, K.E., Lilly, C.L., 2017. Childhood obesity and adult cardiovascular disease risk factors: a systematic review with meta-analysis. BMC Public Health 17.

Van Der Linden, D., Keijsers, G.P.J., Eling, P., Van Schaijk, R., 2005. Work stress and attentional difficulties: an initial study on burnout and cognitive failures. Work Stress 19, 23–36.

Verschueren, S., Eskes, A.M., Maaskant, J.M., Roest, A.M., Latour, C.H.M., Reimer, W.S., 2018. The effect of exercise therapy on depressive and anxious symptoms in patients with ischemic heart disease: a systematic review. J. Psychosom. Res. 105, 80–91.

Virtanen, M., Jokela, M., Nyberg, S.T., Madsen, I.E.H., Lallukka, T., Ahola, K., Alfredsson, L., Batty, G.D., Bjorner, J.B., Borritz, M., Burr, H., Casini, A., Clays, E., De Bacquer, D., Dragano, N., Erbel, R., Ferrie, J.E., Fransson, E.I., Hamer, M., Heikkilä, K., Jöckel, K.H., Kittel, F., Knutsson, A., Koskenvuo, M., Ladwig, K.H., Lunau, T., Nielsen, M.L., Nordin, M., Oksanen, T., Pejtersen, J.H., Pentti, J., Rugulies, R., Salo, P., Schupp, J., Siegrist, J., Singh-Manoux, A., Steptoe, A., Suominen, S.B., Theorell, T., Vahtera, J., Wagner, G.G., Westerholm, P.J.M., Westerlund, H., Kivimäki, M., 2015. Long working hours and alcohol use: systematic review and meta-analysis of published studies and unpublished individual participant data. BMJ, 350.

Wang, T.W., Gentzke, A., Sharapova, S., Cullen, K.A., Ambrose, B.K., Jamal, A., 2018. Tobacco product use among middle and high school students: United States, 2011–2017. Morb. Mortal. Wkly Rep. 67, 629–633.

Warburton, D.E.R., Bredin, S.S.D., 2018. Lost in translation: what does the physical activity and health evidence actually tell us? In: Lifestyle in Heart Health Disease, pp. 175–186.

Warburton, D.E.R., Nicol, C.W., Bredin, S.S.D., 2006. Health benefits of physical activity: The evidence. CMAJ 174, 801.

Warren, J.R., Falster, M.O., Fox, D., Jorm, L., 2013. Factors influencing adherence in long-term use of statins. Pharmacoepidemiol. Drug Saf. 22, 1298–1307.

Whooley, M.A., Vittinghoff, E., Carney, R.M., Ali, S., Feldman, M.D., Schiller, N.B., Browner, W.S., 2008. Depressive symptoms, health behaviors, and risk of cardiovascular events in patients with coronary heart disease. J. Am. Med. Assoc. 300, 2379–2388.

Wiernik, E., Airagnes, G., Lequy, E., Gomajee, R., Melchior, M., Le Faou, A.L., Limosin, F., Goldberg, M., Zins, M., Lemogne, C., 2019. Electronic cigarette use is associated with depressive symptoms among smokers and former smokers: cross-sectional and longitudinal findings from the constances cohort. Addict. Behav. 90, 85–91.

Willcox, B.J., He, Q., Chen, R., Yano, K., Masaki, K.H., Grove, J.S., Donlon, T.A., Willcox, D.C., Curb, J.D., 2006. Midlife risk factors and healthy survival in men. J. Am. Med. Assoc. 296, 2343–2350.

Wilson, F.A., Wang, Y., 2017. Recent findings on the prevalence of e-cigarette use among adults in the U.S. Am. J. Prev. Med. 52, 385–390.

Wing, R., Phelan, S., Tate, D., 2002. The role of adherence in mediating the relationship between depression and health outcomes. J. Psychosom. Res. 53, 877–881.

Wood, A.M., Kaptoge, S., Butterworth, A.S., Willeit, P., Warnakula, S., Bolton, T., Paige, E., Paul, D.S., Sweeting, M., Burgess, S., Bell, S., Astle, W., Stevens, D., Koulman, A., Selmer, R.M., Verschuren, W.M.M., Sato, S., Njølstad, I., Woodward, M., Salomaa, V., Nordestgaard, B.G., Yeap, B.B., Fletcher, A., Melander, O., Kuller, L.H., Balkau, B., Marmot, M., Koenig, W., Casiglia, E., Cooper, C., Arndt, V., Franco, O.H., Wennberg, P., Gallacher, J., de la Cámara, A.G., Völzke, H., Dahm, C.C., Dale, C.E., Bergmann, M.M., Crespo, C.J., van der Schouw, Y.T., Kaaks, R., Simons, L.A., Lagiou, P., Schoufour, J.D., Boer, J.M.A., Key, T.J., Rodriguez, B., Moreno-Iribas, C., Davidson, K.W., Taylor, J.O., Sacerdote, C., Wallace, R.B., Quiros, J.R., Tumino, R., Blazer, D.G., Linneberg, A., Daimon, M., Panico, S., Howard, B., Skeie, G., Strandberg, T., Weiderpass, E., Psaty, B.M., Kromhout, D., Salamanca-Fernandez, E., Kiechl, S., Krumholz, H.M., Grioni, S., Palli, D., Huerta, J.M., Price, J., Sundström, J., Arriola, L., Arima, H., Travis, R.C., Panagiotakos, D.B., Karakatsani, A., Trichopoulou, A., Kühn, T., Grobbee, D.E., Barrett-Connor, E., van Schoor, N., Boeing, H., Overvad, K., Kauhanen, J., Wareham, N., Langenberg, C., Forouhi, N., Wennberg, M., Després, J.P., Cushman, M., Cooper, J.A., Rodriguez, C.J., Sakurai, M., Shaw, J.E., Knuiman, M., Voortman, T., Meisinger, C., Tjønneland, A., Brenner, H., Palmieri, L., Dallongeville, J., Brunner, E.J., Assmann, G., Trevisan, M., Gillum, R.F., Ford, I., Sattar, N., Lazo, M., Thompson, S.G., Ferrari, P., Leon, D.A., Smith, G.D., Peto, R., Jackson, R., Banks, E., Di Angelantonio, E., Danesh, J., Butterworth, A., Verschuren, M., Veikko, S., Gómez de la Cámara, A., Bergmann, M., Crespo, C., Rimm, E.B., Strandberg, T., Nietert, P.J., Dallongeville, J.P., Gillumn, R.F., Ford, I.F., Thompson, S., Davey Smith, G., 2018. Risk thresholds for alcohol consumption: Combined analysis of individual-participant data for 599,912 current drinkers in 83 prospective studies. Lancet 391, 1513–1523.

World Health Organization, 2018a. Healthy Diet—Fact Sheet No. 394. World Health Organization Official Website. Available from: https://www.who.int/nutrition/publications/nutrientrequirements/healthydiet_factsheet/en. (Accessed 1 February 2019).

World Health Organization, 2018b. Global Status Report on Alcohol and Health 2018. World Health Organization Official Website. Available from: https://www.who.int/substance_abuse/publications/global_alcohol_report/en/. (Accessed 1 February 2019).

Further reading

Centers for Disease Control and Prevention, 2008. Prevalence of regular physical activity among adults: United States, 2001 and 2005. J. Am. Med. Assoc. 299, 30.

Maslach, C., Schaufeli, W.B., 2017. Burnout: A Multidimensional Perspective. In: Professional Burnout: Recent developments in theory and Research, pp. 19–32.

McEwen, B.S., 1998. Protective and damaging effects of stress mediators: central role of the brain. N. Engl. J. Med. 338, 171–179.
Rogers, P., Bullman, W., 1995. Prescription medicine compliance: review of the baseline of knowledge: report of the National Council on patient information and education. J. Pharmacoepidemiol. 3, 3–36.
Sackett, D.L., Haynes, R.B., 1976. Compliance with therapeutic regimens. Johns Hopkins Univ. Press, Baltimore, MD, pp. 1–6.
Sarris, J., Logan, A.C., Akbaraly, T.N., Amminger, G.P., Balanzá-Martínez, V., Freeman, M.P., Hibbeln, J., Matsuoka, Y., Mischoulon, D., Mizoue, T., Nanri, A., Nishi, D., Ramsey, D., Rucklidge, J.J., Sanchez-Villegas, A., Scholey, A., Su, K.-P., Jacka, F.N., 2015a. Nutritional medicine as mainstream in psychiatry. Lancet Psychiatry 2, 271–274.
Sarris, J., Murphy, J., Mischoulon, D., Papakostas, G.I., Fava, M., Berk, M., Ng, C.H., 2016. Adjunctive nutraceuticals for depression: a systematic review and meta-analyses. Am. J. Psychiatry 173, 575–587.
World Health Organization, 2015. Chronic Disease: A Vital Investment. World Health Organization Press, Geneva.

CHAPTER 11

The microbiome-gut-brain axis: The missing link in depression

Christine L. Foxx, Christopher A. Lowry
Department of Integrative Physiology and Center for Microbial Exploration, University of Colorado Boulder, Boulder, CO, United States

Contents

Introduction

In this chapter, we describe a newly-characterized player in human health called the *gut microbiome* and the emerging role that this community of microorganisms has in linking stress to the risk of major depressive disorder (MDD) and cardiovascular disease (CVD). In order to do so, we briefly review literature on this teeming ecosystem that exists on and in our own bodies. This chapter presents the bidirectional *microbiome-gut-brain* (MGB) *axis* as a

https://doi.org/10.1016/B978-0-12-815015-3.00011-8

major communication highway that dictates how the human body develops adaptive and maladaptive responses to stress. We inspect the link between the gut microbiome and behaviors relevant to stress-related psychiatric disorders, with a focus on MDD, and additionally review the literature regarding the link between the gut microbiome and CVD. These data suggest that microbial constituents of the gut microbiome are major influencers of the risks of MDD and CVD. We introduce the *Old Friends hypothesis* as a potential mechanism underlying observed increases in the prevalence of MDD and CVD in modern urban societies, and close with potential microbiome-based strategies for preventing and treating the co-occurrence of MDD and CVD.

The gut microbiome: A critical component of the human superorganism

The human gastrointestinal (GI) tract, also known as the gut, anatomically spans the portion of the digestive system from the stomach at its most proximal end to the sigmoid colon at its most distal end. Each region of the GI tract harbors a unique concentration and population of prokaryotes (e.g., bacteria and fungi), viruses, and eukaryotic microorganisms (e.g., yeasts) that collectively form the *gut microbiota* (Dave et al., 2012). Until recently, most scientists have accepted that the gut microbiota overshadows human cells by a factor of nine to one. More recent estimates affirm that the total biomass of the gut microbiota is about 3.8×10^{13} cells in an adult reference man and roughly equals that of the entire host human body in a 1.3 to 1 ratio (Sender et al., 2016).

For most adults, 90% of the gut microbiome falls within two phyla, Bacteroidetes and Firmicutes, and is capable of expressing over 3.3 million distinct, non-redundant genes (Qin et al., 2010). Large-scale consortium efforts, such as the Human Microbiome Project (HMP), have demonstrated that it is possible to conduct a broad survey of bacterial cohabitants living on and in our bodies using targeted molecular analysis of the 16S small subunit ribosomal ribonucleic acid (16S rRNA) gene (Turnbaugh et al., 2007). Alignment of detailed sample metadata with high-throughput sequencing approaches, such as *16S rRNA gene* amplification, provide the scientific community with valuable information on where we might be able to find these bacterial communities and how they differ from one another in relative abundance of phyla and general community composition.

The vocabulary of microbial genomics

The diverse collection of unique and non-redundant genes that the gut microbiota possesses and is capable of collectively expressing over an individual's lifespan is known as the *gut microbiome* (Dave et al., 2012). The gut microbiome characterizes the vast enzymatic and metabolic repertoire of the gut microbiota, a small subset of which functions to ferment indigestible dietary fibers. *Metagenomic* studies are broadly defined as experimental designs and molecular tools that allow scientific researchers to compare multiple genomes against one another. Evolutionary biologists have demonstrated through structural analysis of the *human metagenome* that the gut microbiota has co-evolved symbiotically with humans over several millennia (Hooper et al., 2002).

Some examples of mutually beneficial metabolic functions that the gut microbiota collectively possesses include the ability to produce essential nutrients, such as cobalamin (vitamin B_{12}) and folate (vitamin B_9), de novo, and the ability to synthesize fermentation products, such as *short-chain fatty acids* (SCFAs), from indigestible plant-based materials (Kiela and Ghishan, 2016). These two symbiotic functions alone impart us, as humans, with the ability to extract nutrients from food items that otherwise would not possess substantive metabolic value (Kiela and Ghishan, 2016). In short, having a gut microbiota—and the genetic abilities of a gut microbiome as a direct consequence—imparts humans with increased adaptability to changes in diet and nutrient availability (Hooper et al., 2002).

The complex ecological relationships that play out within and between bacterial communities in each body region are subjects of much interest today and demonstrate just how limited our understanding of the processes that take place in our own bodies may be. In this chapter, we will review how human health is shaped by interactions among the gut microbiota, host epithelium, the mucosal immune system, and the enteric nervous system, with a focus on a potential role for the gut microbiome in determining either resilience or vulnerability to stress-related outcomes such as MDD or CVD. Today, scientific researchers are increasingly convinced that the human gut microbiota plays a critical role in neuronal and immune development and has been overlooked in human health and disease.

Microbiome research: Then and now

In the mid-19th century, Dr. Theodor Escherich was one of the first scientists to develop anaerobic culture techniques in order to characterize the structural characteristics, fermentation substrates, and end-products of bacterial isolates in infant stools. Along with many other physician-scientists of his time, Escherich argued that accumulating waste and bacteria in the colon triggered a state of auto-intoxication, a condition that was widely linked to fever, depression, and anxiety (Chen and Chen, 1989). Given his research background and the outstanding contributions he made in understanding the etiology of several childhood diseases, it is perhaps unsurprising that Escherich did not view the gut microbiota as a symbiotic organ, but rather as a source of potential pathogens.

Thankfully, the detection, identification, and characterization of the vast community of bacterial species known as the gut microbiota has improved immensely since Escherich's time. A series of technological developments in the 1970s allowed scientific researchers to isolate, amplify, and perform targeted sequencing on DNA and RNA. This kickstarted the development of molecular medicine such that we can now positively identify and sequence hundreds of thousands of microbial genomes in a matter of hours. Drs. Carl Woese and George Fox at the University of Illinois were the first to describe a molecular approach that would specifically target and sequence the genomes of bacteria (Woese and Fox, 1977). Woese identified a highly conserved gene present in all bacteria that encoded for 16S rRNA, a discovery that was used to generate the three-domain system of phylogenetic classification now widely accepted, used, and taught broadly in classrooms today.

The recent completion of international consortium efforts to catalogue entire genomes, such as the Human Genome Project (HGP) in 2001, has led to renewed scientific interest in understanding the role that the genes we possess play in human health. Additionally, the development of massively parallel high-throughput sequencing methods in the last 50 years, such as ion torrent sequencing, pyrosequencing, sequencing by synthesis, and chain-termination (Sanger) sequencing, have increased the speed and accuracy of identifying the DNA composition of biological samples, while simultaneously decreasing the cost of analysis to less than \$0.13 per 1 million bases (Hamady and Knight, 2009).

The increase in studies utilizing 16S rRNA gene sequencing to target the gut microbiota and characterize the microbiome was fueled in part by Sanger sequencing of the 16S rRNA gene in the early 2000s. This technology allowed molecular researchers to obtain continuous long reads of approximately 800 base pairs and identify distinct bacteria at a much higher phylogenetic resolution (genus and species) than ever before. Commercialization of high-throughput sequencing platforms, such as the PyroSeq 454® platform (Roche Diagnostics Corp., Indianapolis, IN, United States) and MiSeq/HiSeq® platforms (Illumina Inc., San Diego, CA, United States), have allowed widespread use of microbiome sequencing technology in laboratories all over the world today.

Today, data from high-throughput sequencing platforms are typically processed using two publicly-available and open-source bioinformatics analysis software packages: Quantitative Insights Into Microbial Ecology (QIIME) (Caporaso et al., 2010) and mothur (Schloss et al., 2009). Both QIIME and mothur take raw short read sequences of 50 to 400 base pairs in length from multiple high-throughput sequencing platforms and match them to sample metadata in a process known as demultiplexing (Goodrich et al., 2014). Sequence reads are quality-filtered and matched to *operational taxonomic units* (OTUs), a taxonomic code assignment used to identify closely related organisms, such as bacteria, within the same genus. Positive matches of multiple OTUs to a reference genome sequence leads to phylogenetic identification, which can then be utilized to inform researchers of which bacterial taxa or community profiles are associated with a wide variety of disease conditions, locations on the body, treatments and/or interventions.

The gut microbiome is diverse and reflects the fact that multiple bacterial constituents interact with one another to maintain ecological homeostasis and typically confer some level of functional redundancy (Dave et al., 2012). Phylogenetic data alone offer little to no insight regarding which species are working together to metabolize a particular sugar, utilize a certain SCFA, or adapt to a sudden and specific change in dietary or immune status. Microbiome researchers interested in answering such questions about microbial activity and functionality within the gut can select from a variety of analytical tools to profile the metabolic activities, or the *metabolome*, of the gut microbiota (Vernocchi et al., 2016). Fourier transform infrared (FTIR) spectroscopy and liquid chromatography with mass spectrometry (LC-MS) are capable of rapidly identifying the chemical composition of solids, liquids, and gases in an unknown material. These analytical techniques allow thousands of individual molecular entities to be identified with extremely high specificity and sensitivity.

The bidirectional microbiome-gut-brain axis

The brain and gut microbiome communicate with one another bidirectionally through the MGB axis using both direct and indirect pathways (Petra et al., 2015). While the extent of interactions between the brain and digestive system have been recognized for many years, the mechanisms by which the brain and gut microbiota interact with one another were only recently elucidated thanks to the advent and commercial availability of high-throughput 16S rRNA gene sequencing tools and the development of *germ-free* (GF) rodent models. Many leading researchers now think of humans not just as self-sufficient organisms, but as hosts to complex ecosystems of trillions of microbial species that are constantly competing for space and nutritional accessibility within the gut lumen. Neural, immune, and endocrine signaling mechanisms ensure constant communication between the central nervous system (CNS) and the gut microbiome and, conversely, between the gut microbiome and the CNS.

Communication pathways from the brain to the gut microbiome

Neuronal signals from the CNS act on GI functions such as production of mucus and motility, in part through activation of central autonomic control systems. In addition, aversive events directly affect the hypothalamic-pituitary-adrenal (HPA) axis, resulting in increases in circulating glucocorticoid hormones, which can impact gut physiology, as well as the composition of the microbiome (Guo et al., 2009). Neuronal signals from higher order brain structures involved in cognitive and affective function, such as the medial prefrontal cortex (mPFC), provide input to subcortical brain regions (e.g., the amygdala and hypothalamus) controlling complex homeostatic bodily states, including gut homeostasis, food intake, and visceral pain (Mayer, 2011). Brainstem monoaminergic neurons are part of central autonomic control systems with descending spinal pathways that regulate a wide variety of GI functions (Fields, 2004). These neural systems are engaged by strong behavioral and motivational states, such as vigilance and fear, providing a neuroanatomical context for the colloquial phrase "gut feeling."

Activation of post-ganglionic sympathetic neurons of the GI tract stimulates vasoconstriction, modulates immune cell activity within Peyer's patches and the intestinal mucosa, inhibits mucosal secretion, and slows motility throughout the GI tract (Lyte et al., 2010). Parasympathetic inputs to the GI tract, traveling in the vagus nerve, mediate gastric acid secretion in the cephalic phase of digestion and stimulate enteroendocrine cells in the gut epithelium to release gastrin and somatostatin in the bloodstream. A small subset of enteroendocrine cells known as enterochromaffin (EC) cells also release histamine and serotonin (5-hydroxytryptamine, 5-HT) (Stephens and Tache, 1989). All told, a wide variety of neural, endocrine, and immune signals influence the functions of the GI tract, with broad implications for a number of potential interactions with the gut microbiome.

Communication pathways from the gut microbiome to the brain

The gut microbiome can signal to the brain through diverse mechanisms, including but not limited to the production of neuroactive metabolites, modulation of immune signaling, and modulation of afferent neural signaling through vagal and spinal afferents. For example, several members of the gut microbiota synthesize neurotransmitters, such as γ-aminobutyric acid (GABA) and dopamine (DA) de novo, and are capable of stimulating EC cells to produce 5-HT through the increased expression of tryptophan hydroxylase type 1 (TPH1) (Petra et al., 2015). The gut microbiome profoundly influences the production of anti-inflammatory cytokines, such as interleukin (IL)-10, which can act directly on the CNS and immune systems alongside SCFAs like butyrate (Petra et al., 2015). Some molecular microbiome-derived signals can cross the blood-brain barrier (BBB) and affect its permeability, antigenically stimulate regulatory T (Treg) cells, macrophages, and dendritic cells that traffic to the CNS to direct proinflammatory or anti-inflammatory responses, and activate neural afferents, such as the intestinal branches of the vagus nerve (Petra et al., 2015).

Studies demonstrate that vagal afferents are responsive to cytokines, histamine, tryptases, and 5-HT released by EC cells, mast cells, dendritic cells, macrophages, T cells, and B cells that are localized to the intestinal mucosa (Keita and Soderholm, 2010). Vagal afferents from the GI tract project to the nucleus tractus solitarii (NTS) in the brainstem, where sensory information pertaining to the cardiovascular system is also integrated (Craig, 2009). Signals in the NTS are then integrated in the parabrachial nucleus and relayed to forebrain regions, such as the hypothalamus and amygdala, to control behavioral and emotional states in a neurological signal feedback loop (Craig, 2009). Although less well-studied, afferent fibers innervating the gut mucosa travel in sympathetic nerve bundles and terminate in lamina I of the spinal cord (Mayer, 2011). From there, signals are relayed to the brain through spinoparabrachial and spinothalamic pathways (Mayer, 2011). Data also demonstrate that bidirectional interactions of the MGB axis affect the HPA axis (Sudo et al., 2004), immune system (Keita and Soderholm, 2010), and a wide variety of metabolic pathways (Bryleva and Brundin, 2017).

Many of the studies establishing the MGB axis were initially correlative. However, studies utilizing GF models and fecal microbiota transfer (FMT) in *gnotobiotic* animals have provided important mechanistic insights.

Relationships between stress and the gut microbiome

Effects of stress on the gut microbiome

Preclinical studies in mice demonstrate that psychosocial stressors, such as social disruption with an aggressor (Bailey et al., 2011), chronic subordinate colony housing (Reber et al., 2016a,b), and maternal separation (Desbonnet et al., 2010), significantly alter the gut microbiota. Bailey et al. (2011) demonstrated that the cecal microbiome of mice subjected to social disruption for 2 hours each day, 6 days total, severely reduced the relative abundance of bacteria in the genus *Bacteroides* while members of the genus *Clostridium* proliferated in its place.

Chronic stress is a recognized factor in increased intestinal permeability (Øines et al., 2012), which may allow enteric pathogens to evade the gut microbiota biofilm and easily translocate through the gut epithelium into the bloodstream. Stress also affects the activation of sympathetic efferent nerves discussed above in this chapter's overview of the MGB axis. This, in turn, directly affects gut motility and the secretion of mucin, gastric acids, and bile salts (Lyte et al., 2010). Gut motility, of all the factors above, is most strongly associated with changes in microbial diversity and abundance within the gut (Malan-Muller et al., 2018).

Chronic psychosocial stress has been shown to result in proliferation of pathobionts, such as *Helicobacter* spp. (Langgartner et al., 2017; Reber et al., 2016a,b), which are known to induce colitis in the absence of adequate immunoregulation, such as that found in IL-10-deficient mice (Kullberg et al., 1998). Stress-induced expansion of *Helicobacter* spp. is associated with spontaneous colitis, exaggerated release of proinflammatory cytokines, such as interferon (IFN)-γ and IL-6, from mesenteric lymph node cells stimulated ex vivo, exaggerated chemically-induced colitis in a model of inflammatory bowel disease, and increased anxiety (Langgartner et al., 2017; Reber et al., 2016a,b). Other studies have shown that inescapable stressors can also induce neuroinflammatory responses, which are characterized by microglial priming and exaggerated release of proinflammatory cytokines (Fonken et al., 2018; Frank et al., 2018). Microbiome-dependent, stress-induced induction of systemic inflammatory and neuroinflammatory responses is one mechanism through which stress exposure, including psychosocial stress exposure, may influence risk of MDD and CVD.

Effects of the gut microbiome on physiological and behavioral responses to stress

GF animal models are great tools that examine the effects of the gut microbiota (or lack thereof) on behavior in order to determine causation (Malan-Muller et al., 2018). One of the first GF studies was conducted by Sudo et al. (2004), who demonstrated that GF mice exhibited unusually high HPA axis reactivity to restraint stress, a psychological stressor. Sudo et al. (2004) noted that elevated plasma adrenocorticotropic hormone and corticosterone (CORT) levels observed in restrained ex-GF mice were significantly improved by prenatal oral immunization with 5×10^4 colony-forming units (CFUs) of the *probiotic* species *Bifidobacterium infantis*. In this way, the authors demonstrated that HPA axis reactivity to restraint stress could be ameliorated to normal levels following inoculation with a single, potentially beneficial bacterium early in development (Sudo et al., 2004).

Cruymeyrolle-Arias et al. (2014) demonstrated that GF and SPF rats in a stress-sensitive F344 rat strain also behaved differently in social interaction and open-field (OF) tests. GF F344 rats spent much less time sniffing unknown partners and displayed *anxiogenic* or anxiety-like defensive behavioral responses in the OF test, increasing latency to enter the center zone and spending less time in the center zone (Cruymeyrolle-Arias et al., 2014). GF rats also exhibited 2.8-fold higher serum CORT concentrations after OF test exposure relative to SPF rats, in findings very similar to that of Sudo et al. (2004).

Relationships between depression and the gut microbiome

One mechanism through which the gut microbiome may influence the risk of MDD, persistence of MDD symptoms, or antidepressant response is through its effects on L-tryptophan metabolism, including serotonergic pathways. 5-HT, also known as serotonin, is a monoamine neurotransmitter named for its abundant presence in serum platelets and its vasoconstrictive effects on the cardiovascular system (Sodhi and Sanders-Bush, 2004). Dysregulation of 5-HT is implicated in mood and anxiety disorders and selective serotonin reuptake inhibitors (SSRIs), such as citalopram and fluoxetine, are well-characterized as pharmacological treatments for affective disorders such as MDD.

The gut microbiome has a critical role in stimulating the proliferation and development of EC cells in the gut epithelium that secrete over 95%

of 5-HT in the body (Yano et al., 2015). Peripherally released 5-HT interacts with 5-HT receptors on terminal afferent vagal endings that project directly to the CNS (Yano et al., 2015). Platelets circulating through the bloodstream also gather 5-HT released by EC cells in the gut epithelium and transport it to a variety of body sites to promote homeostasis, including immunoregulatory responses associated with Treg cells (Baganz and Blakely, 2013). Wikoff et al. (2009) demonstrated that basal concentrations of serum 5-HT from GF male Swiss-Webster mice are nearly 2.8 times lower than 5-HT detected in SPF controls. Decreases in serum 5-HT concentrations were accompanied by increases in L-tryptophan and *N*-acetyltryptophan concentrations, data suggesting that 5-HT synthesis was severely impaired in GF mice (Wikoff et al., 2009). Although 5-HT does not cross the BBB, these findings highlight how the gut microbiome influences peripheral serotonergic signaling, which may directly influence physiological responses relevant to CVD (Musselman et al., 1998) and indirectly influence brain functions relevant to MDD.

Other studies show that the kynurenine metabolic pathway of L-tryptophan is involved in both neuroinflammation and suicidality in depressive patients (Sublette et al., 2011; Bradley et al., 2015). Production of neuroactive compounds derived from the kynurenine pathway of L-tryptophan metabolism is chiefly mediated by indoleamine-2,3-dioxygenase (IDO), an enzyme expressed primarily in extrahepatic tissues (Maes, 2011). Proinflammatory cytokines, such as IFN-γ, IL-1β, and IL-6, induce IDO and are capable of activating the kynurenine pathway, reducing the bioavailability of L-tryptophan and potentially that of 5-HT (Bryleva and Brundin, 2017). Neuroactive products of kynurenine metabolism, such as quinolinic acid, can bypass the BBB and activate *N*-methyl-D-aspartic acid receptors (NMDAR) expressed primarily in the hippocampus, dorsal striatum, and neocortex (Guillemin, 2012).

Many scientists believe that depression can be caused by chronic inflammation in the brain, broadly mediated by soluble molecular signals known as *cytokines* that are secreted by immune cells and travel to a diverse array of organs and tissues to have broadly pro- or anti-inflammatory effects (Raison et al., 2006). Blood samples taken from patients with MDD who are otherwise medically healthy demonstrate that MDD is associated with increased proinflammatory cytokine levels, including but not limited to IL-1, tumor necrosis factor (TNF), the acute phase protein, C-reactive protein (CRP), and intercellular adhesion molecule 1 (ICAM-1) (Raison et al., 2006). Several studies have shown that plasma concentrations of several proinflammatory cytokines previously mentioned in this chapter are associated with the severity of depressive symptoms, and that these associations are apparent throughout the adult lifespan. Studies suggest that the pleiotropic cytokine IL-6, which is often associated with inflammatory responses, is elevated in persons with a diagnosis of MDD (Pace et al., 2006) and causally induces depression-like behavioral responses in rodent models of stress (Hodes et al., 2014).

Evidence for a role of the gut microbiome in risk of MDD

Preclinical experiments utilizing chronic psychosocial stress paradigms, particularly prolonged restraint stress (Bailey et al., 2010), chronic social defeat (Bailey et al., 2011; Bharwani et al., 2016; Aoki-Yoshida et al., 2016), and grid floor-induced stress (Bangsgaard Bendtsen et al., 2012), demonstrate that significant compositional changes in the gut microbiome and metabolome occur in association with stress-induced depressive-like behaviors in mouse models. Although it is not yet possible to indicate a causal role of the gut microbiome in the development of MDD, FMT from MDD patients to humanized GF mice results in depression-like behavioral responses relative to behaviors in GF mice colonized with microbiota derived from healthy control individuals (Zheng et al., 2016). Additional data indicate that behavioral and immune effects of chronic subordinate colony housing, a preclinical model for chronic psychosocial stress, can be ameliorated by the transplantation of feces from single-housed control mice (Langgartner et al., 2018).

Evidence for effects of MDD on the gut microbiome

It is conceivable that, rather than the gut microbiome playing a causal role in the etiology of MDD, MDD or behavioral changes associated with MDD may alter the gut microbiome. This possibility highlights the need for robust microbiome analyses that take into account all potential confounders, such as medication and diet (Malan-Muller et al., 2018).

Early-life stress and depression

Several epidemiological and preclinical animal studies have demonstrated that early-life stress has a profound effect on the risk of developing MDD and severe depressive symptoms. *Maternal separation* is a well-characterized model to induce early-life stress in preclinical studies. Preclinical models of maternal separation reduce the potential for confounding factors to obscure the etiology of MDD and depressive-like behaviors, and allow researchers interested in investigating novel therapies for MDD to evaluate their efficacy.

Desbonnet et al. (2010) demonstrated this in a study of maternal separation in which rats were exposed to maternal separation during early life, then given either *Bifidobacterium infantis* (strain 35624, Alimentary Health Ltd., University College Cork, Cork, Ireland) or citalopram in drinking water from postnatal day 50 (i.e., late adolescence) until euthanasia in adulthood. They found that maternal separation significantly affected immobility in the forced-swim test (FST), similar to other studies measuring depressive-like behaviors in rodents, and that oral administration of *Bifidobacterium infantis* ameliorated the effects of maternal separation in the FST (Desbonnet et al., 2010). Clinical studies in humans confirm that maternal stress is an important determinant of the gut microbiome of the offspring (Zijlmans et al., 2015), and it is conceivable that these changes in the gut microbiome alter subsequent risk of MDD and other stress-related psychiatric disorders.

Relationships between cardiovascular disease and the microbiome

Evidence for a role of the gut microbiome in risk of CVD

Evidence suggests that the gut microbiota is involved in the metabolism of several compounds that ultimately generate the proatherosclerotic compound trimethylamine-*N*-oxide (TMAO) (Jie et al. 2017). Compounds such as choline, phosphatidylcholine, and L-carnitine (Wang et al., 2011; Koeth et al., 2013) are metabolized by the gut microbiome to produce trimethylamine (TMA), which can then be easily oxidized by the human liver into TMAO (Jie et al., 2017). Bacterial DNA has been detected within atherosclerotic plaques in some studies (Ott et al., 2006; Koren et al., 2011), although whether that is a consequence of bacterial translocation from the gut lumen into the bloodstream or an indication of direct residential associations to CVD remains to be seen.

Fermentation of complex indigestible carbohydrates by gut microbiota produces various metabolites, such as butyrate, acetate, and propionate known as SCFAs (Morrison and Preston, 2016). These microbial-derived metabolites are known to induce Treg cell proliferation and activity, thereby calibrating immune function to potentially prevent inflammatory diseases (Galisteo et al., 2008). Metagenomic analysis of the gut microbiomes of 218 individuals with atherosclerotic CVD demonstrated that alterations to the gut metabolome reflected reductions in fermentative capability and increases in inflammatory phenotypes (Jie et al., 2017). Diets high in certain fats and sugars are known to deplete anti-inflammatory bacteria, allowing potentially harmful opportunist species, such as *Bilophila wadsworthia* and *Helicobacter pylori,* to proliferate in their place within the GI tract (Turnbaugh et al., 2009a,b). These species thin the protective lining of the GI tract's mucous layer, creating the potential for systemic inflammation to occur (Turnbaugh et al., 2009a,b).

Evidence for effects of CVD on the gut microbiome

As with MDD discussed earlier, it is conceivable that CVD or behavioral changes associated with CVD alter the gut microbiome. Again, this possibility highlights the need for robust microbiome analyses that take into account all potential confounders, such as medication and diet.

Evidence that gut microbiome responses to stress could be a mechanism through which stress leads to MDD and CVD

Although we have largely focused on how stress-induced changes in the gut microbiome may induce inflammatory responses and negative outcomes, specific microorganisms have been shown to prime immunoregulatory circuits and suppress pathological inflammation (Rook et al., 2004). Microorganisms that have evolved to prime immunoregulatory circuits include: (i) commensal microbiota that have been altered by the Western lifestyle, including a diet low in microbiota-accessible carbohydrates (Sonnenburg and Sonnenburg, 2014; von Hertzen et al., 2015); (ii) infectious pathogens associated with the hunter-gatherer period of human evolution (Atherton and Blaser, 2009); and (iii) organisms from the natural environment, including saprophytic species such as *Mycobacterium vaccae,* with which humans were inevitably in daily contact (and so had to be tolerated by the immune system).

However, evidence demonstrates that radical shifts in the human microbiome as a result of living a modern urban lifestyle, where entire classes of microorganisms are now either absent or much-reduced, are associated with a substantive increase in inflammatory and autoimmune diseases (Hooper et al., 2002). The *Disappearing Microbes hypothesis* (Blaser and Falkow, 2009), *Microbial Biodiversity hypothesis* (von Hertzen et al., 2011), and *Old Friends hypothesis* (Rook et al., 2004) each provide potential mechanisms by which changes in microbiome communities and functions may cause this trend. It has been argued that urban lifestyle-associated changes in microbiome composition alter the manner in which the peripheral immune system responds to challenge, resulting in a shift towards imbalanced immunoregulation (as indicated by deficits in Treg development and proliferation) (Rook et al., 2004). This reasoning has led to the development of strategies to prevent or treat psychiatric disorders by restoring some of the lost "Old Friends" through microbial-based interventions, thereby shifting immune signaling back towards immunoregulation (Lowry et al., 2016).

Based on the weight of the evidence discussed above, it is possible that the gut microbiome's responses to stress could be a mechanism through which psychosocial stress leads to MDD and CVD. A particularly attractive hypothesis is that psychosocial stress-induced changes in the gut microbiome cause increases in systemic inflammation and neuroinflammation that, together, increase the risk of developing CVD and MDD. The next frontier in stress microbiome research lies in understanding the physiological and behavioral consequences of chronic adverse experience in the context of the gut microbiome's role in *allostasis*, which collectively represents the adaptive behavioral, immune (i.e., inflammatory), autonomic, and neuroendocrine mediator responses (i.e., those involved in growth, reproduction, and metabolism) to stress (McEwen, 1998; Miller and Raison, 2016; Tsigos and Chrousos, 2002).

Conclusions and future directions

The gut microbiome plays an important role in human development from the first few moments of birth to the last breath that an individual takes, and there is much more research not discussed here that is published or currently underway to characterize its functions in health and disease. Evidence suggests that the MGB axis serves as a link between the gut microbiome and the brain and plays a formative role in modulating stress-related physiological and behavioral responses. The published scientific literature mentioned in this chapter touches on just a few studies that have been pivotal in defining the MGB axis and demonstrating potential mechanisms through which the gut microbiota may influence the risk of development of stress-related disorders, including CVD and MDD.

Epidemiological data demonstrate that sex differences exist in the clinical presentation of cardiac risk factors and mortality due to CVD (Bairey Merz et al., 2010), despite unclear explanations as to why these differences occur. Presentation of atherogenic CVD risk factors, such as hypertriglyceridemia and diabetes in women, are associated with poorer clinical outcomes and cluster together during and after menopause (Bairey Merz et al., 2010), making the search for potential therapies and treatments for CVD especially challenging without taking sex and age into consideration. The literature also demonstrates that incidence rates and morbidity risk due to depressive disorders, including MDD, are persistently much higher in women than in men across the lifespan, with a strong sex-dependent correlation to adverse childhood experiences (Piccinelli and Wilkinson, 2000). Given these data, it is perhaps not unsurprising that recent forays into potential sex differences in MGB axis development and contributions to the etiology of several diseases have been proposed (Jašarević et al., 2016). It is important that future studies take sex differences into account when modeling the MGB axis in MDD and CVD risk.

Data from the University College Cork in Ireland by Dr. John Cryan's lab suggest that probiotic treatments can be as effective, if not more effective, than pharmaceutical drugs in the reduction of depression- and anxiety-like behavioral responses in rodent models (Savignac et al., 2014). The development of probiotics to confer mental health benefits has led to a new interest in the potential for microbial *psychobiotics* to treat and potentially prevent stress-related psychiatric disorders, including mood disorders, such as MDD. However, much more work remains to be done, including large-scale randomized, placebo-controlled, double-blind clinical trials, in order to translate these findings for use in the prevention or treatment of stress-related outcomes including CVD and MDD.

Disclosures

C.A.L. serves on the Scientific Advisory Board of Immodulon Therapeutics, Ltd.

References

Aoki-Yoshida, A., Aoki, R., Moriya, N., Goto, T., Kubota, Y., Toyoda, A., Takayama, Y., Suzuki, C., 2016. Omics studies of the murine intestinal ecosystem exposure to subchronic and mild social defeat stress. J. Proteome Res. 15 (9), 3126–3138.

Atherton, J.C., Blaser, M.J., 2009. Coadaptation of *Helicobacter pylori* and humans: ancient history, modern implications. J. Clin. Invest. 119 (9), 2475–2487.

Baganz, N.L., Blakely, R.D., 2013. A dialogue between the immune system and brain, spoken in the language of serotonin. ACS Chem. Neurosci. 4 (1), 48–63.

Bailey, M.T., Dowd, S.E., Parry, N.M.A., Galley, J.D., Schauer, D.B., Lyte, M., 2010. Stressor exposure disrupts commensal microbial populations in the intestines and leads to increased colonization by *Citrobacter rodentium*. Infect. Immun. 78 (4), 1509–1519.

Bailey, M.T., Dowd, S.E., Galley, J.D., Hufnagle, A.R., Allen, R.G., Lyte, M., 2011. Exposure to a social stressor alters the structure of the intestinal microbiota: implications for stressor-induced immunomodulation. Brain Behav. Immun. 25 (3), 397–407.

Bairey Merz, C.N., Mark, S., Boyan, B.D., Jacobs, A.K., Shah, P.K., Shaw, L.J., Taylor, D., Marbán, E., 2010. Proceedings from the scientific symposium: sex differences in cardiovascular disease and implications for therapies. J. Women's Health 19 (6), 1059–1072.
Bangsgaard Bendtsen, K.M., Krych, L., Sørensen, D.B., Pang, W., Nielsen, D.S., Josefsen, K., Hansen, L.H., Sørensen, S.J., Hansen, A.K., 2012. Gut microbiota composition is correlated to grid floor-induced stress and behavior in the BALB/c mouse. PLoS ONE 7 (10), e46231.
Bharwani, A., Mian, M.F., Foster, J.A., Surette, M.G., Bienenstock, J., Forsythe, P., 2016. Structural & functional consequences of chronic psychosocial stress on the microbiome & host. Psychoneuroendocrinology 63, 217–227.
Blaser, M.J., Falkow, S., 2009. What are the consequences of the disappearing human microbiota? Nat. Rev. Microbiol. 7 (12), 887–894.
Bradley, K.A.L., Case, J.A.C., Khan, O., Ricart, T., Hanna, A., Alonso, C.M., Gabbay, V., 2015. The role of the kynurenine pathway in suicidality in adolescent major depressive disorder. Psychiatry Res. 227 (2–3), 206–212.
Bryleva, E.Y., Brundin, L., 2017. Kynurenine pathway metabolites and suicidality. Neuropharmacology 112, 324–330.
Caporaso, J.G., Kuczynski, J., Stombaugh, J., Bittinger, K., Bushman, F.D., Costello, E.K., Fierer, N., Peña, A.G., Goodrich, J.K., Gordon, J.I., Huttley, G.A., Kelley, S.T., Knights, D., Koenig, J.E., Ley, R.E., Lozupone, C.A., McDonald, D., Muegge, B.D., Pirrung, M., Reeder, J., Sevinsky, J.R., Turnbaugh, P.J., Walters, W.A., Widmann, J., Yatsunenko, T., Zaneveld, J., Knight, R., 2010. QIIME allows analysis of high-throughput community sequencing data. Nat. Methods 7 (5), 335–336.
Chen, T.S., Chen, P.S., 1989. Intestinal autointoxication: a medical leitmotif. J. Clin. Gastroenterol. 11 (4), 434–441.
Craig, A.D., 2009. How do you feel—now? The anterior insula and human awareness. Nat. Rev. Neurosci. 10 (1), 59–70.
Cruymeyrolle-Arias, M., Jaglin, M., Bruneau, A., Vancassel, S., Cardona, A., Daugé, V., Naudon, L., Rabot, S., 2014. Absence of the gut microbiota enhances anxiety-like behavior and neuroendocrine response to acute stress in rats. Psychoneuroendocrinology 42, 207–217.
Dave, M., Higgins, P.D., Middha, S., Rioux, K.P., 2012. The human gut microbiome: current knowledge, challenges, and future directions. Transl. Res. 160 (4), 246–257.
Desbonnet, L., Garrett, L., Clarke, G., Kiely, B., Cryan, J.F., Dinan, T.G., 2010. Effects of the probiotic *Bifidobacterium infantis* in the maternal separation model of depression. Neuroscience 170 (4), 1179–1188.
Fields, H., 2004. State-dependent opioid control of pain. Nat. Rev. Neurosci. 5 (7), 565–575.
Fonken, L.K., Frank, M.G., Gaudet, A.D., Maier, S.F., 2018. Stress and aging act through common mechanisms to elicit neuroinflammatory priming. Brain Behav. Immun. 73, 133–148.
Frank, M.G., Fonken, L.K., Dolzani, S.D., Annis, J.L., Siebler, P.H., Schmidt, D., Watkins, L.R., Maier, S.F., Lowry, C.A., 2018. Immunization with *Mycobacterium vaccae* induces an anti-inflammatory milieu in the CNS: attenuation of stress-induced microglial priming, alarmins, and anxiety-like behavior. Brain Behav. Immun. 73, 352–363.
Galisteo, M., Duarte, J., Zarzuelo, A., 2008. Effects of dietary fibers on disturbances clustered in the metabolic syndrome. J. Nutr. Biochem. 19 (2), 71–84.
Goodrich, J.K., Di Rienzi, S.C., Poole, A.C., Koren, O., Walters, W.A., Caporaso, J.G., Knight, R., Ley, R.E., 2014. Conducting a microbiome study. Cell 158 (2), 250–262.
Guillemin, G.J., 2012. Quinolinic acid, the inescapable neurotoxin. FEBS J. 279 (8), 1356–1365.
Guo, G., Jia, K.R., Shi, Y., Liu, X.F., Liu, K.Y., Qi, W., Guo, Y., Zhang, W.J., Wang, T., Xiao, B., Zou, Q.M., 2009. Psychological stress enhances the colonization of the stomach by *Helicobacter pylori* in the BALB/c mouse. Stress 12 (6), 478–485.

Hamady, M., Knight, R., 2009. Microbial community profiling for human microbiome projects: tools, techniques, and challenges. Genome Res. 19 (7), 1141–1152.

Hodes, G.E., Pfau, M.L., Leboeuf, M., Golden, S.A., Christoffel, D.J., Bregman, D., Rebusi, N., Heshmati, M., Aleyasin, H., Warren, B.L., Labonté, B., Horn, S., Lapidus, K.A., Stelzhammer, V., Wong, E.H.F., Bahn, S., Krishnan, V., Bolaños-Guzman, C.A., Murrough, J.W., Merad, M., Russo, S.J., 2014. Individual differences in the peripheral immune system promote resilience versus susceptibility to social stress. Proc. Natl Acad. Sci. U.S.A. 111 (45), 16136–16141.

Hooper, L.V., Midtvedt, T., Gordon, J.I., 2002. How host-microbial interactions shape the nutrient environment of the mammalian intestine. Annu. Rev. Nutr. 22, 283–307.

Jašarević, E., Morrison, K.E., Bale, T.L., 2016. Sex differences in the gut microbiome–brain axis across the lifespan. Philos. Trans. R. Soc. Lond. B: Biol. Sci. 371 (1688), 20150122.

Jie, Z., Xia, H., Zhong, S.L., Feng, Q., Li, S., Liang, S., Zhong, H., Liu, Z., Gao, Y., Zhao, H., Zhang, D., Su, Z., Fang, Z., Lan, Z., Li, J., Xiao, L., Li, J., Li, R., Li, X., Li, F., Ren, H., Huang, Y., Peng, Y., Li, G., Wen, B., Dong, B., Chen, J.Y., Geng, Q.S., Zhang, Z.W., Yang, H., Wang, J., Wang, J., Zhang, X., Madsen, L., Brix, S., Ning, G., Xu, X., Liu, X., Hou, Y., Jia, H., He, K., Kristiansen, K., 2017. The gut microbiome in atherosclerotic cardiovascular disease. Nat. Commun. 8 (1), 845.

Keita, Å.V., Söderholm, J.D., 2010. The intestinal barrier and its regulation by neuroimmune factors. Neurogastroenterol. Motil. 22 (7), 718–733.

Kiela, P.R., Ghishan, F.K., 2016. Physiology of intestinal absorption and secretion. Best Pract. Res. Clin. Gastroenterol. 30 (2), 145–159.

Koeth, R.A., Wang, Z., Levison, B.S., Buffa, J.A., Org, E., Sheehy, B.T., Britt, E.B., Fu, X., Wu, Y., Li, L., Smith, J.D., DiDonato, J.A., Chen, J., Li, H., Wu, G.D., Lewis, J.D., Warrier, M., Brown, J.M., Krauss, R.M., Tang, W.H.W., Bushman, F.D., Lusis, A.J., Hazen, S., 2013. Intestinal microbiota metabolism of L-carnitine, a nutrient in red meat, promotes atherosclerosis. Nat. Med. 19 (5), 576–585.

Koren, O., Spor, A., Felin, J., Fåk, F., Stombaugh, J., Tremaroli, V., Behre, C.J., Knight, R., Fagerberg, B., Ley, R.E., Bäckhed, F., 2011. Human oral, gut, and plaque microbiota in patients with atherosclerosis. Proc. Natl. Acad. Sci. U.S.A. 108, 4592–4598.

Kullberg, M.C., Ward, J.M., Gorelick, P.L., Caspar, P., Hieny, S., Cheever, A., Jankovic, D., Sher, A., 1998. *Helicobacter hepaticus* triggers colitis in specific-pathogen-free interleukin-10 (IL-10)-deficient mice through an IL-12- and gamma interferon-dependent mechanism. Infect. Immun. 66 (11), 5157–5166.

Langgartner, D., Peterlik, D., Foertsch, S., Füchsl, A.M., Brokmann, P., Flor, P.J., Shen, Z., Fox, J.G., Uschold-Schmidt, N., Lowry, C.A., Reber, S.O., 2017. Individual differences in stress vulnerability: the role of gut pathobionts in stress-induced colitis. Brain Behav. Immun. 64, 23–32.

Langgartner, D., Vaihinger, C.A., Haffner-Luntzer, M., Kunze, J.F., Weiss, A.L.J., Foertsch, S., Bergdolt, S., Ignatius, A., Reber, S.O., 2018. The role of the intestinal microbiome in chronic psychosocial stress-induced pathologies in male mice. Front. Behav. Neurosci. 12, 252.

Lowry, C.A., Smith, D.G., Siebler, P.H., Schmidt, D., Stamper, C.E., Hassell Jr., J.E., Yamashita, P.S., Fox, J.H., Reber, S.O., Brenner, L.A., Hoisington, A.J., Postolache, T.T., Kinney, K.A., Marciani, D., Hernandez, M., Hemmings, S.M.J., Malan-Muller, S., Wright, K.P., Knight, R., Raison, C.L., Rook, G.A., 2016. The microbiota, immunoregulation and mental health: implications for public health. Curr. Environ. Health Rep. 3 (3), 270–286.

Lyte, M., Vulchanova, L., Brown, D.R., 2010. Stress at the intestinal surface: catecholamines and mucosa–bacteria interactions. Cell Tissue Res. 343 (1), 23–32.

Maes, M., 2011. Depression is an inflammatory disease, but cell-mediated immune activation is the key component of depression. Prog. Neuro-Psychopharmacol. Biol. Psychiatry 35 (3), 664–675.

Malan-Muller, S., Valles-Colomer, M., Raes, J., Lowry, C.A., Seedat, S., Hemmings, S.M.J., 2018. The gut microbiome and mental health: implications for anxiety- and trauma-related disorders. OMICS 22 (2), 90–107.

Mayer, E.A., 2011. Gut feelings: the emerging biology of gut–brain communication. Nat. Rev. Neurosci. 12 (8), 453–466.

McEwen, B.S., 1998. Stress, adaptation, and disease: allostasis and allostatic load. Ann. N.Y. Acad. Sci. 840 (1), 33–44.

Miller, A.H., Raison, C.L., 2016. The role of inflammation in depression: from evolutionary imperative to modern treatment target. Nat. Rev. Immunol. 16 (1), 22–34.

Morrison, D.J., Preston, T., 2016. Formation of short chain fatty acids by the gut microbiota and their impact on human metabolism. Gut Microbes 7 (3), 189–200.

Musselman, D.L., Evans, D.L., Nemeroff, C.B., 1998. The relationship of depression to cardiovascular disease: epidemiology, biology, and treatment. Arch. Gen. Psychiatry 55 (7), 580–592.

Øines, E., Murison, R., Mrdalj, J., Grønli, J., Milde, A.M., 2012. Neonatal maternal separation in male rats increases intestinal permeability and affects behavior after chronic social stress. Physiol. Behav. 105 (4), 1058–1066.

Ott, S.J., El Mokhtari, N.E., Musfeldt, M., Hellmig, S., Freitag, S., Rehman, A., Kühbacher, T., Nikolaus, S., Namsolleck, P., Blaut, M., Hampe, J., Sahly, H., Reinecke, A., Haake, N., Günther, R., Krüger, D., Lins, M., Herrmann, G., Fölsch, U.R., Simon, R., Schreiber, S., 2006. Detection of diverse bacterial signatures in atherosclerotic lesions of patients with coronary heart disease. Circulation 113 (7), 929–937.

Pace, T.W.W., Mletzko, T.C., Alagbe, O., Musselman, D.L., Nemeroff, C.B., Miller, A.H., Heim, C.M., 2006. Increased stress-induced inflammatory responses in male patients with major depression and increased early life stress. Am. J. Psychiatr. 163 (9), 1630–1633.

Petra, A.I., Panagiotidou, S., Hatziagelaki, E., Stewart, J.M., Conti, P., Theoharides, T.C., 2015. Gut-microbiota-brain axis and its effect on neuropsychiatric disorders with suspected immune dysregulation. Clin. Ther. 37 (5), 984–995.

Piccinelli, M., Wilkinson, G., 2000. Gender differences in depression: critical review. Br. J. Psychiatry 177 (6), 486–492.

Qin, J., Li, R., Raes, J., Arumugam, M., Burgdorf, K.S., Manichanh, C., Nielsen, T., Pons, N., Levenez, F., Yamada, T., Mende, D.R., Li, J., Xu, J., Li, S., Li, D., Cao, J., Wang, B., Liang, H., Zheng, H., Xie, Y., Tap, J., Lepage, P., Bertalan, M., Batto, J.M., Hansen, T., Le Paslier, D., Linneberg, A., Nielsen, H.B., Pelletier, E., Renault, P., Sicheritz-Ponten, T., Turner, K., Zhu, H., Yu, C., Li, S., Jian, M., Zhou, Y., Li, Y., Zhang, X., Li, S., Qin, N., Yang, H., Wang, J., Brunak, S., Doré, J., Guamer, F., Kristiansen, K., Pedersen, O., Parkhill, J., Weissenbach, J., MetaHIT Consortium, Bork, P., Ehrlich, S.D., Wang, J., 2010. A human gut microbial gene catalogue established by metagenomic sequencing. Nature 464 (7285), 59–65.

Raison, C.L., Capuron, L., Miller, A.H., 2006. Cytokines sing the blues: inflammation and the pathogenesis of depression. Trends Immunol. 27 (1), 24–31.

Reber, S.O., Langgartner, D., Foertsch, S., Postolache, T.T., Brenner, L.A., Guendel, H., Lowry, C.A., 2016a. Chronic subordinate colony housing paradigm: a mouse model for mechanisms of PTSD vulnerability, targeted prevention, and treatment—2016 Curt Richter Award Paper. Psychoneuroendocrinology 74, 221–230.

Reber, S.O., Siebler, P.H., Donner, N.C., Morton, J.T., Smith, D.G., Kopelman, J.M., Lowe, K.R., Wheeler, K.J., Fox, J.H., Hassell Jr., J.E., Greenwood, B.N., Jansch, C., Lechner, A., Schmidt, D., Uschold-Schmidt, N., Füchsl, A.M., Langgartner, D., Walker, F.R., Hale, M.W., Perez, G.L., van Treuren, W., González, A., Halweg-Edwards, A.L., Fleshner, M., Raison, C.L., Rook, G.A., Peddada, S.D., Knight, R., Lowry, C.A., 2016b. Immunization with a heat-killed preparation of the environmental bacterium *Mycobacterium vaccae* promotes stress resilience in mice. Proc. Natl. Acad. Sci. U.S.A. 113 (22), E3130–E3139.

Rook, G.A., Adams, V., Hunt, J., Palmer, R., Martinelli, R., Brunet, L.R., 2004. Mycobacteria and other environmental organisms as immunomodulators for immunoregulatory disorders. Springer Semin. Immunopathol. 25 (3–4), 237–255.

Savignac, H.M., Kiely, B., Dinan, T.G., Cryan, J.F., 2014. *Bifidobacteria* exert strain-specific effects on stress-related behavior and physiology in BALB/c mice. Neurogastroenterol. Motil. 26 (11), 1615–1627.

Schloss, P.D., Westcott, S.L., Ryabin, T., Hall, J.R., Hartmann, M., Hollister, E.B., Lesniewski, R.A., Oakley, B.B., Parks, D.H., Robinson, C.J., Sahl, J.W., Stres, B., Thallinger, G.G., Van Horn, D.J., Weber, C.F., 2009. Introducing mothur: open-source, platform-independent, community-supported software for describing and comparing microbial communities. Appl. Environ. Microbiol. 75 (23), 7537–7541.

Sender, R., Fuchs, S., Milo, R., 2016. Revised estimates for the number of human and bacteria cells in the body. PLoS Biol. 14 (8), e1002533.

Sodhi, M.S.K., Sanders-Bush, E., 2004. Serotonin and brain development. Int. Rev. Neurobiol. 59, 111–174.

Sonnenburg, E.D., Sonnenburg, J.L., 2014. Starving our microbial self: the deleterious consequences of a diet deficient in microbiota-accessible carbohydrates. Cell Metab. 20 (5), 779–786.

Stephens, R.L., Tache, Y., 1989. Intracisternal injection of a TRH analogue stimulates gastric luminal serotonin release in rats. Am. J. Physiol. 256 (2 Pt 1), G377–G383.

Sublette, M.E., Galfalvy, H.C., Fuchs, D., Lapidus, M., Grunebaum, M.F., Oquendo, M.A., Mann, J.J., Postolache, T.T., 2011. Plasma kynurenine levels are elevated in suicide attempters with major depressive disorder. Brain Behav. Immun. 25 (6), 1272–1278.

Sudo, N., Chida, Y., Aiba, Y., Sonoda, J., Oyama, N., Yu, X.N., Kubo, C., Koga, Y., 2004. Postnatal microbial colonization programs the hypothalamic-pituitary-adrenal system for stress response in mice. J. Physiol. 558 (1), 263–275.

Tsigos, C., Chrousos, G.P., 2002. Hypothalamic–pituitary–adrenal axis, neuroendocrine factors and stress. J. Psychosom. Res. 53 (4), 865–871.

Turnbaugh, P.J., Ley, R.E., Hamady, M., Fraser-Liggett, C.M., Knight, R., Gordon, J.I., 2007. The human microbiome project. Nature 449 (7164), 804–810.

Turnbaugh, P.J., Hamady, M., Yatsunenko, T., Cantarel, B.L., Duncan, A., Ley, R.E., Sogin, M.L., Jones, W.J., Roe, B.A., Affourtit, J.P., Egholm, M., Henrissat, B., Heath, A.C., Knight, R., Gordon, J.I., 2009a. A core gut microbiome in obese and lean twins. Nature 457 (7228), 480–484.

Turnbaugh, P.J., Ridaura, V.K., Faith, J.J., Rey, F.E., Knight, R., Gordon, J.I., 2009b. The effect of diet on the human gut microbiome: a metagenomic analysis in humanized gnotobiotic mice. Sci. Transl. Med. 1 (6), 6ra14.

Vernocchi, P., Del Chierico, F., Putignani, L., 2016. Gut microbiota profiling: metabolomics based approach to unravel compounds affecting human health. Front. Microbiol. 7, 1144.

von Hertzen, L., Hanski, I., Haahtela, T., 2011. Natural immunity. Biodiversity loss and inflammatory diseases are two global megatrends that might be related. EMBO Rep. 12 (11), 1089–1093.

von Hertzen, L., Beutler, B., Bienenstock, J., Blaser, M., Cani, P.D., Eriksson, J., Färkkilä, M., Haahtela, T., Hanski, I., Jenmalm, M.C., Kere, J., Knip, M., Kontula, K., Koskenvuo, M., Ling, C., Mandrup-Poulsen, T., von Mutius, E., Mäkelä, M.J., Paunio, T., Pershagen, G., Renz, H., Rook, G., Saarela, M., Vaarala, O., Veldhoen, M., de Vos, W.M., 2015. Helsinki alert of biodiversity and health. Ann. Med. 47 (3), 218–225.

Wang, Z., Klipfell, E., Bennett, B.J., Koeth, R., Levison, B.S., DuGar, B., Feldstein, A.E., Britt, E.B., Fu, X., Chung, Y.M., Wu, Y., Schauer, P., Smith, J.D., Allayee, H., Tang, W.H.W., DiDonato, J.A., Lusis, A.J., Hazen, S.L., 2011. Gut flora metabolism of phosphatidylcholine promotes cardiovascular disease. Nature 472 (7341), 57–63.

Wikoff, W.R., Anfora, A.T., Liu, J., Schultz, P.G., Lesley, S.A., Peters, E.C., Siuzdak, G., 2009. Metabolomics analysis reveals large effects of gut microflora on mammalian blood metabolites. Proc. Natl. Acad. Sci. U.S.A. 106 (10), 3698–3703.

Woese, C.R., Fox, G.E., 1977. Phylogenetic structure of the prokaryotic domain: the primary kingdoms. Proc. Natl. Acad. Sci. U.S.A. 74 (11), 5088–5090.
Yano, J.M., Yu, K., Donaldson, G.P., Shastri, G.G., Ann, P., Ma, L., Nagler, C.R., Ismagilov, R., Mazmanian, S.K., Hsiao, E.Y., 2015. Indigenous bacteria from the gut microbiota regulate host serotonin biosynthesis. Cell 161 (2), 264–276.
Zheng, P., Zeng, B., Zhou, C., Liu, M., Fang, Z., Xu, X., Zeng, L., Chen, J., Fan, S., Du, X., Zhang, X., Yang, D., Yang, Y., Meng, H., Li, W., Melgiri, N.D., Licinio, J., Wei, H., Xie, P., 2016. Gut microbiome remodeling induces depressive-like behaviors through a pathway mediated by the host's metabolism. Mol. Psychiatry 21 (6), 786–796.
Zijlmans, M.A.C., Korpela, K., Riksen-Walraven, J.M., de Vos, W.M., de Weerth, C., 2015. Maternal prenatal stress is associated with the infant intestinal microbiota. Psychoneuroendocrinology 53, 233–245.

CHAPTER 12

Future directions and concluding remarks

Paul D. Chantler[a], **Kevin T. Larkin**[b]
[a]Division of Exercise Physiology, Robert C. Byrd Health Sciences Center, School of Medicine, West Virginia University, Morgantown, WV, United States
[b]Department of Psychology, West Virginia University, Morgantown, WV, United States

Contents

Scientists have long searched for behavioral risk factors for cardiovascular disease (CVD) that extend our ability to predict disease onset beyond what we can learn from one's family history of heart disease, tobacco use, and cholesterol, glucose, and blood pressure levels (Centers for Disease Control and Prevention, National Center for Health Statistics, 2017). Such efforts

Cardiovascular Implications of Stress and Depression
https://doi.org/10.1016/B978-0-12-815015-3.00012-X

have commonly examined the role of exposure to environmental stress and have occasionally ignited interest in identifiable behavioral patterns shown to be associated with increased risk for CVD, like the Type A Behavior Pattern (e.g., Friedman and Rosenman, 1959) or presence of cynical hostility (e.g., Siegler et al., 1992). In providing the foundation for this book, diagnoses of depressive disorders and/or the presence of depressive symptoms now have over two decades of solid empirical work demonstrating strong associations with onset of CVD among initially healthy adults and subsequent cardiac events among patients diagnosed with coronary heart disease (for meta-analytic reviews, see Rugulies, 2002; Barth et al., 2004, respectively). As noted in Chapter 1, following an exhaustive review of the association between depression and CVD, the American Heart Association issued a statement recommending that depression be considered an established risk factor for poor prognosis among patients with CVD (Lichtman et al., 2014).

Despite evidence linking depression to CVD, it has yet to earn the designated status as a risk factor for CVD and is not routinely screened in many medical settings. There are two factors that have prolonged the adoption of depression as an established risk factor for CVD: (1) a limited comprehension of whether the association between depression and CVD is correlational or causal; and (2) a lack of understanding of the physiological mechanisms through which a presumably mental disease, like depression, results in perturbations within the circulatory system that promote atherosclerosis etc. The purpose of this book is to evaluate the theoretical models for how environmental stress gets "under the skin" and leads to conditions like depression and CVD, and most importantly, to examine evidence for a causal relation between depression and subsequent CVD. In so doing, we considered viable behavioral and biological mechanisms responsible for this association, many of which have been highlighted in previous reviews of this literature (e.g., Carney et al., 1995; Lett et al., 2004; Skala et al., 2006). In this final chapter, we will summarize the primary findings of the literature cited in the preceding works of scholarship and aim to integrate them into a comprehensive model of how stress, depression, and CVD are inter-related. We will also identify key work that has yet to be done and outline potential strategies for carrying out this type of empirical work. Naturally, considerable research is still required in order to build a more comprehensive understanding of how these biopsychosocial variables interact with one another. Hopefully, this book will serve as a blueprint for future work in this area so that behavioral risk factors can be recognized as important factors that contribute to the onset of CVD and be considered as routine assessments in all contemporary health care settings.

Causal relations among stress, depression, and cardiovascular disease

Several traditional risk factors for CVD have biologically-plausible pathways that promote the atherosclerotic process. For example, various toxins contained in tobacco smoke have been shown to result in endothelial injury (e.g., Guarino et al., 2011; Pittilo, 2000), which in turn elicits the body's inflammatory reaction that leads to the development of atherosclerotic plaques. Similarly, elevated lipids in the blood provide the raw materials to create fatty streaks that accumulate beneath the endothelium and serve as the foundation for subsequent plaque development (Insull, 2009). In contrast, although exposure to stress or experiencing symptoms of depression may be as highly predictive of CVD onset as other traditional risk factors, the biologically-plausible mechanisms that promote atherosclerosis are not as clear, leading some to question the causality of these relations as well as the importance of evaluating psychological risk factors for CVD.

To examine potential causal relations among stress, depression, and CVD, findings from epidemiological research on humans, translational research using animal models, and intervention research were reviewed in Chapters 2–4. Each body of literature represents a different approach to examining these associations and any causal pathways that might exist among them.

Epidemiologic research

Epidemiologic trials have the advantage of gathering large numbers of humans existing in real life settings using temporal measurement sequences. Rutledge and Gould covered the vast amount of research demonstrating strong direct associations between both stress and depression in predicting onset of CVD in healthy populations and disease progression among patients with diagnosed CVD (see Chapter 2). These relations are remarkably stable across numerous types of samples, methods of measuring both psychological predictor and CVD outcome variables, and duration of follow-up. The fact that psychological factors are measured years before CVD outcome variables in these studies lends some credence to interpreting these findings to support a causal pathway from stress and/or depression to CVD. Despite the appeal of this conclusion, it does not preclude alternative pathways from being considered. It is well known, for example, that experiencing a cardiac event and subsequent hospitalization leads to increased stress in one's life (e.g., Krantz, 1980). It is also possible that while measures of stress

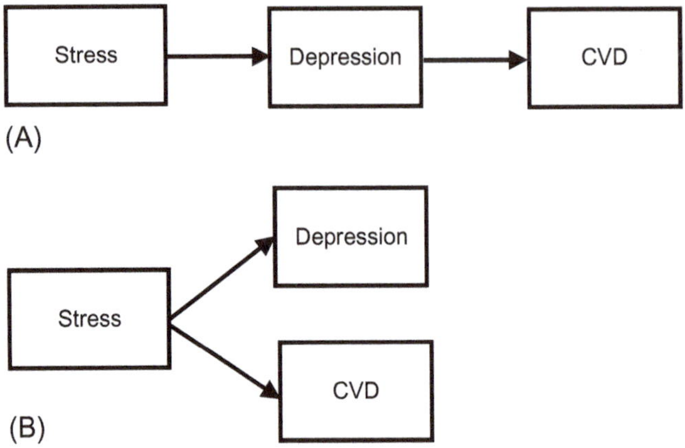

Fig. 12.1 Causal and correlational models linking stress with depression and CVD. (A) Depression mediates the relation between stress and CVD. (B) Depression as a corollary of the relation between stress and CVD.

or depression are obtained long before CVD health outcome parameters are assessed that these psychological variables serve as reliable predictors of CVD disease onset or exacerbation, but do not cause these phenomena. In this case, a shared underlying physiological vulnerability may lead to both onset of depression and CVD, with one not causing the other. In brief, findings from the epidemiologic literature clearly support strong associations among stress, depression, and CVD, but cannot address the issue of causation fully nor do they enable us to draw conclusions whether the corollary or mediational models displayed in Fig. 12.1 (redrawn from Chapter 1 here) best reflect the actual state of these associations.

Translational animal research

In contrast to the epidemiologic approach, translational research employing animal models can employ experimental methods that permit causal inferences to be made. In Chapter 3, Grippo reviewed the animal research linking exposure to stress with both measures of depressive behaviors as well as cardiovascular consequences. Consistent with the prior chapter, strong associations between exposure to stress and both depressive behaviors (e.g., anhedonia, helplessness, passive social behaviors, reduced locomotion) and altered circulatory system functioning have been observed. These relations have been detected in studies using numerous experimental paradigms, including exposure of organisms to unpredictable chronic stress, social defeat or isolation, and direct stimulation of bodily processes through surgical or pharmacologic means. Of interest is the fact that some research has shown that the depressive symptoms detected in these paradigms can be normalized by using anti-depressant medications or electroconvulsive approaches, similar to treating depressive disorders in humans (e.g., Katz, 1981). Additionally, these models have also shown that animals exposed to stress exhibit dysregulation of many of the physiological systems hypothesized to mediate the linkage between stress, depression, and CVD in humans (e.g., autonomic nervous system, vascular modeling, inflammation, etc.). In this regard, we have learned a great deal about the associations among stress, depression, and CVD from this approach.

Unfortunately, most research using animal models to date has failed to illuminate the causal connections between stress, depression, and CVD completely. By devising careful control conditions with animals from the same genetic strains and ensuring that all extraneous environmental factors other than those being studied are equivalent across groups (e.g., diet, temperature, housing, exposure to light), experimenters can be assured that any group differences in behavior or physiological functioning are caused by the stress manipulation. In this regard, studies of this type permit us to conclude with some confidence that exposure to stress can lead to depressive behaviors and physiological disturbances to the cardiovascular system. As suggested in Chapter 1, environmental stress is the stimulus and depression and/or CVD are responses. Although experimental approaches like those surveyed by Grippo permit us to make these sorts of causal statements, they have done very little to determine whether depression, too, is part of a causal pathway leading to CVD. For the most part, animal studies have measured depressive behaviors concurrent with assessments of the physiological phenomena that disrupt cardiovascular functioning. Consequently, these studies have not allowed us to determine whether depression mediates the relation between stress and CVD or whether it reflects a concurrent bodily response, as shown in Fig. 12.1. The next generation of translational animal works might aim to tackle this question by devising longitudinal designs that contain continuous assessments of both behavior and physiology across the lifespan of organisms exposed to environmental stress by contrasting them with animals in non-stressful living environments. Through such methods, we could test whether differential CVD outcomes are observed among organisms that display greater or lesser behavioral displays of depressive symptoms in response to environmental stress.

Clinical intervention trials

In Chapter 4, Cavanagh, Gaffey, Rosman, and Burg reviewed clinical trials aimed at treating either depression or CVD using pharmacotherapeutic agents or psychological interventions, including stress management, cognitive behavior therapy, and nutritional and exercise interventions. Given the strong associations detected between depression and CVD, clinical scientists began to consider what impact on CVD outcomes would be evidenced if patients with CVD who were depressed were treated using one of the evidence-based pharmacologic or non-pharmacologic interventions known to reduce depression. In Cavanagh et al.'s appraisal of this work, it is clear that mood symptoms are improved with these treatments when applied to CVD patients, but they very rarely translate into improvements with respect to CVD events, CVD mortality rates, or traditional CVD risk factors. There are limitations to making causal inferences from clinical trials research that were duly noted in this chapter, of course, but it is safe to conclude that evidence that treatment of depression with established intervention approaches results in improved CVD outcomes is limited. This could be due to the lack of a causal relation between depression and CVD, but also could reflect the state of our knowledge regarding the mechanisms through which our current treatments for depression operate.

Although we have convincing data that demonstrate the efficacy and effectiveness of many pharmacologic and non-pharmacologic interventions for treating depression (Gartlehner et al., 2017), the underlying physiological mechanisms for their salubrious effects still remain a mystery. Regarding the use of selective serotonin reuptake inhibitors (SSRIs), one can assume that the availability of serotonin in the synaptic cleft or the density of serotonergic receptors could be responsible for the efficacy of this intervention approach. To date, however, we have only limited evidence that either of these physiological processes is responsible for the anti-depressant effects of SSRIs. Likewise, one can assume that modification of belief systems could be the mechanism through which cognitive behavior therapy (CBT) exerts its influence on negative mood states. Again, we have only limited evidence that alteration of such belief systems is responsible for the mood-enhancing effects of CBT. In a study using positron-emission tomography, Goldapple et al. (2004) demonstrated that different brain regions were modulated when treating depression with an SSRI versus treating depression with CBT, suggesting that changes in different brain regions are responsible for explaining how each of these efficacious treatment approaches for depression operates.

Summary of evidence regarding causality

Given the limitations of epidemiologic, animal, and intervention research for establishing causal inferences, findings from studies in each of these areas are likely not sufficient to convince practitioners that measurement of depression should become routine in health care settings. In brief, support for causal relations between exposure to stress and onset of both depressive disorders and CVD is on much more solid footing than evidence that depression causes CVD. It could be argued, however, that demonstration of causality is not necessarily a pre-requisite for adopting a construct like depressed mood as a risk factor. If a test could be constructed that possessed adequate psychometric characteristics and it predicted unique variance in disease onset above and beyond that of other traditional risk factors, the assessment could have clinical utility as a credible marker for CVD even though it may not be linked to the disease process causally.

Based on our current understanding of this body of literature, interventionists might stress the importance of engaging in behavior change among those factors known to reduce risk for CVD more for patients presenting with high depression scores or being diagnosed with a major mood disorder. In this regard, validated screening tools could be quite effective in guiding the treatment plans of health care professionals regardless of whether depression causes CVD. Although differentiating whether depression is a causal psychological phenomenon or a corollary phenomenon is not important for establishing depression as a behavioral risk factor for CVD, it is critical for developing intervention and prevention programs. If depression is not part of the causal pathway linking stress with CVD, which may well be the case, evidence-based interventions for treating depression will improve mood symptoms but not influence subsequent incidence of CVD. Although treating depression has many positive benefits in terms of improving CVD patient's psychological health and daily functioning, interventionists will have a tougher time citing evidence that treating their depression will have positive effects on the functioning of their circulatory systems.

Mechanisms linking stress with depression and cardiovascular disease

Although future studies are needed to confirm the role of depressed affect and behavior in CVD onset and progression, there is insufficient evidence at this point in time to suggest that depression causes CVD. However, because it is clear that exposure to stress is associated with both increased prevalence and incidence of CVD, it is critical to examine the biologically-plausible mechanisms linking stress with both onset of depression as well as CVD. The chapters that comprise the second part of this volume examined potential candidates for explaining how stress infects organisms leading to the dramatic changes in the cardio-vasculature that lead to onset of CVD. As mentioned in Chapter 1, this is not a new question, as several prior reviews have considered a plethora of mechanisms linking stress, depression, and CVD (e.g., Carney et al., 1995; Lett et al., 2004; Skala et al., 2006). However, given that over a decade had passed since these reviews were conducted and results of many new and exciting discoveries have been published since that time, the second aim of this book was to examine the progress made in understanding how stress influences the circulatory system and leads to atherosclerotic complications, including the magnitude and patterning of autonomic and neuroendocrine responses to stress (Chapter 5), cardiac changes (e.g., electrical conductance, cardiac remodeling) and vascular alterations (peripheral arterial disease; arterial stiffness and atherosclerosis) that occur in response to stress (Chapter 6), central nervous system involvement (e.g., brain signaling; cerebrovascular functioning) in response to stress (Chapter 7), inflammatory system responses to stress (Chapter 8), physiological influences of social support and lack thereof (Chapter 9), alterations in health behaviors, including dietary and exercise behaviors, sleep, smoking tobacco and use of other substances, and engagement with health care providers (Chapter 10), and alterations in the microbiome of the gut in response to stress (Chapter 11).

A summary of the information presented in Chapters 5–11 is presented in Fig. 12.2 and Table 12.1. In brief, the various mechanisms considered in this book can be described as stimuli that exist in the environment outside of the body, covert responses that occur within the body, and overt behavioral responses. Using Fig. 12.2 and Table 12.1 as a guide, we summarize the body of literature that has examined these various mechanisms through which stress infects the body and the numerous feedback systems that attest to the complexity of the interconnections that exist among these parameters.

External stimuli

As noted in the work of Selye (1955) and Cannon (1915), the bodily response to stress requires exposure to some type of antecedent stimuli detected in the environment. Without such exposures, there is no need for a bodily adaptation, and consequently, homeostatic processes will function normally to properly sustain life (e.g., heart rate regulation, digestion,

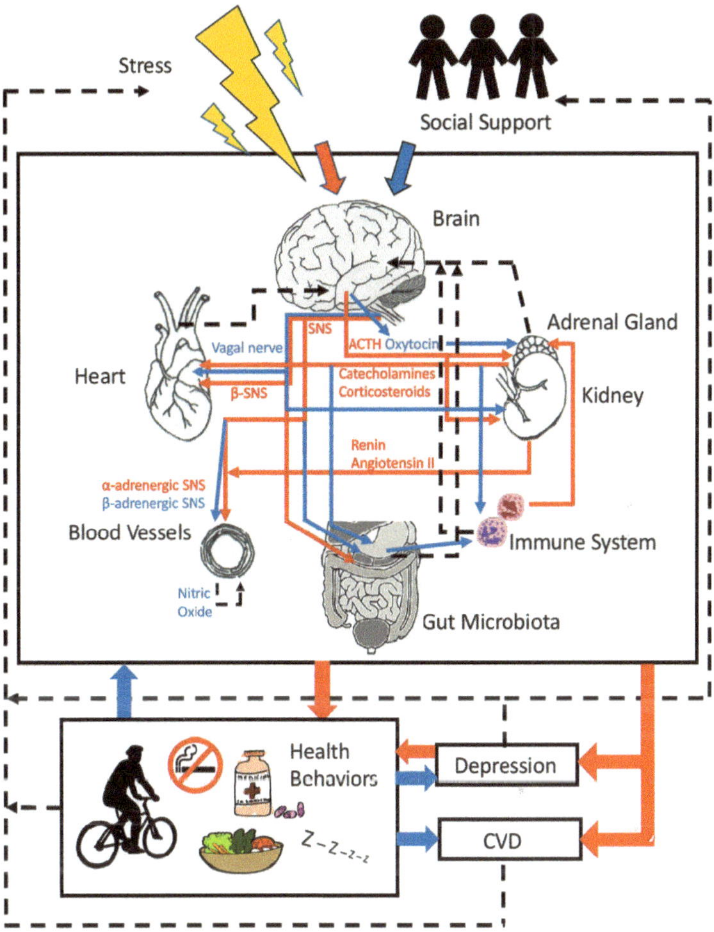

Fig. 12.2 Recognized pathways linking stress with both depression and CVD. External stimuli (Stress, Social Support) at the top of the figure provide both excitatory *(red arrow)* and inhibitory *(blue arrow)* effects on the subsequent bodily response to stress, respectively. Physiological processes known to be involved in the stress response are depicted in the middle box. *Red arrows* indicate excitatory connections and *blue arrows* indicate inhibitory connections. *Black dotted arrows* reflect negative feedback loops. The bottom frame displays observable outcomes, including engagement in health behaviors and disease processes. Please note that not all associations are displayed in order to highlight the primary relations supported by empirical evidence. Some images in the public domain were obtained from openclipart.org.

temperature control). From their work, however, we know that organisms are physiologically wired to respond to certain stimuli interpreted as threatening quite differently from stimuli interpreted as being either benign or in some cases approachable. Organisms rely on the sensory systems of the brain to assist in detecting these environmental stimuli and orchestrating the

Table 12.1 Physiological mechanisms altered by exposure to stress.

Cortex	– Amygdala activation – Medial prefrontal lobe activation – Activation of anterior and posterior cingulate cortex – Activation of insula – Activation of thalamus – Reduced activity in dorsolateral prefrontal cortex – Reduced activity in basal ganglia and striatum – Reduced activity in nucleus accumbens
Autonomic nervous system	– Increased or blunted beta-adrenergic activity – Increased or blunted alpha-adrenergic activity – Reduced vagal tone – Delayed autonomic recovery from stress
Neuroendocrine system	– Increased ACTH from the pituitary gland – Increased or blunted cortisol/corticosterone secretion from adrenal gland – Increased catecholamine secretion from adrenal gland – Delayed cortisol recovery from stress
Renin-angiotensin-aldosterone system	– Increased renin secretion from kidney – Increased angiotensin II from kidney
Vasculature	– Endothelial dysfunction and injury – Reduced nitric oxide bioavailability – Increased vascular wall remodeling – Arterial stiffness/atherosclerosis – Microvascular rarefaction – Elevated blood pressure
Cardiac	– Cardiac ischemia and arrhythmias – Increased resting HR and cardiac output – Cardiac remodeling/fibrosis
Immune system	– Change in adaptive and innate immune system – Shift in pro-inflammatory phenotype (M2 to M1) – Increased cytokines (e.g., interleukin-1, TNF, and interleukin-6 ....) – Pro-oxidative environment
Microbiome in the gut	– Change in the gut microbiome: proliferation of pathobionts – Exaggerated release of proinflammatory cytokines

appropriate physiological responses needed to sustain life. It is not surprising then, that the brain is the first organ where stimulus information enters the body and that launches the physiological stress responses designed to enhance survival (see location of the brain in Fig. 12.2).

It should be noted, however, that not all environmental stimuli are threatening to the organism. As shown in Fig. 12.2, exposure to nurturing, supportive, or comforting stimuli (labeled Social Support), in contrast to stimuli perceived as threatening (labeled Stress), results in different physiological and behavioral responses (e.g., to approach rather than to avoid). Presumably, this is the avenue through which social support influences the body's stress response as well as its vulnerability for becoming depressed or developing CVD. Based on the data on this parameter, Uchino and colleagues (Chapter 9) provide evidence that shows the importance of developing socially supportive relationships in ameliorating risk for both depression and CVD. Presumably, the presence of social support in the surrounding environment alters the entire sequence of stress responding that follows, including the autonomic nervous system, the neuroendocrine system, and immunological parameters. Uniquely associated with presence of social support, Uchino and colleagues also discuss the role of the hormone, oxytocin, that is secreted by the pituitary gland and has been shown to dampen the magnitude of many components of the stress response, particularly in the adrenal gland (see oxytocin link in Fig. 12.2).

Uchino and colleagues also emphasize the importance of the distinction between *received* and *perceived* social support in conducting studies examining this important parameter. Clearly, being in close proximal contact with a spouse who is rejecting and hyper-critical will trigger a different physiological response than interacting with a loving and nurturing spouse. In this regard, it is not surprising that stronger health benefits have been observed with the perception of support provided by others rather than simply assessing the number of people available in one's social network.

Cortical response to stress

Among all organ systems in the body, the brain is responsible for transducing environmental stimuli (e.g., detection of threatening, benign, or nurturing stimuli) into neural phenomena that enable the body to develop appropriate responses to these stimuli. In brief, detection of threat orchestrates a different physiological and behavioral response than detection of nurturing stimuli. As reviewed by Brefczynski-Lewis in Chapter 7, there are several cortical regions influenced by the detection of threat, including those in the frontal, temporal, and parietal lobes as well as important regions of the limbic system (see Table 12.1).

The brain is a complicated organ, continuously receiving sensory inputs from not only the external environment, but also all of the other organ systems of the body. In this regard, while the brain is involved in detecting threat (or lack thereof) in the external environment, it is also receiving information from the body via numerous neural pathways from peripheral organs and systems (see dashed feedback loops depicted in Fig. 12.2). For example, the vagal nerve sends efferent messages to peripheral organs, but also receives afferent input from them (Berthoud and Neuhuber, 2000). Baroreceptors detect alterations in blood pressure in the carotid arteries and provide input to the brainstem to modulate heart rate (Lawrence and Jarrott, 1996). Circulating cortisol levels are detected in the brain and corticotropin-releasing factor secretion is modified based on this input (Tasker and Herman, 2011). Likewise, circulating inflammatory cytokines are detected by the brain, which in turn affect neurotransmitters involved in the regulation of mood, emotion, and the physiological stress response (Miller et al., 2013). The brain accumulates all of this information in determining the appropriate physiological response to emit. In this regard, it is critical to assess the influence of these feedback loops when measuring the magnitude and patterning of the body's physiological stress response.

Accumulating all of this information, the brain then orchestrates a physiological response that is carried out by the entire body. Responses to the detection of threat versus safety and support in the surrounding environment are quite distinctive and carried out by different physiological messengers and systems in the body. Beginning with the initial sensory neurons involved in stimulus detection, Brefczynski-Lewis identifies many brain regions that function together to coordinate the immediate bodily response that impacts functioning of virtually all other organ systems in the body. As noted above, the brain then monitors and secures information from the bodily systems involved in the response through various feedback loops and directs the ongoing response with the aim of sustaining, intensifying, or dampening the response. Because brains are quite capable of responding to environmental stressors very rapidly, it becomes very difficult for scientists to ascertain whether the functional brain changes they are detecting following exposure to test stimuli represent the initial detection and appraisal of the stimuli or the brain's reactions to the information it is receiving from the body via feedback from the periphery as the brain appraises whether an optimal physiological response is occurring. This is quite likely why so many different brain regions have been identified as being influenced by exposure to environmental stressors (see Table 12.1). In this regard, the stress response is not simply a brain phenomenon, but involves the entire body. Extending this logic further, one could also say that CVD is not purely a disorder of the cardiovascular system and depression is not purely a brain or mental disorder. Because of the complexity of the stress response, it is clear that the entire body is involved in the etiology of both of these diseases as they present in modern health care settings.

One remaining feature regarding the empirical findings reported by Brefczynski-Lewis is that some evidence exists to suggest that the dysregulation of patterns of cortical activity that occurs in response to stress is malleable and capable of at least partially resuming normal levels of functioning (e.g., Liston et al., 2009; Savic et al., 2017). The extent to which existing interventions promote the normalization of brain activity is far from understood, but she illuminates several that are worthy candidates for future study (e.g., modifications in dietary and exercise habits and various stress management programs).

Autonomic nervous system response to stress

The autonomic nervous system (ANS) comprises one of the first body systems launched by the brain once it detects threat in the surrounding environment. In Chapter 5, Larkin, Brown, and Tiani reviewed the extensive literature showing that ANS functioning is altered by exposure to stress, experiencing depression, and being diagnosed with CVD. Congruent with prior reviews of this literature (e.g., Carney et al., 1995), they concluded that convincing evidence exists to link dysfunctional patterns of ANS responding with exposure to stress as well as to diagnoses of depression and CVD. The nature of dysfunctional ANS arousal, however, has shifted slightly over time. Rather than focusing entirely on exaggerated heart rate and blood pressure responses to stress commonly reported in the earlier literature (Krantz and Manuck, 1984), there is now strong evidence that two other dysfunctional response patterns exist that are associated with risk for subsequent health problems: the delayed physiologic recovery from stress and the blunted or dampened physiologic response to stress.

As depicted in Fig. 12.2, the ANS, comprised of the vagal nerve (see blue arrows linking the brain with the heart and kidney in Fig. 12.2) and the alpha- (see red arrows linking the brain with blood vessels in Fig. 12.2) and beta-adrenergic branches (see red arrow linking the brain with the heart and blue arrow linking the brain with blood vessels in Fig. 12.2) of the sympathetic nervous system, provides the neural connection between the brain and most peripheral organs and structures of the body. As noted above, feedback mechanisms exist between these peripheral organs and the brain to assist the body in constructing the optimal response to the type of environmental stimulus detected. One of the best examples of this regulatory response is the baroreceptor feedback loop (see dashed black arrow linking the heart with the brain in Fig. 12.2). These pressure receptors located in the carotid arteries monitor arterial pressure and send feedback to the nucleus tractus solitarius in the brainstem to modify tonic heart rate via the vagal nerve to protect the brain when rapid increases in arterial pressure are sensed. The vagal nerve functions as an afferent neural fiber as well as an efferent fiber (Berthoud and Neuhuber, 2000), providing another example of how the body transmits information to the brain to assist in formatting the optimal stress response.

Although it was initially perplexing to read reports that exposure to stress was associated with both *exaggerated* and *blunted* cardiovascular responses, it can be understood given the nature of these intricate feedback loops involved in regulating the ANS's contribution to the stress response. Although acute stress presentations might elicit an exaggerated response, as exposure to stress persists, a hyperactive feedback process that accompanies an exaggerated response may actually result in a dampened heart rate and blood pressure responses to stress. Thus, for participants in studies examining ANS response to stress who are experiencing a good bit of stress in their lives, blunted responses to stress may be more likely observed. Consequently, it becomes important for future studies of this type to assess the amount of stress study participants are experiencing on a daily basis because this may influence the pattern of ANS stress responses observed.

Neuroendocrine response to stress

Larkin, Brown, and Tiani also reviewed the literature examining the neuroendocrine (NE) response to stress in Chapter 5, primarily studies that explored the role of the cortisol response to stress. As they noted in their chapter, in contrast to the ANS, the NE response takes a little longer to detect due to the fact that it uses the bloodstream to transmit information about stress detection to peripheral organs. Similarly, the feedback mechanisms involved in the regulation of cortisol take longer to occur because they too rely on the bloodstream as their mode of transit to the brain.

Like the ANS, there is evidence that exposure to stress and experiencing depressed moods are associated with both exaggerated and blunted cortisol responses as well as delayed recovery from stress. As shown in Fig. 12.2, the glucocorticoid and catecholamine responses to stress are excitatory in nature (see red arrow linking the brain to the kidney and adrenal gland in Fig. 12.2), although with prolonged exposure to stress, levels of these hormones are dampened, presumably through an overactive feedback system between adrenal and cortical functioning (see black dashed arrow linking the corticosteroid action from the adrenal gland with the brain in Fig. 12.2).

Investigations exploring the role of both ANS and NE systems in the stress response are largely based on laboratory work during which participants are exposed to acute cognitive or interpersonal challenges. This approach enables presentations of a standard series of stressors to study participants and is a valid strategy for examining the initial phases of a relatively time-limited environmental stressor. However, these studies provide very little information about the physiological responses to exposures to prolonged environmental stress (e.g., job stress, living in poverty, repetitive adverse living conditions). Consequently, we know more about the acute stress response than the chronic stress response.

In addition to evaluating differences in response patterns based on the chronicity of the stress stimuli used in empirical work, Larkin and colleagues also noted the importance of gathering normative data on the optimal response to a designated challenge in order to ascertain whether a participant's responses are exaggerated or blunted. Unless we know the optimal increase of ANS parameters in response to a stressful task (i.e., arousal level that optimizes task performance without wasting any energy), it will be impossible to determine whether a given response is adaptive or maladaptive.

Immune system response to stress

A relation exists between the neuroendocrine and immune system, whereby stress activates the HPA axis which normally suppresses the immune response through the release of glucocorticoids from the adrenals (see blue arrow linking the adrenal gland with the immune system in Fig. 12.2). The activation of the immune system also influences NE activity, presumably through detection of cytokines by the adrenal glands (see red arrow linking immune system and the adrenal gland in Fig. 12.2), which provides a feedback loop between the immune system and brain via detection of cytokines in the bloodstream (see black dashed arrow in Fig. 12.2).

In Chapter 8, Wood describes the immune and oxidative responses to stress, depression, and CVD. In both pre-clinical and clinical models, stress triggers the release of immune cells (Interleukin-1, TNF, and Interleukin-6), which can disrupt the integrity of the blood brain barrier and contribute to increased neuroinflammation. Microglia are highly adaptive innate immune cells that are capable of transitioning between pro-inflammatory (M1) and anti-inflammatory (M2) states. Under chronic stressful conditions and with the activation of glucocorticoid receptors, the microglia morphology shifts from an M2 to M1 phenotype, reflecting a pro-inflammatory response (Tang et al., 2018). In addition, the microglia become sensitized and respond to a subsequent stress event with a faster and more robust cytokine and chemokine release. The nature of this response accumulates with the degradation of proteins, mitochondria dysfunction, and impaired axonal transport, which can lead to neuronal dysfunction and cell death (Waisman et al., 2015), and the development of cerebrovascular diseases and cognitive impairment.

Wood goes onto describe that stress is a catalyst for the accumulation of reactive oxygen species (ROS). Although ROS play a role in several critical neuronal functions such as neuronal plasticity and learning and memory, an overproduction and accumulation of ROS can be detrimental. Using pre-clinical models, stress decreases mitochondrial function which is the major producer of ROS. Interestingly, a bi-directional relation exists between inflammation and oxidative stress, whereby an increase in proinflammatory cytokines also directly increase ROS, thus leading to a vicious cycle of a negative neurovascular environment.

With regard to depression and the presence of CVD, similar increased pro-inflammatory responses, including a shift in the microglia phenotype, and an increase in ROS accumulation has been documented (Sacher et al., 2015; Scapagnini et al., 2012). Evidence also suggests that the level of neuroinflammation may be correlated with severity of depressive symptoms. (Sasayama et al., 2013). Such data perhaps suggests a link between inflammation, stress, depression, and the development of CVD. Because of the complexity, future research is required tease out these relations and determine mechanisms of cause and effect.

Cardiac and vascular response to stress

The cardiovascular system consists of the heart, blood vessels, and blood. The function of the cardiovascular system involves the transportation of nutrients, oxygen, and hormones to cells throughout the body, and removal of metabolic wastes (carbon dioxide, nitrogenous wastes). In Chapter 6, Chantler and Frisbee describe the cardiac and vascular adaptations to stress, depression, and CVD, and their interplay. Evidence suggests that chronic stress and depression elicit multifaceted functional impairments to the cardiovascular system, which play a critical role in the health of the brain, heart, and vasculature, and the pathogenesis of stress-related cognitive impairment and cerebrovascular and cardiovascular events. Cardiac complications include stress-induced cardiac ischemia and arrhythmias, likely via the increased activation of the sympathetic nervous system and increased circulating hormones. Disrupted autonomic function has been observed in the development of depression and its association with CVD. Indeed, depressed individuals display higher resting heart rates, leading to increased cardiac output, and altered basal and stressor-associated blood pressure, higher basal levels of plasma norepinephrine, and greater sympathetic nervous system reactivity to stressors. The effects of chronic stress on cardiac remodeling reflect an increased accumulation of fibrous tissue in the heart, maladaptive cardiac hypertrophy, and changes in electrical conduction system of the heart. Such maladaptive changes place extra strain on the heart and increase the risk of cardiovascular events.

Chantler and Frisbee also describe the effects of chronic stress on vascular health. Importantly, the effects of chronic stress on vascular dysfunction are noted in a variety of caliber vessels (i.e., aorta, gracilis, and cerebral arterioles), suggesting a systemic effect of stress. Pre-clinical and clinical models have shown that the inner lining of arteries and arterioles (i.e., endothelial cells) are damaged with chronic stress and in patients with depression. The endothelium (see blood vessels in Fig. 12.2) plays a critical role in the initiation and progression of atherosclerosis. Thus, it is not surprising that in addition to impaired arterial function, chronic stress leads to arterial stiffening and the formation of plaque. Such adverse changes likely play a significant role linking chronic stress to CVD. However, evidence also suggests that arterial stiffness may accelerate the development of depressive disorders through altering brain structure (i.e., small vessel lesions and endothelial cell permeability).

Although the mechanisms through which chronic stress leads to endothelial dysfunction and vascular remodeling remains a process of exploration, many of the studies noted significant increases in the vascular production of inflammatory cytokines, ROS, and a reduction in nitric oxide (NO) bioavailability (see feedback loop indicated by a black dashed arrow through which NO regulates vasoconstriction and vasodilation in Fig. 12.2). However, stress hormones are also regulators of vascular function and likely contribute to the dysfunction noted. Another pathway that leads to vascular dysfunction and accelerates the development of atherosclerosis, is the over-activation of the renin-angiotensin-aldosterone system (see red arrow linking the kidney to blood vessels through the renin-aldosterone-angiotensin II system in Fig. 12.2).

Although both exposure to stress and presence of diagnoses of depression have been associated with identifiable changes in cardiac and vascular responses, very little effort has focused on how they interact in influencing cardiac and vascular processes. Further, evidence suggests that the stress response on the vasculature is mediated by pre-existing CVD and may be sex-dependent. Future studies need to tease out these complex relations that likely interact with one another to determine the overall effects of chronic stress on cardiovascular health.

Microbiota and gut response to stress

In Chapter 11, Cho and Lowry describe the bidirectional microbiome-gut-brain (MGB) axis as a major communication highway that dictates how the human body develops adaptive and maladaptive responses to stress. Given that depression is not simply a mental disorder or brain disease, but a systemic disease, evidence has shown that depressed patients often have gut appetite and metabolic disturbances, functional gastrointestinal disorders, and gut microbiota abnormalities (Liang et al., 2018). Cho and Lowry explain how the gut may impact the brain through several pathways including the production of neuroactive metabolites, modulation of immune signaling, and modulation of afferent neural signaling through vagal and spinal afferents. Both the ANS and NE systems influence digestive activity, namely that activation of the sympathetic nervous system and secretion of cortisol is associated with decreased gastric motility (see blue arrows linking the sympathetic nervous system and corticosteroids to the gut in Fig. 12.2). In contrast, the activation of the vagal nerve (see red arrow from the vagal nerve to the gut in Fig. 12.2) enhances digestive processes. Interestingly, healthy gut microbiota promotes the healthy development of the immune system responses (see blue arrow from the gut to the immune system in Fig. 12.2). Whereas chronic stress results in proliferation of pathobionts and the subsequent exaggerated release of proinflammatory cytokines, evidence demonstrates that compositional changes in the gut microbiome occur with an increase in intestinal permeability in association with stress-induced depressive-like behaviors. Such a response may allow enteric pathogens to evade the gut microbiota biofilm and easily bypass the gut epithelium to translocate into the bloodstream and travel to end organs (i.e., brain; see dashed feedback loop from the gut to the brain in Fig. 12.2).

Much research over the last decade has supported the hypothesis that the gut microbiota is a key mediator of depression (Bruce-Keller et al., 2018). Two striking studies supports this hypothesis: (1) microbiome transplants from healthy control mice into stressed mice ameliorates the behavioral and immune effects associated with the depressive state (Langgartner et al., 2018); and (2) probiotic treatment was effective in reducing the depressive- and anxiety-like behavior in rodent models (Savignac et al., 2014). Such research has led to a new interest in the potential for microbial psychobiotics to treat and potentially prevent stress-related psychiatric disorders.

Health behaviors in response to stress

Exposure to stress results in the physiologic response depicted in the center of Fig. 12.2 and listed in more detail in Table 12.1. However, it is well known that individuals engage in observable behavioral responses as well. Some develop habits of smoking tobacco to handle stress, some drink alcohol or use other substances, some consume too much food and fail to engage in an adequate amount of physical activity, some develop poor sleeping habits, and some fail to seek health care and/or follow the recommendations of their health care providers. Goodie, Wilfong, and Kroke provided a comprehensive review of the literature examining many of these health behaviors in Chapter 10. Across the full range of health behaviors, they provided evidence that stress is associated with adoption of fewer healthy behaviors (see red arrow in Fig. 12.2) and that engagement in healthy behaviors is associated with reductions in stress (see blue arrow in Fig. 12.2). Furthermore, these same associations were linked with depression, in that engagement in healthy living is associated with substantial improvements in mood (see blue arrow linking health behaviors with depression in Fig. 12.2). As depicted in Fig. 12.2, health behaviors also have direct positive health benefits for CVD and comprise a substantial part of most health care professional's recommendations to reduce risk for CVD (see blue arrow linking health behaviors with CVD). Notably, a feedback loop also exists between the various types of health behaviors and environmental stress (see dashed feedback loop in Fig. 12.2), in that engagement in few health behaviors involving ingestion of foods high in fat and salt and low in fiber, alcohol, other substances, inhaling tobacco smoke, and failing to obtain proper health care, regular exercise, or a good night's sleep constitute additional stressors to which the body reacts.

Summary

Based upon the empirical evidence shown in Fig. 12.2 and presented here, it is clear that depression and CVD are both body (systemic) disorders, not simply nervous system and circulatory system diseases, respectively. Pathways between environmental stimuli and bodily response systems are complex, bi-directional, and interactive. As depicted in Fig. 12.2, not only does stress launch complicated physiological and behavioral responses that culminate in depression and/or CVD, but these diseases contribute to new sources of stress or reductions in social support that modulate the complete bodily response (see dashed feedback loops around the perimeter in Fig. 12.2). If markers of inflammation increase in response to stress, for example, the brain will detect this alteration and orchestrate a sickness response (e.g., social withdrawal; sleep more; eating disruption; anhedonia). What if depression is an inflammatory disorder? Treating mood disturbances using interventions that fail to correct the abnormal inflammatory response would not alter its cause, nor would it impact CVD risk! Testing of novel interventions (e.g., anti-inflammatory agents) to target the physiological mechanisms directly responsible for the relation between stress and CVD would likely yield greater effects. Considerable work needs to be done, however, to provide the scientific foundation to support these types of approaches. Consideration of recommended next steps for pursuing this direction follow in the final section of this chapter.

Future research directions

Chronic versus acute stress

To date, we have a rather good understanding of how acute stress affects the cardiovascular system, how the endocrine and inflammatory systems respond to the acute stress stimulus. We also have excellent epidemiological data showing strong associations between stress, depression, and CVD. However, more longitudinal studies are needed to track the integrative responses to continued stress. Although pre-clinical models provide insight into this, most of these studies expose a stressful stimulus to a rodent model over days and weeks rather than months and years, and then examine the cardiovascular or hormonal consequences. It would also be important to understand if the stress-induced cardiovascular adaptations are permanent or can be resolved once the chronic stress has ended. In addition, not all individuals respond to stress in a similar way; indeed, there are a number of animal models that are resilient to chronic stress (e.g., DBA/2 mice, Flinders Resistant Line rats), and there are also heterogenetic within-stress responding species. Understanding why certain species are resilient to exposure to stress may provide important insights to future therapeutic approaches.

Individual difference moderators

There is evidence that the magnitude and pattern of one's physiologic response to stress is moderated by several individual difference variables, including an individual's sex, race, and disease status. Although these variables are not depicted in Fig. 12.1, they should provide an ample source of hypotheses for future work exploring how stress gets "under the skin" for various constellations of people. Although men are diagnosed with CVD at higher rates than women earlier in life, women are diagnosed at much higher rates of CVD following menopause (Lorenzo et al., 2007). Conversely, women are diagnosed with higher rates of depressive disorders than men throughout the entire lifespan (Albert, 2015). The role of sex differences in ANS responding to stress are highlighted in Chapter 5 and in vascular responding to stress in Chapter 6. These findings suggest that the hormonal influences on both the stress response and the onset of CVD and depression warrant continued investigation. A comparable set of hypotheses can direct future investigations into racial differences in physiologic stress responding, in particular, given that many minority racial groups are exposed to more stressful life circumstances than those from the majority culture.

Data do indicate that less healthy persons have different physiologic stress responses than their healthy counterparts. In this regard, being diagnosed with either depression or CVD only adds to life stress and perpetuates the physiologic sequelae that follow. In this regard, future research should continue to explore how various diseases create added vulnerabilities for subsequent disease consequences as well as dysfunctional physiologic responses to stress.

Multi-method multi-system research

Although all of the proposed mechanisms depicted in Fig. 12.2 that show reactivity to stress play an important role in explaining how stress leads to depression and CVD, they are not likely the only physiological mechanisms involved. Additionally, researchers do not regularly explore interactions among the various bodily systems involved in orchestrating the stress response. Investigators into ANS responses to stress do not often measure cortical responses simultaneously or immunologic parameters. Certainly, investigations into the role of the gut microbiome in the stress response have only been conducted by a few research teams. We may have contributed to this "siloed" approach by inviting authors to focus on single physiological systems in their chapters. However, that is the state of the art at the present time. That said, we would urge future investigators to use multiple methods of assessing the stress response that cut across multiple organ systems. Although this approach will have costs associated with it, both in terms of the expense for conducting additional assays as well as establishing interdisciplinary collaborations to gain the expertise to interpret these findings, it will be essential in gaining a fuller integrative understanding of how exposure to stress leads to conditions like CVD and depression. Future empirical work aimed at uncovering important connections across physiologic systems and identifying their causal relations will bring us closer to understanding how stress leads to CVD.

Measurement of stress and depression

In contrast to relying on blood levels of circulating messengers from the immune, endocrine, and gastro-intestinal systems, the assessment of stress and depression in humans typically involves completing self-report questionnaires or through clinical interviews. As such, this information is not immune from personal biases or attempts to portray oneself in either a more positive or negative light. Reliance on such self-reported methods of assessment is often considered less valid than the more reliable measures of physiological functioning. That said, most practicing health care providers rely on asking patients whether they smoke cigarettes or not rather than drawing cotinine levels to assess this important risk factor. Conversely, blood levels of stress hormones, microbiota, or immunological parameters are about as stable as measures of blood pressure that can vary significantly over relatively brief durations of time. In this regard, it would seem that several existing measures of depression possess adequate psychometric characteristics for use as assessment tools in health care environments. It would be useful, however, if guidelines were created and endorsed so that health care providers could employ standardized tools for assessing stress and depression that could be translated into numerous languages and adopted for uniform practice. By doing so, data across sites could be compared easily and even integrated as we attempt to learn more about relations among stress, depression, and CVD.

Genetic influences

Over the last decade, researchers have examined how genetics may play an integral role with depression. For example, family and twin studies have provided strong evidence for the contribution of genetic factors to the risk of depression. A meta-analysis of twin research data shows that the heritability rate for depression is 37% (95% CI: 31%–42%), and data from family studies show a two- to threefold increase in the risk of depression in first-degree offspring of patients with depression

(Sullivan et al., 2000). Researchers found that chronic exposure to a stress hormone causes modifications to DNA in the brains of mice, prompting changes in gene expression, possibly through epigenetic modifications (i.e., changes to DNA that don't alter sequences but influence gene expression) (Lee et al., 2010). Such data suggests that environmental stress could lead to epigenetic modifications, and provide a link between stress and depression. As such, important questions need to be answered: Can exposure to an acute stressor influence genomic regulation long after cessation of the stimulus? Are there distinct periods in life (young vs. old; men vs. women, pre vs. post-menopause) during which genomic regulation may be particularly sensitive to the impact of stress exposure? And, may the effects of concurrent multiple or repetitive stressors over time accumulate and sculpt genomic function, and what role does pre-existing CVD contributing to the development of distinct phenotypes? Elucidating the molecular mechanisms that shape this risk could open new opportunities for the prevention and treatment of stress-related pathologies.

Conclusion

Psychological factors, primarily exposure to stress holds considerable promise in advancing our understanding and predicting risk for CVD. Although a considerable amount of evidence has accumulated linking stress with CVD, health care professionals do not regularly assess life stress when evaluating overall risk for initial CVD events or subsequent events among patients already diagnosed with CVD. As shown in this volume, the evidence does not currently support a causal relation between depression and CVD, but its status as a risk factor cannot be argued. In this regard, we would advocate for adopting the recommendations of the American Heart Association to screen for depression routinely in health care settings (Lichtman et al., 2014).

Through examination of the current state of the literature on potential mechanisms that explain how exposure to stress leads to conditions of depression and CVD, it is clear that exposure to stress propagates a sequence of physiological responses that influences the circulatory system in less than positive ways. Continued research aimed at understanding the biologically plausible mechanisms that link stress with CVD will encourage the development and testing of innovative strategies for modifying them with the goal of reducing the consequences of the disabling array of diseases that affect the cardiovascular system.

References

Albert, P.R., 2015. Why is depression more prevalent in women? J. Psychiatry Neurosci. 40 (4), 219–221.

Barth, J., Schumacher, M., Herrmann-Lingen, C., 2004. Depression as a risk factor for mortality in patients with coronary heart disease: a review and meta-analysis. Psychosom. Med. 66 (6), 802–813.

Berthoud, H.R., Neuhuber, W.L., 2000. Functional and chemical anatomy of the afferent vagal system. Auton. Neurosci. 85 (1–3), 1–17.

Bruce-Keller, A.J., Salbaum, J.M., Berthoud, H.R., 2018. Harnessing gut microbes for mental health: getting from here to there. Biol. Psychiatry 83 (3), 214–223.

Cannon, W.B., 1915. Bodily Changes in Pain, Hunger, Fear and Rage: An Account of Recent Researches Into the Function of Emotional Excitement. Appleton.

Carney, R.M., Freedland, K.E., Rich, M.W., Jaffe, A.S., 1995. Depression as a risk factor for cardiac events in established coronary heart disease: a review of possible mechanisms. Ann. Behav. Med. 17 (2), 142–149.

Centers for Disease Control and Prevention, National Center for Health Statistics, 2017. Underlying Cause of Death 1999–2016 on CDC WONDER Online Database, released December, 2017. Data are from the Multiple Cause of Death Files, 1999–2016, as compiled from data provided by the 57 vital statistics jurisdictions through the Vital Statistics Cooperative Program.

Friedman, M., Rosenman, R.H., 1959. Association of specific overt behavior pattern with blood and cardiovascular findings; blood cholesterol level, blood clotting time, incidence of arcus senilis, and clinical coronary artery disease. J. Am. Med. Assoc. 169 (12), 1286–1296.

Gartlehner, G., Wagner, G., Matyas, N., Titscher, V., Greimel, J., Lux, L., Gaynes, B.N., Viswanathan, M., Patel, S., Lohr, K.N., 2017. Pharmacological and non-pharmacological treatments for major depressive disorder: review of systemic reviews. Br. Med. J. 7 (6), e014912.

Goldapple, K., Segal, Z., Garson, C., Lau, M., Bieling, P., Kennedy, S., Mayberg, H., 2004. Modulation of cortical-limbic pathways in major depression: treatment-specific effects of cognitive behavior therapy. Arch. Gen. Psychiatry 61 (1), 34–41.

Guarino, F., Cantarella, G., Caruso, M., Russo, C., Mancuso, S., Arcidiacono, G., Cacciola, R.R., Bernardini, R., Polosa, R., 2011. Endothelial activation and injury by cigarette smoke exposure. J. Biol. Regul. Homeost. Agents 25 (2), 259–268.

Insull Jr., W., 2009. The pathology of atherosclerosis: plaque development and plaque responses to medical treatment. Am. J. Med. 122 (1 Suppl), S3–S14.

Katz, R.J., 1981. Animal model of depression: effects of electroconvulsive shock therapy. Neurosci. Biobehav. Rev. 5 (2), 273–277.

Krantz, D.S., 1980. Cognitive processes and recovery from heart attack: a review and theoretical analysis. J. Hum. Stress. 6 (3), 27–38.

Krantz, D.S., Manuck, S.B., 1984. Acute psychophysiologic reactivity and risk of cardiovascular disease: a review and methodologic critique. Psychol. Bull. 96 (3), 435–464.

Langgartner, D., Vaihinger, C.A., Haffner-Luntzer, M., Kunze, J.F., Weiss, A.J., Foertsch, S., Bergdolt, S., Ignatius, A., Reber, S.O., 2018. The role of the intestinal microbiome in chronic psychosocial stress-induced pathologies in male mice. Front. Behav. Neurosci. 12, 252.

Lawrence, A.J., Jarrott, B., 1996. Neurochemical modulation of cardiovascular control in the nucleus tractus solitarius. Prog. Neurobiol. 48 (1), 21–53.

Lee, R.S., Tamashiro, K.L., Yang, X., Purcell, R.H., Harvey, A., Willour, V.L., Huo, Y., Rongione, M., Wand, G.S., Potash, J.B., 2010. Chronic corticosterone exposure increases expression and decreases deoxyribonucleic acid methylation of Fkbp5 in mice. Endocrinology 151 (9), 4332–4343.

Lett, H.S., Blumenthal, J.A., Babyak, M.A., Sherwood, A., Strauman, T., Robins, C., Newman, M.F., 2004. Depression as a risk factor for coronary artery disease: evidence, mechanisms, and treatment. Psychosom. Med. 66 (3), 305–315.
Liang, S., Wu, X., Hu, X., Wang, T., Jin, F., 2018. Recognizing depression from the microbiota-gut-brain axis. Int. J. Mol. Sci. 19 (6), E1592.
Lichtman, J.H., Froelicher, E.S., Blumenthal, J.A., Carney, R.M., Doering, L.V., Frasure-Smith, N., Freedland, K.E., Jaffe, A.S., Leifheit-Limson, E.C., Sheps, D.S., Vaccarino, V., Wulsin, L., On Behalf of the American Heart Association Statistics Committee of the Council on Epidemiology and Prevention and the Council on Cardiovascular and Stroke Nursing, 2014. Depression as a risk factor for poor prognosis among patients with acute coronary syndrome: systematic review and recommendations. Circulation 129 (12), 1350–1369.
Liston, C., McEwen, B.S., Casey, B.J., 2009. Psychosocial stress reversibly disrupts prefrontal processing and attentional control. Proc. Natl. Acad. Sci. 106, 912–917.
Lorenzo, C., Williams, K., Hunt, K.J., Haffner, S.M., 2007. The National Cholesterol Education Program—Adult Treatment Panel III, International Diabetes Federation, and World Health Organization definitions of the metabolic syndrome as predictors of incident cardiovascular disease and diabetes. Diabetes Care 30 (1), 8–13.
Miller, A.H., Haroon, E., Raison, C.L., Felger, J.C., 2013. Cytokine targets in the brain: impact on neurotransmitters and neurocircuits. Depress. Anxiety 30 (4), 297–306.
Pittilo, R.M., 2000. Cigarette smoking, endothelial injury and cardiovascular disease. Int. J. Exp. Pathol. 81 (4), 219–230.
Rugulies, R., 2002. Depression as a predictor for coronary heart disease: a review and meta-analysis. Am. J. Prev. Med. 23 (1), 51–61.
Sacher, J., Rekkas, P.V., Wilson, A.A., Houle, S., Romano, L., Hamidi, J., Rusjan, P., Fan, I., Stewart, D.E., Meyer, J.H., 2015. Relationship of monoamine oxidase-A distribution volume to postpartum depression and postpartum crying. Neuropsychopharmacology 40, 429–435.
Sasayama, D., Hattori, K., Wakabayashi, C., Teraishi, T., Hori, H., Ota, M., Yoshida, S., Arima, K., Higuchi, T., Amano, N., Kunugi, H., 2013. Increased cerebrospinal fluid interleukin-6 levels in patients with schizophrenia and those with major depressive disorder. J. Psychiatr. Res. 47, 401–406.
Savic, I., Perski, A., Osika, W., 2017. MRI shows that exhaustion syndrome due to chronic occupational stress is associated with partially reversible cerebral changes. Cereb. Cortex 28, 894–906.
Savignac, H.M., Kiely, B., Dinan, T.G., Cryan, J.F., 2014. Bifidobacteria exert strain-specific effects on stress-related behavior and physiology in BALB/c mice. Neurogastroenterol. Motil. 26 (11), 1615–1627.
Scapagnini, G., Davinelli, S., Drago, F., De Lorenzo, A., Oriani, G., 2012. Antioxidants as antidepressants: fact or fiction? CNS Drugs 26, 477–490.
Selye, H., 1955. Stress and disease. Science 122 (3171), 625–631.
Siegler, I.C., Peterson, B.L., Barefoot, J.C., Williams, R.B., 1992. Hostility during late adolescence predicts coronary risk factors at mid-life. Am. J. Epidemiol. 136, 146–154.
Skala, J.A., Freedland, K.E., Carney, R.M., 2006. Coronary heart disease and depression: a review of recent mechanistic research. Can. J. Psychiatr. 51 (12), 738–745.
Sullivan, P.F., Neale, M.C., Kendler, K.S., 2000. Genetic epidemiology of major depression: review and meta-analysis. Am. J. Psychiatr. 157 (10), 1552–1562.
Tang, J., Yi, W., Chen, S., Gao, Z., Xiao, B., 2018. Microglia polarization and endoplasmic reticulum stress in chronic social defeat stress induced depression mouse. Neurochem. Res. 43 (5), 985–994.
Tasker, J.G., Herman, J.P., 2011. Mechanisms of rapid glucocorticoid feedback inhibition of the hypothalamic–pituitary–adrenal axis. Stress 14 (4), 398–406.
Waisman, A., Liblau, R.S., Becher, B., 2015. Innate and adaptive immune responses in the CNS. Lancet Neurol. 14 (9), 945–955.

Author index

B

D

E

F

G

H

K

L

O

S

T

U

V

W

X

Y

Z

Subject index

Note: Page numbers followed by *f* indicate figures and *t* indicate tables.

T

U

V

X

FSC
www.fsc.org
MIX
Papier | Fördert
gute Waldnutzung
FSC® C083411

Kolibri 360 GmbH
Ferdinand-Jühlke-Straße 7
99095 Erfurt, Deutschland
produktsicherheit@kolibri360.de